HOW TO BUILD AND USE ELECTRONIC DEVICES WITHOUT FRUSTRATION, PANIC, MOUNTAINS OF MONEY, OR AN ENGINEERING DEGREE

HOW TO BUILD AND USE ELECTRONIC DEVICES WITHOUT FRUSTRATION, PANIC, MOUNTAINS OF MONEY, OR AN ENGINEERING DEGREE

Stuart A. Hoenig, PhD
Professor of Electrical Engineering,
University of Arizona, Tucson

Second Edition

Little, Brown and Company, Boston

Second Edition

Library of Congress Catalog Card No. 80-80590

ISBN 0-316-36808-3

Printed in the United States of America

SEM

Over the years I have seen books dedicated to wives, children, secretaries, and sometimes even students, all of whom played a big part in the writing of the book. However, I haven't seen any books dedicated to Mammon, which is really why most of them are written in the first place. I almost didn't have a dedication at all, except that this seemed a good spot to point out that electronics really is fun *once you get used to it and stop being afraid. The invention of op-amps played a big part in getting one engineer (SAH) over the transistors-are-too-much-for-me hang-up. So perhaps the book should be dedicated to Harold S. Black, who first recognized that a linear DC amplifier could be used for all sorts of useful things and called it a "negative feedback amplifier."*

Contents

Preface		*page ix*
A Little Circuit Theory and an Instrument or Two	1	*page 1*
The Op-Amp and How to Make it Work for You	2	*page 69*
Op-Amp Applications for Fun and Profit	3	*page 105*
Biomedical Applications of Op-Amp Circuits	4	*page 229*
The Computer and Its Applications (or Space War for Fun and Profit)	5	*page 249*
Op-Amp Problems and How to Fix Them	6	*page 283*
Discrete Devices (If You Must Use Them)	7	*page 301*
Conclusions	8	*page 347*
Appendixes		*page 351*
Where to Buy It		*page 353*
Selected Reading List		*page 357*
An Informal Glossary of Technical Terms		*page 361*
Index		*page 379*

Preface

Dear Reader, you are looking at the second edition of this electronics book. The fact that there is a second edition implies that the first edition sold enough copies to justify a second – I must have done something right. This in turn suggests that I keep the same approach that worked the last time.

The book is designed to show you how to build circuits in the minimum length of time without the mathematical juggling that fills most books on electronics. The math may be necessary for electrical engineer types, but my observations have convinced me that practicing engineers seldom use it. Most circuits are designed by scratching on the backs of envelopes, building the damn thing, and testing it out. I urge you to read this book with "soldering-iron in hand."

You can start at a particular chapter, learn the material you need to know, and build circuits using what you have learned. In some cases, the designs will be inelegant; in fact, "quick and dirty" is the best descriptive phrase. Engineering often involves settling for the "third best" of anything. "Second best" takes too long, and the "best" design comes only when the whole process is obsolete. One might paraphrase the poet: "Come learn along with me, third best will work – you'll see."

My method of teaching is somewhat unorthodox. It is based on the system used in the Navy during World War II to teach farm boys to be sailors. The system was called "monkey see, monkey do." First, you do something by brute copying; then, when you know how to do it, you ask about how it works. Then they give you a book called *Operation of Mark IV Torpedo,* and you are on your way to being a Chief Torpedo Mechanic. Why not start with the book first? Well, if you started with the theory first, you would never get to take a torpedo apart; you would be a Seaman II forever.

As a professor or graduate student, you may be insulted by being treated like an 18-year-old. The question is, do you want to learn something fast, or do you want to play games that soothe your ego? If you want to learn, sit down at the laboratory bench and start reading the book!

You might be more justified in asking what I am trying to teach you. First, I am going to try to bring you to the point where you can design your own instruments using op-amps. Second, I want to teach you the nomenclature and jargon of the electronics business so that you can steal useful circuits from the manufacturers and the trade magazines. In the Selected Reading List in the Appendixes several good

"You realize, of course, that this is grounds for a divorce?"

sources for such circuits are listed. Many of these are intended for electrical engineers, but when you have gone halfway through this book, you'll be stealing from them like a veteran. Try to get on the mailing lists of the free magazines such as *Electrical Design News.* Once word gets out that you are in the market for op-amps, you will be deluged with catalogs.

Some of you may be wondering why I emphasize op-amps so much. Are they any better or different than transistors? The answer is yes! Op-amps *are* different, and by the time Hoenig is done with you, you'll be talking about op-amps in your sleep. For the moment, I will simply state that an *op-amp* is a transistorized linear DC amplifier with high gain and good stability. At least five PhD electrical engineering man-years have gone into designing and testing this gadget. You can expect it to be *better* than anything you could learn to build in the next year. The point is that you don't have to reinvent the wheel, electronically speaking; the product of all this effort is ready for you to use.

You might also be wondering just how good op-amps are for building circuits. To answer that question, I could quote the industry trade reports that say the number of op-amp manufacturers has increased from one in 1946 to 40 in 1978. Believe it or not, most of them are making money. What better proof could there be?

The cooperation of the Burr-Brown Research Corp. (Tucson) and the Motorola Semiconductor Products Division (Phoenix) in providing equipment and supplies is gratefully acknowledged. The cartoons in the text were drawn by Ms. Derith Glover of Tucson. Paul R. Stauffer helped with the corrections and improvements for the second edition.

Before turning you loose on the op-amps, I have one more quote that I hope will help when things don't go right. It is sometimes paraphrased "persistence pays," and is attributed to Calvin Coolidge, 30th president of the United States.

Press On

Nothing in the world can take the place of persistence. Talent will not; nothing is more common than unsuccessful men with talent. Genius will not; unrewarded genius is almost a proverb. Education alone will not; the world is full of educated derelicts. Persistence and determination alone are omnipotent.

S.A.H.

HOW TO BUILD AND USE ELECTRONIC DEVICES WITHOUT FRUSTRATION, PANIC, MOUNTAINS OF MONEY, OR AN ENGINEERING DEGREE

A Little Circuit Theory and an Instrument or Two

Before starting to become op-amp experts, you have to understand a little circuit theory. Even if you flunked Physics 1 and are terrified by the toaster, you will find that circuit theory is not an ineffable mystery.

We propose to start at the beginning and teach by our own technique. First we explain in the simplest terms; then you do an experiment to convince yourself that it's all true. Finally – after you have used the device – we explain the theory. It is a little backward, but it seems to work.

In this book you should *not* hesitate to skip over certain dull subjects to get to something interesting. When you get stuck, you can go back and perhaps discover that certain dull topics aren't so dull after all.

You should first arm yourself with a soldering iron, a 45 volt dry-cell battery, a volt-ohmmeter (a Heathkit MM-1 is great*), and some carbon resistors (the cheapest kind). This book is written with the assumption that you will read it with soldering iron in hand. Do the experiments as you read the book – *don't wait, do them now!*

RESISTORS

Resistors are marked with colored stripes. The stripes tell the resistance value and tolerance according to the resistor code. The code itself is given in Table 1-1, and you might want to try to memorize it. If you do, there is a mnemonic to help. The colors go black, brown, red, orange, yellow, green, blue, violet, gray, and white. The mnemonic is "**B**etty **B**rown **R**uns **O**ver **Y**ellow **G**rapes **B**ut **V**iolet **G**oes **W**alking." The application of this color code is shown below. These colored bands are grouped toward one end of the resistor body. Starting with that end of the resistor, the first band represents the first digit of the resistive value; the second band represents the second digit; and the third band represents the power of 10 by which the first two digits are multiplied. A fourth band of gold or silver represents a tolerance of ±5% or ±10% respectively. The absence of a fourth band indicates a tolerance of ±20%.

The physical size of a carbon† resistor is related to its wattage rating: its size

*Now and then you will note that we mention certain commercial products. These are *not* paid-for advertisements (at least we haven't had any offers yet), but they represent the results of our experience.

†Carbon resistors are the ones most commonly used. For high currents, wire-wound resistors are available. Other special types are listed in the various catalogs. Just remember that carbon resistors are the lowest in cost. Some military type resistors will have extra bands indicating reliability – ignore them. The first three bands are what really count.

Table 1-1. Resistor Color Code

Multiplier Color	First Digit	Second Digit	Multiplier
Black	0	0	0
Brown	1	1	10
Red	2	2	100
Orange	3	3	1000
Yellow	4	4	10,000
Green	5	5	100,000
Blue	6	6	1,000,000
Violet	7	7	Not used
Gray	8	8	Not used
White	9	9	Not used
Gold	–	–	.1
Silver	–	–	.01

Tolerance Color	Tolerance
Gold	±5%
Silver	±10%
No band	±20%

increases progressively as the wattage rating is increased. The diameters of 1/4, 1/2, 1, and 2 watt resistors are approximately 3/32, 1/8, 1/4, and 5/16 inch respectively. For now, just get the 1/4 watt size – they will do for most of our experiments. If high wattages are needed, we will say so.

The color code chart (Table 1-1) and examples (Figure 1-1) provide the information required to identify color-coded resistors.

Another thing worth mentioning is the abbreviations used in electrical engineering. Learning them will help you read this book and the associated literature. (In engineering we read the literature to steal cute circuits from one another.)

A = ampere or amp
Å = Angstrom unit (10^{-10} m)
F = farad (unit of capacitance, C)
f = frequency
Hz = hertz (unit of frequency or cycles per second)
I = symbol for current
Ω = ohm
R = symbol for resistance
V = volt
W = symbol for energy, watt
Z = impedance

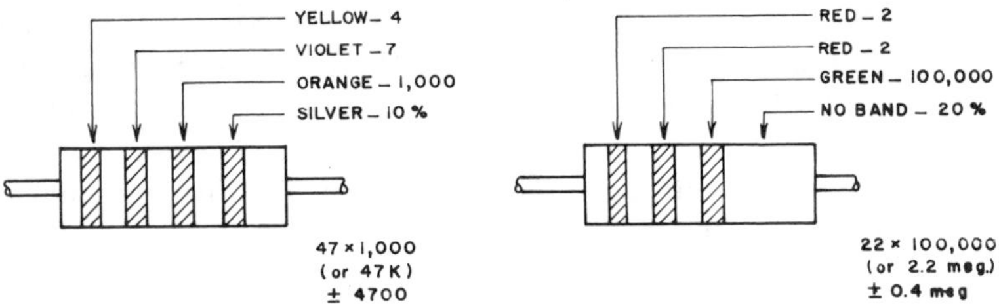

Figure 1-1. Color-coded resistors.

ρ = resistivity = rho
σ = conductivity = sigma
$\omega = 2\pi f$, angular frequency = omega

p = pico- = 0.000000000001 = 10^{-12} (pF = picofarad)
n = nano- = 0.000000001 = 10^{-9} (nm = nanometer)
μ = micro = 0.000001 = 10^{-6} (μV = microvolt)
m = milli- = 0.001 = 10^{-3} (mA = milliamp)
k = kilo- = 1000 = 10^{3} (k = kilohm)
M = mega- = 1,000,000 = 10^{6} (MHz = megahertz)

Chances are that you probably don't want to memorize the code anyway, so run down to your friendly radio shop and buy a resistor guide for 25¢.* You should also start getting catalogs from the following companies (the catalogs are free – you will end up paying for them in the stuff you buy):

Allied Electronics
3160 Alfred Street
Santa Clara, CA 95050

Edmund Scientific Co.
101 E. Gloucester Pike
Barrington, NJ 08007

EICO Instrument Co.
283-T Malta Street
Brooklyn, NY 11207

Heath Company
Benton Harbor, MI 49022

Lafayette Radio
111 Jericho Turnpike
Syosett, NY 11791

Newark Electronics
500-T N. Pulaski Road
Chicago, IL 60624

Olson Electronics
260 S. Forge Street
Akron, OH 44308

Poly Paks
P.O. Box 942
South Lynnfield, MA 01940

Each of these companies specializes in some particular line of items. Many other

*Please remember that when we mention prices, they are for the era 1979–1980 and are subject to inflation. However, we think that it will help if you have some "order-of-magnitude" prices to go by.

companies will be listed later in the book; these are just some good ones to start out with. Edmund sells primarily optical equipment; Heath sells electronic instruments; Allied and Newark sell new electronic gear. The others sell both new and old parts, as well as reclaimed military or manufacturing surplus. Buying used or surplus stuff can be great – but watch out: there may be little or no operating data and replacement may be impossible. But don't let this warning scare you. Such problems provide part of the fun of electronics. As part of the fun and games, you may want to start reading things like *Popular Electronics.*

Popular Electronics
Ziff-Davis Publishing Co.
1 Park Avenue
New York, NY 10016

You will find all sorts of useful circuits, many of which will be simple enough for beginners, but it's hoped you won't be a beginner very long.

CIRCUIT SYMBOLS

Part of beginning is learning circuit language. We will now introduce you to the most common circuit symbols (Figure 1-2). Some of them you won't see again for many pages, but there are a few of them that we use all the time.

VOLT-OHMMETERS

The only instrument you will need right now is the volt-ohmmeter (VOM). Later, we will discuss this device in detail, but for now, just look at the meter scales. You

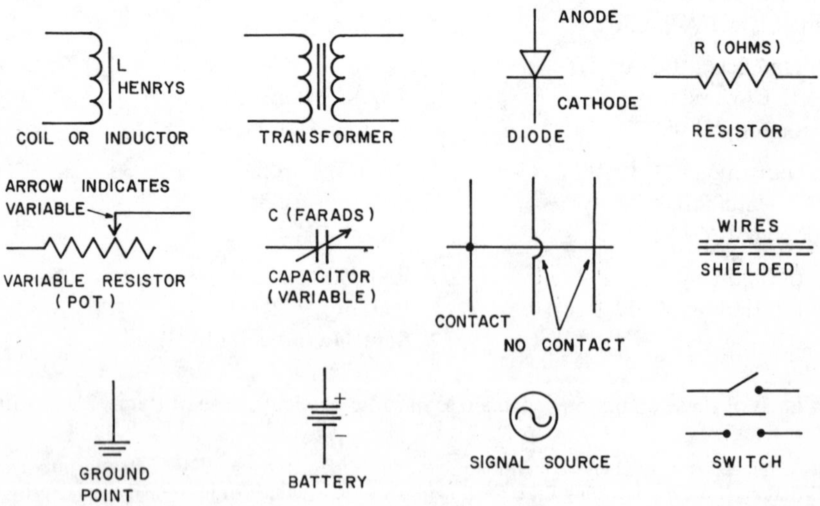

Figure 1-2. Circuit symbols.

have a series of ranges, usually 0-50 and 0-15. These are voltage ranges. By means of a switch, the 0-50 range can cover 0-5, 0-50, 0-500, and 0-5000 volts. The 0-15 scale can be used for measuring 0-1.5, 0-15, 0-150, and 0-1500 volts. The meter will read AC or DC voltages depending on your use of input jacks or switches. The "ohms" scale is nonlinear and reads backward (don't worry about why for now). By means of decade switches, you can read 0-2 kΩ, 0-20 kΩ, 0-200 kΩ, and 0-20 MΩ. You get from the ohms to volts scales by means of a switch on the meter. Some VOMs have a series of current scales or use the voltage scales to measure amperage, i.e., 0-150 mA, 0-15 mA, and so on. Again, a switch is used to get on to the current scales.

VOMs are rugged instruments and *almost* student proof: don't use the 0-1.5 volt scale on 110 volts or the ohms scale on any circuit that has current in it. If you have already bought a Heath VOM at this time (as suggested on page 1), *read the instructions.*

POWER RECEPTACLES

At this point we should say a few words about the AC power receptacles found in our homes and laboratories. The old-fashioned, and still very common, type is the two-prong system shown in Figure 1-3A. In this case, one side is "hot," the other is "neutral"; the electricity comes out of the hot side at a "high" electrical pressure (voltage), goes through the appliance or whatever, and returns to the power plant at

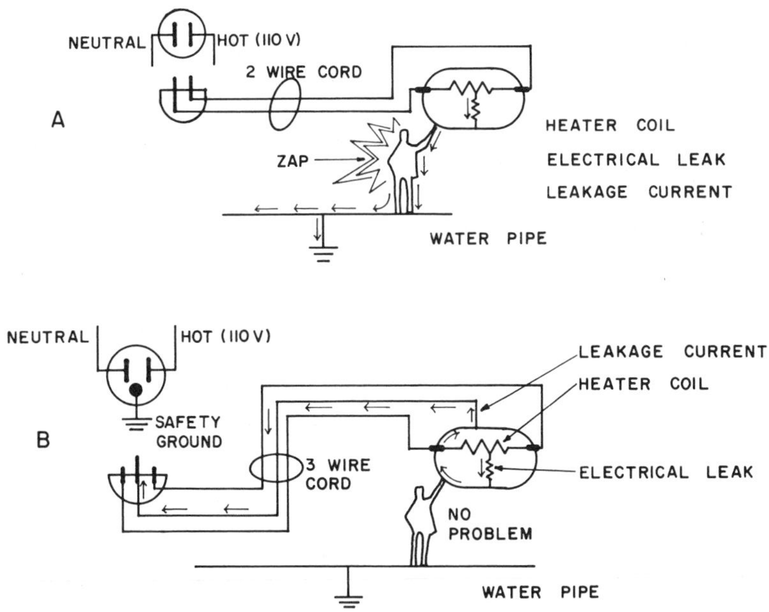

Figure 1-3. Conventional power systems. *A*. Two-wire system. *B*. Three-wire system with safety ground.

low pressure via the neutral lead. It is a real mistake to refer to the neutral lead as "ground," and those who do may be setting themselves up for a "shocking surprise" (sorry, we couldn't resist it).

To appreciate the advantages of the three-wire or grounded system and the dangers of the two-wire system, we show Figure 1-3 where both systems are used with a "defective" appliance. Make sure you understand why in one case (A) the little figure is "shocked." In the other drawing (B) the ground wire carries off the leaking electricity so that it does not go through the person in the circuit.

What do you do if your house or laboratory only has two-wire outlets? This is like the question, how do porcupines make love? To which the answer is, *very carefully.* All we can suggest is (1) be careful, (2) ground every electrical appliance that has a metal case to a water pipe, and (3) keep your fingers dry. (Actually, if you're a college student, you are in more danger from auto accidents than from everything else put together, including the military.) We will try to warn you as much as possible, but there is *no substitute for common sense!*

At this point we could consider the matter closed but one of our student readers pointed out that "common sense" is usually the product of experience and that is what students don't have. "Besides," he continued, "I still don't understand why the term 'ground neutral' is ever used." We heard this man ask for help and promptly wrote the expanded discussion which follows. The student in question said that "it helped a lot."

To understand where the term *ground neutral* came from let's start right at the local power plant. The local power company delivers what is called *three-phase power* (Figure 1-4). Note in this figure that there are four wires: A, B, C, and ground. Ground is connected to a big copper plate that is buried in the earth at the power plant and can at this point be correctly called *ground.* Wires A, B, and C are called *hot* because they can deliver a painful electrical burn if you are careless. All four wires are carried on power poles throughout the city and countryside. Ground is connected to the earth ground at every pole to reduce the damage from accidental lightning strokes.

For home service, the high voltages are reduced by transformers. The usable voltages, which are nominally 110 and 220 volts, are delivered to the home or laboratory as shown in Figure 1-5. Notice that now the two "hot leads" and the ground are used to provide power to various areas of the house. The point of this three-wire arrangement should be clear to you. If something goes wrong and the case of an appliance receives some electrical pressure due to leakage, the electricity goes back to the power plant via the ground lead instead of to the body of someone who might be using the device.

Looking at Figure 1-5, you can appreciate where the neutral lead appears and that it is only THERE that neutral and ground are the same. Inside the house the neutral lead carries the normal current returning to the power plant while ground serves only as a safety wire in case something goes wrong. Outside the house the neutral lead does not exist.

Returning to the question of neutral versus ground, we leave you with the

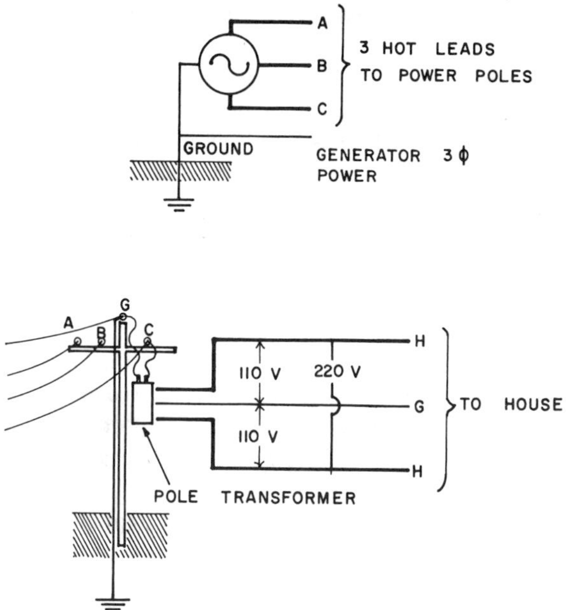

Figure 1-4. Electrical generator and transmission lines.

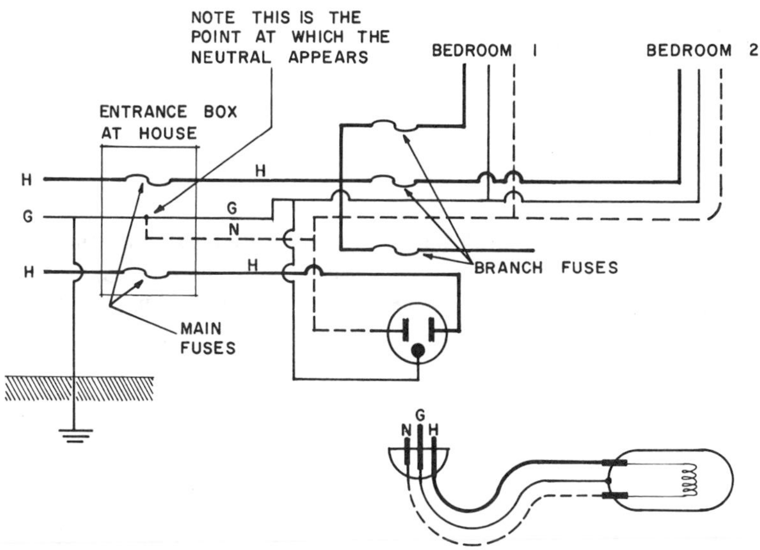

Figure 1-5. House wiring.

knowledge that *neutral is not, is not, ground inside a building! Never, Never, Never!!!*

This concludes our discussion of house wiring and how it works. For more details you will have to consult another book or wait until we write one. Our next step will be to take you through some simple circuits and instruments that you will need to make use of op-amps. Some of them you may be familiar with, and in that case skip over the material. However, if you haven't seen something before, read the discussion over carefully. We didn't write it into the book just to take up space.

A VOLTAGE DIVIDER

For our first experiment, let's take the VOM and adjust it for DC volts on a range that is greater than the battery voltage. Now measure the battery voltage. If it reads 45 volts, great! If you have noticed that the VOM goes off scale, the leads are reversed. But this is typical when working with DC; you can expect to do it wrong the first time (the chances are 50–50).

Step two is to make up a voltage divider. Set up a circuit as shown in Figure 1-6. Use 5.1 kΩ and 10 kΩ 1/4 watt resistors.

Ohm's law states – remember Physics 1? If not, see the next section. But do the experiment first! – that V (volts) = I (amps) X R (ohms). In this case

$$I = \frac{V}{R}$$

$$I = \frac{45 \text{ volts}}{15.1 \text{ k}\Omega}$$

$$I = 3 \times 10^{-3} \text{ amps}$$

$$I = 3 \text{ mA}$$

You can convince yourself of this, if you have any doubts about Ohm's law, by putting the VOM on the milliamp scale (full scale must be more than 3 mA) and connecting it in series with R_1 and R_2 at points A and B by opening the circuit at A

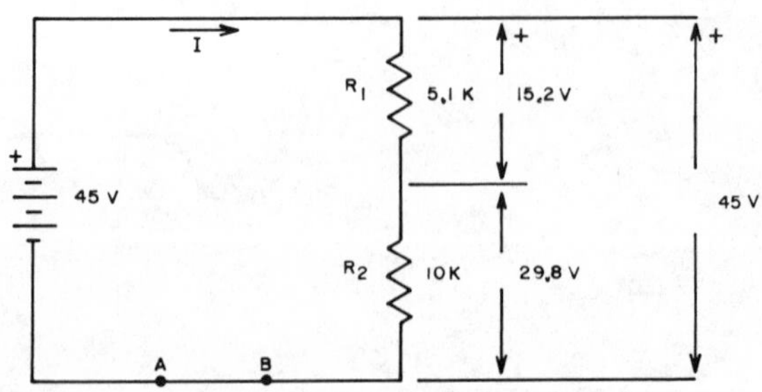

Figure 1-6. Voltage divider.

and B and connecting in the VOM. Now the voltage across R_2(10 kΩ), from Ohm's law, should be $V_2 = 3 \times 10^{-3}$ amps (10^4 ohms) = 30 volts. Across R_1 (5.1 kΩ), the voltage should be $V_1 = 3 \times 10^{-3}$ (5×10^3) = 15 volts. You have built a *voltage divider.* This is a useful gadget for obtaining any voltage between 0 and 45 volts from a 45 volt battery (all calculations are only slide-rule-accurate).

How much power are we dissipating in the resistors? The power dissipation law is P(watts) = I^2 (amps) $\times$ R (ohms). In the 10 kΩ resistor, we dissipate $P = (3 \times 10^{-3})^2 \times 10^4 = 9 \times 10^{-2}$ watt; in the 5.1 kΩ resistor, the dissipation, by a similar calculation, is 0.045 watt. The use of 1/4 watt resistors thus provides an adequate margin of safety; besides, they are easily available.

After having performed this simple experiment with the voltage divider, we are now ready to get into some physical concepts of resistance and some of the details of circuit theory, which are not so simple. As we said before, if you want to skip the details of circuit theory, go ahead. You *should* read the sections on capacitors, inductors, filters, power supplies, and oscilloscopes. Of course, if you want to go on to Chapter 2 now and come back to this chapter as you need it, that is your decision. Who are we to tell you how to learn?

OHM'S LAW AND RESISTIVITY

An important property of all materials is called *conductivity,* i.e., the ability to pass electricity. Suppose we have a bar of length L and cross-sectional area A, and we apply a voltage V between the ends of the bar. If V is in volts and L is in meters, we can define the *voltage gradient,* E, as

$$E = V/L \text{ (volts/meter)}$$

Now if this voltage V forces a current I (in amps) through the bar of area A (meters^2), we can define the *current density,* J, as

$$J = I/A \text{ (amps/meter}^2\text{)}$$

The conductivity, σ, is defined as the current density per unit voltage gradient, so

$$\sigma = \frac{J}{E} = \frac{I/A}{V/L} \left(\frac{\text{amps}}{\text{volts} \times \text{meters}}\right)$$

The nice thing about σ is that it is *independent* of the shape of the conductor. Sometimes we use $\rho = \frac{1}{\sigma}$, where ρ is called the resistivity. The fact that resistivity (or conductivity) is a natural property of a certain material leads us to Ohm's law.

Consider our bar of length L and area A. If it has a resistivity ρ, then its *resistance* (which *is* shape-dependent) is

$$R = \rho \text{ (L/A)}$$

The resistance R is expressed in *ohms.* Since the resistivity, ρ, is the reciprocal of conductivity, then

$$\rho = \frac{V/L}{I/A}$$

when this is substituted into the equation for R, we obtain

$$R = \frac{V/L}{I/A}\left(\frac{L}{A}\right) = \frac{V}{I}$$

The relation $R = V/I$ (or $V = IR$) is known as *Ohm's law.*

The fact that we used Ohm's law earlier without deriving it doesn't bother us a bit. *First* we show you how to do something useful; *then* we explain it. (The old Navy system again!)

Ohm's law as stated above assumes that the resistance is *linear,* i.e., if the voltage across the resistance is doubled, the current through it is also doubled. The resistance of materials like carbon, aluminum, copper, silver, gold, and iron is linear. These materials are also *bilateral,* that is, a wire made of one of these metals will conduct electricity equally well in either direction. Many devices that conduct electricity, however, are both nonlinear and unilateral. Such devices include rectifiers and transistors. Have patience, we will get to them faster than you think.

While we are on this topic, we should throw in another useful law (though without proof) that we mentioned previously. The current through a resistor R (ohms) is I (amps) when we apply a voltage of V (volts). The heat produced is given by P (watts) $= I^2R = I\,V$ (this is called Joulean Heating or Joule's law heat because Joule first proposed that $P = I^2R$). Usually P is called *power,* the rate at which heat is produced. The power rating or wattage rating of a resistor tells you how much heat (power) you can dissipate in that resistor without destroying it. For example, a 1000 ohm, 1/4 watt resistor will be at its maximum temperature when a current of about 16 mA (0.016 amp) is passing through it. If you want to pass, say, 40 mA (0.040 amp) through a 1000 ohm resistor, it had better be rated at 1.6 to 2 watts to be safe.

FACTORS AFFECTING RESISTANCE

Some materials are better conductors of electricity than others, and the conductivity varies with temperature. Metals have a *positive coefficient of resistivity,* i.e., their resistance increases with temperature. Table 1-2 shows the effect of temperature on the resistance of some metals. The table lists the ratio R/R_0, which is the ratio of the resistance of a given piece of wire at the given temperature to its resistance at zero degrees Centigrade.

The point of providing you with Table 1-2 is to impress you with the fact that

Table 1-2. Variation of Resistance with Temperature

Temp. (°C)	R/R_0 Copper	R/R_0 Silver	R/R_0 Nickel	R/R_0 Iron	R/R_0 Platinum
-200	0.117	0.176	....	....	0.177
-100	0.557	0.596	....	....	0.599
0	1.000	1.000	1.000	1.000	1.000
+100	1.431	1.408	1.663	1.650	1.392
+200	1.862	1.827	2.501	2.464	1.773

the resistance of a wire goes up with its temperature. If you're going to be running experiments using high currents in hot environments, you might have to take account of the change in wire resistance. Also, you might note the nice linear change in the resistance of platinum with temperature. Platinum is great for resistance thermometers and is surprisingly low in cost (contact Minco Products, 7300 Commerce Lane, Minneapolis, MN 55432).

A useful relationship between resistance and temperature is:

$$R = R_0\ [1 + a(T - T_0)]$$

In this equation, R is the resistance of a material at temperature T, and R_0 is the resistance at some reference temperature, T_0. The coefficient α is almost a constant for some materials like platinum, which allows us to build super-accurate devices called *resistance thermometers.* Their output isn't very high ($\alpha \approx 0.003$ for platinum), but we will show you later how to make big signals out of little ones. Now, on to circuit theory!

KIRCHOFF'S LAWS

An electric circuit must consist of a complete closed loop in order for current to flow. The simplest closed circuit possible is shown in Figure 1-7. The current I flowing in R causes a voltage drop V. According to Ohm's law, $V = RI$. Knowing V and R, we can solve for I.

If there are two resistors in series connected across V as shown in Figure 1-8, then the sum of the two voltage drops, $R_1I + R_2I$, is equal to V. (Note that the current is shown as flowing *out* of the *positive* terminal of the battery and into its negative terminal, and that the current *enters* the *positive* terminal of the resistor and leaves the negative terminal.* The battery is supplying energy, the resistor is absorbing energy).

*Those of you who have had physics know that current *really* flows from the negative side of the battery. We know it, too, but EEs do it the other way, and we have to teach you their jargon.

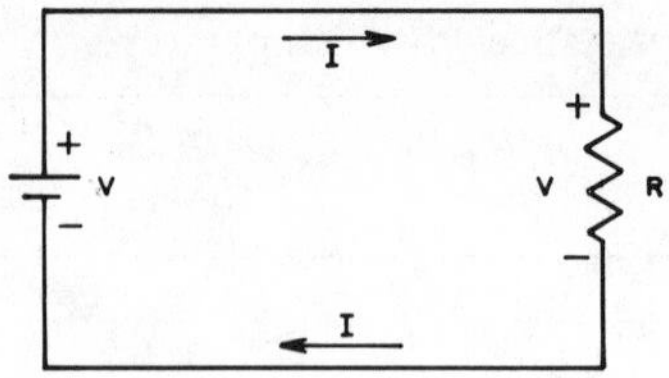

Figure 1-7. Simplest closed-loop circuit.

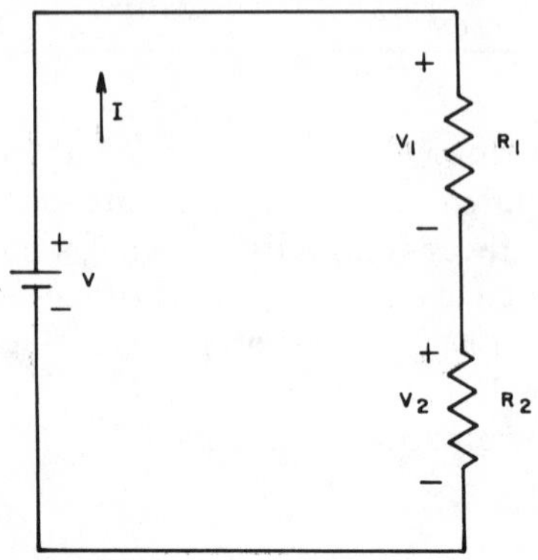

Figure 1-8. Two-resistor series loop.

Kirchoff's Loop Voltage Law

Figure 1-8 illustrates *Kirchoff's loop law.* This law simply states that *the sum of the voltage drops around any closed loop is equal to zero,* a voltage rise being considered as a negative voltage drop. In other words, the voltage drops in any closed loop are equal to the voltage rises in the same loop. Applying this law to Figure 1-8:

$$R_1 I + R_2 I = V$$

From Kirchoff's loop law, we can deduce a very useful relation showing how voltage divides across two (or more) resistors in series. For the circuit shown in Figure 1-8, Kirchoff's loop law states that

$$V = V_1 + V_2$$

in which V_1 is the voltage across R_1 and V_2 is the voltage across R_2. However,

$$V_1 = R_1 I$$

and since resistance in series is additive,

$$I = \frac{V}{R_1 + R_2}$$

Substituting for *I:*

$$V_1 = R_1\left(\frac{V}{R_1 + R_2}\right) = V\left(\frac{R_1}{R_1 + R_2}\right)$$

Similarly,

$$V_2 = V\left(\frac{R_2}{R_1 + R_2}\right)$$

This concept of voltage division in a series circuit is very useful in solving circuit problems; for now, you can take our word for it.

Kirchoff's loop law applies even though a circuit may consist of many closed loops with many batteries, but care must be taken to consider all currents through any one circuit element. The procedure is to assign a current to each closed loop and then write the voltage drops around each loop realizing that, because of the arbitrary assignment of a current to each loop, more than one of these currents may be flowing in any one resistor. Since in a complicated circuit it is not possible to reason out intuitively the direction of current flow in all parts of the circuit, it is best to assign all currents in a clockwise direction.

Kirchoff's Node Current Law

For circuits containing many loops, the above method becomes very laborious, and it may be better to use *Kirchoff's node law,* which states that *the sum of all the currents entering a node is equal to the sum of all the currents leaving that node.*

Consider the simple circuit shown in Figure 1-9. There are two current loops, so Kirchoff's node law states that:

$$I = I_1 + I_2 \tag{1-1}$$

This is true unless there is a storage of electric charge at point *a.* From Kirchoff's node law, we can deduce a very useful relationship showing how current divides in two resistors connected in parallel. The voltage drop across R_1 must be equal to the voltage drop across R_2 since the two resistors are connected together at their terminals. From Ohm's law, the voltage drop V_a across R_1 is

$$V_a = R_1 I_1 \tag{1-2}$$

in which V_a is the voltage at point *a* if the voltage at point *b* is considered to be zero.

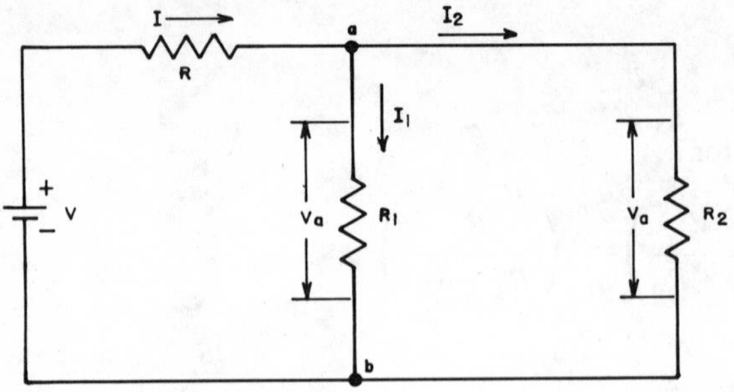

Figure 1-9. Kirchoff's node currents.

V_a, then, is the voltage across R_1. Also:

$$V_a = R_2 I_2 \tag{1-3}$$

From equations (1-2) and (1-3):

$$R_1 I_1 = R_2 I_2$$

Solving for I_2:

$$I_2 = \left(\frac{R_1}{R_2}\right) I_1$$

Substitution in equation (1-1) yields

$$I = I_1 + \left(\frac{R_1}{R_2}\right) I_1 = I_1 \left(1 + \frac{R_1}{R_2}\right) = I_1 \left(\frac{R_2 + R_1}{R_2}\right)$$

Thus:

$$I_1 = I\left(\frac{R_2}{R_1 + R_2}\right)$$

And similarly:

$$I_2 = I\left(\frac{R_1}{R_1 + R_2}\right)$$

This is often called the *current divider equation.*

Equations (1-2) and (1-3) can be solved for I_1 and I_2 respectively:

$$I_1 = \frac{V_a}{R_1} \tag{1-4}$$

$$I_2 = \frac{V_a}{R_2} \tag{1-5}$$

However, there is a complication with regard to writing an expression for I. This complication is that V_a is not the voltage across R; the voltage across R is $V - V_a$. I, then, must be expressed as:

$$\frac{V - V_a}{R} = I \tag{1-6}$$

Substituting equations (1-4), (1-5), and (1-6) into equation (1-1) gives

$$\frac{V_a}{R_1} + \frac{V_a}{R_2} = \frac{V - V_a}{R}$$

from which V_a can be determined if V, R, R_1, and R_2 are known. The values of I, I_1, and I_2 can be calculated from equations (1-6), (1-4), and (1-5) respectively, and the performance of the circuit shown in Figure 1-9 is completely known.

IDEAL AND REAL DC POWER SOURCES

In DC circuits the sources are batteries. A new battery consists of an *electromotive force (emf)* in series with a very low resistance; for our purposes, emf can be expressed in volts. A new 1.5 volt dry-cell battery, for example, consists of an emf of approximately 1.55 volts in series with a resistance of approximately 0.01 ohm. As this dry-cell battery ages, its emf remains approximately the same, but its equivalent series resistance increases and may become as high as 100 ohms or more. (Of course, as the battery becomes very old and its chemical reaction ceases, its emf becomes zero.) For this reason, testing a dry-cell battery with a high resistance voltmeter *gives very little information concerning its condition.* The battery must be tested under load. A good test is to measure the voltage across the terminals under no load with a high resistance voltmeter and then measure the terminal voltage when 0.01 amp is being delivered to an external resistance. If the battery terminal voltage decreases appreciably, the battery's internal resistance has increased to the point where it should be replaced.

In solving DC circuit problems, the source is usually separated diagrammatically into its emf and its internal resistance, as shown in Figure 1-10. The terminal volt-

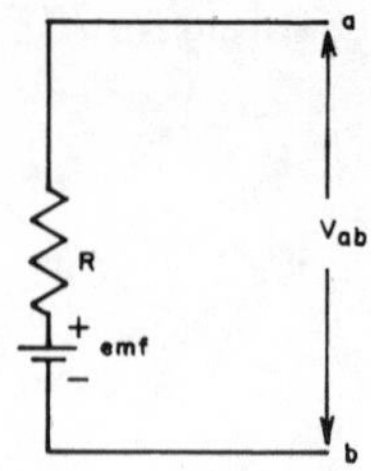

Figure 1-10. A real voltage source.

age V_{ab} is that voltage which is available for the *external* circuit. If *zero current* is being drawn from the circuit, this terminal voltage is equal to the emf. If a current I is being drawn from this source, the terminal voltage is the emf minus the drop $R\,I$ across the internal resistance R. A new battery can usually be considered as an emf with zero internal resistance. As such, it is called an *ideal source.* If the internal resistance is considered, then it is called a *real source.* The difference between "ideal" and "real" batteries should explain why a dry cell which reads the full 1.5 volts on a good voltmeter (which draws very little current) won't light your flashlight bulb (which takes about 300 mA.)

AC POWER

Up to this point we have only discussed DC circuits. In electronics, however, AC circuits are more often used. The reasons for this will become apparent as we move into this topic.

A typical DC voltage-versus-time plot is shown in Figure 1-11A and an AC voltage-time plot is shown in Figure 1-11B for comparison. When dealing with DC signals we use *peak voltage* (V_{peak}), and it is important to realize that it *may* vary with time. With AC we use *root-mean-square* (rms) *voltage,* where $V_{rms} = 0.707\ V_{peak}$ (in this book all AC signals are pure sine waves).

Whenever we talk about AC voltage, current, or power, *we mean rms unless otherwise specified.* You might wonder why this is done. The answer is that when a current is forced through a resistor, power is dissipated: $P = I^2R$. For DC circuits,

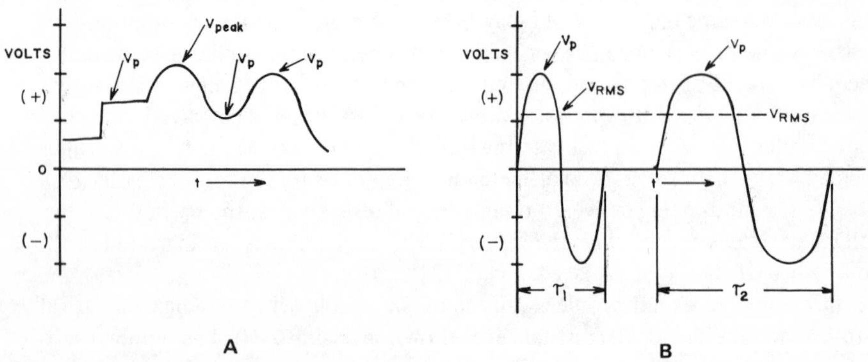

Figure 1-11. *A.* DC voltage. *B.* AC voltage.

there wasn't much doubt about what numbers to put in this equation. We just use the actual current at that instant of time, so for DC the power is $P = (I_{peak})^2R$. For AC circuits, we want to use the same formula ($P = I^2R$) and get the same result whether I is AC or DC amperage. To do this, we use I (root-mean-square) if it is AC and I (peak) if it is DC.

How do we define rms? Well, the AC signal shown in Figure 1-11 is described by a formula of the form

$$V = A \sin 2\pi f t$$

Here f is frequency in cycles per second or hertz, t is time in seconds, and π and A are constants. Given V, we define V_{rms} as

$$V_{rms} = \left[\frac{1}{\tau} \int_0^{\tau} V^2 \, dt \right]^{\frac{1}{2}} = 0.707 \; V_{peak}$$

where τ is the period of time that we integrate over. I_{rms} is defined in a similar manner, *but you don't really have to learn integral calculus to use these definitions.* For now, remember that if we talk about AC voltages and currents, we *always* mean *rms.* If we talk about DC voltages and currents, they are always *peak* unless defined otherwise.

FILTERS (HOW TO SEPARATE SIGNALS OF DIFFERENT FREQUENCIES)

Having defined the basic AC terminology, we can do something useful almost immediately and build some filters. In the first part of this section, we will cheat a little bit by leaving out some complex and confusing material in order to provide you with the *concept* of a filter and an idea of why filters act the way they do. (The real EEs who read this book will be screaming, "No! The numbers won't come out right!") If you build filters from the formulas we give, the circuits will *work,* but the numerical values won't quite be correct. We feel, though, that it is more important right now for you to build a filter that *works,* however badly, than to be buried and turned off by a "correct" mathematical analysis. The "correct" analysis is given in a later part of this section on page 24. You math freaks who dig complex numbers can start reading that.

Basic Filter Design

If we get a little sloppy (in the semantic sense), we can define the resistance of a resistor as a *reactance,* X_R. Note that X_R is still expressed in ohms, and that for a resistor, it does *not* vary with frequency, at least not at the frequencies we will be dealing with.

Now let's introduce two other devices: *capacitors* (which are measured in farads) and *inductors* (which are measured in henrys). The number of *farads* (or microfarads) is denoted by *C* and the number of *henrys* (or millihenrys) by *L;* the abbreviations for farads and henrys are "F" and "H," respectively. The interesting thing about these devices is that their reactance *is* a function of frequency, *f,* as shown in the formulas below. For inductors,

$$X_L = 2\pi fL$$

For capacitors,

$$X_C = \frac{1}{2\pi fC}$$

These devices can be used to design circuits that stop certain frequencies and allow signals of other frequencies to pass. A circuit that passes low freqnencies and stops high frequencies is called a *low-pass filter.* Conversely, a filter that passes high frequencies and stops low ones is called a *high-pass filter.* Since you will be needing circuits of these types, we will teach you how to design simple but useful filters as the next step.

Let's assume that you want to measure some DC signal but the measurement is made difficult by some 1000 Hz electrical "noise"* from a local industrial plant. You need a *low-pass filter* that will pass the DC signal and stop the 1 kHz noise. The 1 kHz signal looks like a sine wave of the form $A \sin 2\pi f_1 t$, where $f_1 = 1000$ Hz. The DC signal looks like a constant of value *B,* so the *total* signal looks like $V = A \sin 2\pi f_1 t + B$. This is shown in Figure 1-12, where we picked $A = 1$, $B = 3$.

To design our filter, we assume that we want to pass at least 90% of all signals having a frequency of 10 Hz or less and that we want to stop at least 90% of all signals at the noise frequency of 1 kHz. A simple filter circuit to perform this function is shown in Figure 1-13 – an approximate analysis follows.

In the circuit of Figure 1-13, the signal input voltage *V* is applied across the series combination of R and C. The reactance of R is just X_R, and the reactance of C is $X_C = \frac{1}{2\pi fC}$. The current *I* is determined by the total reactance of R plus C:

$$I = \frac{V}{X_R + X_C}$$

For our filter, we want to pass a 10 Hz signal with only 10% loss (there has to be some loss) to the output. The 1000 Hz signal must lose 90% of its value before it

**Electrical noise* is simply an unwanted messy signal that makes our desired signal come out messy, too; it's like static on your radio.

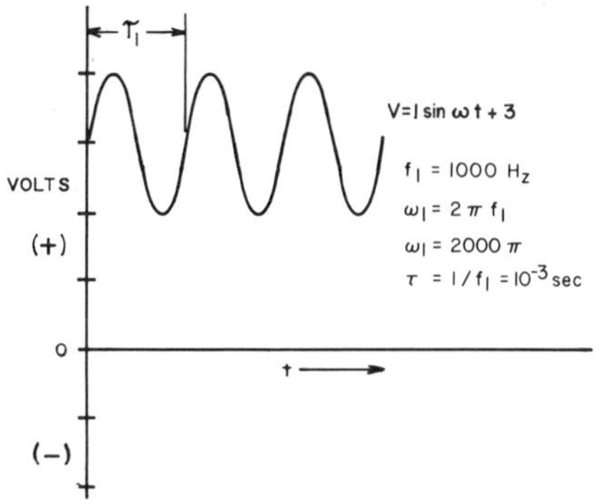

Figure 1-12. AC-plus-DC voltage versus time.

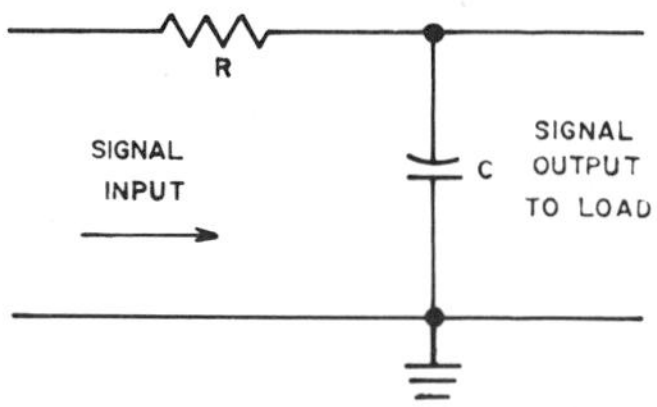

Figure 1-13. Low-pass filter.

gets to the output. This means that at 10 Hz, the voltage across the capacitor, C, should be 90% of the input value, i.e.,

$$\frac{V_C}{V} = 0.9$$

and at 1000 Hz, this voltage must be

$$\frac{V_C}{V} = 0.1$$

The output signal (if we ignore any current going to the load, which is the usual assumption in designing this type of filter) is the current times the reactance of C. (Notice that we can use Ohm's law and treat capacitors and inductors just like resistors as long as we use the reactance concept.)

$$V_C \text{ (output)} = I\,X_C = \frac{V X_C}{X_R + X_C}$$

Rewriting the above and recalling that $X_C = 1/(2\pi fC)$, we obtain

$$\frac{V_C}{V} = \frac{1}{2\pi f RC + 1}$$

Using the values we assumed ($f = 10$ Hz, $V_C/V = 0.9$), then

$$\frac{1}{20\pi\, RC + 1} = 0.9$$

So $18\pi RC = 0.1$, or $RC = 0.006/\pi$. At 1000 Hz,

$$\frac{V_C}{V} = \frac{1}{12 + 1} = 0.08$$

which more than satisfies our requirement that $V_C/V = 0.1$ at 1000 Hz. Now to pick R and C. Any RC value will work, but the bigger the R, the bigger our $I^2 R$ losses. If we pick $R = 100\ \text{k}\Omega$, then

$$C = \frac{0.006}{10^5 \pi} = 1.59 \times 10^{-8}\ \text{F} = 16\ \text{n F}$$

(There is one possible point of confusion here: when we say "the bigger the R is, the bigger the $I^2 R$ losses," we have a special situation in mind. Suppose you want to pass a given current I_L to the load. Then the loss $I_L{}^2 R$ will go up with R, assuming that the applied voltage is big enough to keep I_L constant.)

At this point we might note that it is *not* possible to write

$$0.9 = \frac{1}{20\pi\, RC + 1} \quad \text{and} \quad 0.1 = \frac{1}{200\pi\, RC + 1}$$

and solve the equations simultaneously, or the answer is "over-determined." You have to pick RC to satisfy *one* equation and check that it *more* than satisfies the other equation. In this case the equations $18\pi\, RC + 0.9 = 1$ and $200\pi\, RC + 0.1 = 1$ require that

$$RC \leqslant \frac{0.1}{18\pi} = 1.77 \times 10^{-3}$$

and

$$RC \geqslant \frac{0.9}{200\pi} = 1.43 \times 10^{-3}$$

so

$$1.43 \times 10^{-3} \leqslant RC \leqslant 1.77 \times 10^{-3}$$

Turning back to the design of filters, suppose we interchanged the capacitor and resistor in the circuit shown in Figure 1-13. The current to ground would still be

$$I = \frac{V}{X_R + X_C}$$

But the output voltage would be taken across the resistor and we can assume $X_R = R$, so

$$V_R = IR = \frac{VR}{R + X_C}$$

or

$$\frac{V_R}{V} = \frac{2\pi fCR}{2\pi fCR + 1}$$

Now as f approaches zero, V_R/V approaches zero. However, as f approaches infinity, V_R/V approaches unity. This is a *high-pass filter* and again our analysis is only an approximation.

[At this point, we do have to sound a warning note because of a problem we have seen students run into when they try to design a filter network for a hi-fi system. A typical 20 watt amplifier driving a 4 ohm loudspeaker will deliver some 2.24 amps; if you don't believe it, try the formula $W = I^2R$. If you design a filter with 100 ohms of resistance into the circuit, the voltage drop (if the amplifier could produce it) would be 224 volts. Since the usual output voltage is about 10 volts from an amplifier of this type, you don't want to design high-value resistors into the circuit. Inductor capacitor systems waste far less power than systems using resistors. Another point to remember – we never quit – is that the resistance heating of the resistors controls the wattage rating. A 100 ohm resistor carrying 2.24 amps will liberate 500 watts of heat and must be designed accordingly.]

We suggest that you stop right here and build a high-pass and a low-pass filter. To test them, you can skip to the sections on signal generators and oscilloscopes. Hook a signal generator to the input, the oscilloscope to the output, and go to it. If you are chicken, keep reading and do the experiment when you have finished the chapter. You are the best judge of how and when to learn.

You can build even better filters by replacing the resistor shown in the circuit of Figure 1-13 with an *inductor.* You can see why by recalling that $X_L = 2\pi fL$, where L is in henrys. To pick the right inductor, you can replace R with $X_L = 2\pi fL$ in our formula

$$V_C = \frac{V X_C}{R + X_C}$$

and calculate it from there.

We should warn you that inductors are more expensive than resistors, and they are seldom sold with large power-handling capability. You can wind them yourself by winding 100 turns of #28 insulated wire on an iron nail. Use the inductor in a filter with the nail *left in;* then pull the nail out. You should see a drop in L unless you are using a very high frequency.

To conclude our discussion of filters, we have to introduce two new concepts: *roll-off* and the *Bode plot.* Roll-off is a measure of the ability of a filter to separate the signal frequency you want from the noise signal you don't want. The faster the roll-off, the better you can separate two signals that differ in frequency by only a small number of cycles. A typical roll-off curve is shown in the Bode plot of Figure 1-14. The ratio of output to input voltage is some value V_{out}/V_{in}, The corresponding decibel (dB) value is

$$\text{dB} = 20 \log_{10}(V_{out}/V_{in})$$

Decibels are a convenient way of expressing large number ratios, e.g., $10^5 = 100$ dB, $10^4 = 80$ dB, and so on. However, the decibel is not just a way of handling large ranges of numbers. It has a basic relationship to the way the human eye and ear respond to changes in signal level. If an observer is asked to listen to two signals (of the same frequency) whose intensity differs by a factor of 2, we can expect the observer to say that one is louder than the other by a factor of 0.3, which is $\log_{10}(2)$. This is not an exact relationship, but it does demonstrate that the response of the ear is more logarithmic than linear.

Another application of the decibel concept is in characterizing the slope of the V_{out}/V_{in} versus frequency curve. Assume that we are testing a low-pass filter at some frequency f_1 above its cut-off value. We go to some higher frequency $2f_1$ (one octave), and the ratio of the output voltages at the two frequencies defines the decibel per octave value:

$$\text{dB/octave} = 20 \log_{10} \, [V_o(f_2)/V_o(f_1)]$$

It happens that most filters roll off in multiples of 6 decibels per octave (6, 12, 18, . . . dB/octave). Due to this general characteristic, filters are referred to as first, second, third, . . . order filters. The higher the order, the more nearly the perfect filter is approached. However, more passive elements (resistors and capacitors) are needed to realize higher order filters. In practice, orders higher than 6 or 7 are rarely

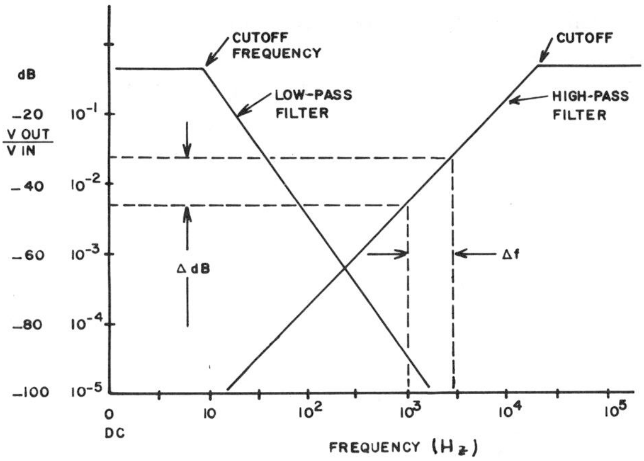

Figure 1-14. Bode plot of filter characteristics.

seen. Actually, first-order filters are often found useful, and second-order filters are used extensively.

In Figure 1-14, output versus input filter characteristics are plotted against frequency in both decibels and powers of 10. The high-pass filter rolls off about 6 decibels per octave below cut-off.

Figure 1-15 shows circuits for various types of filters. Those adventurous souls who want to build a 60 Hz slot filter can follow the diagram given in Figure 1-15C. The combination of a high-pass and a low-pass filter produces a *band-pass* filter; this is shown in Figure 1-15A. You have to pick the values of *L*, *R*, and *C* for the fre-

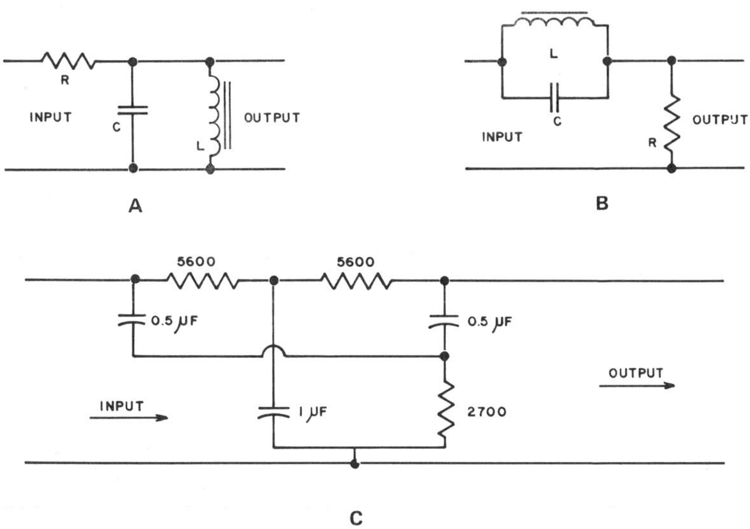

Figure 1-15. Filter circuits. *A*. Band-pass, second order. *B*. Band-attenuation, second order. *C*. 60 Hz slot (hum reduction) filter.

quency bands that you want to stop or pass, which you can do by using the formulas we have given you.

In case you get more deeply into filter design, the best "simple" reference we know of is by R. P. Sallen and E. L. Key, *A Practical Method of Designing R. C. Active Filters.* Another excellent reference is J. L. Hilburn's *Manual of Active Filter Design.* (See the Bibliography for the publisher.)

In the next part of this section, we will give the "correct" mathematical analysis of filter circuits. If you don't need it now, feel free to skip it and start reading about phase shift. We are throwing out this sop to those purists who would object to our rough but handy formulations given above. Be warned: we won't always be so generous in the pages to follow.

Mathematical Analysis of Filter Circuits

The correct analysis of a filter circuit requires that we define

$$A e^{i\theta} = A(\cos\theta + i \sin\theta)$$

where $i = \sqrt{-1}$ and $e = 2.7183$. (EEs use lower case "j" instead of the lower case "i" that physicists use. Be warned!) In this system of notation, numbers may have a real part as well as an imaginary part indicated by the "i" factor. Such numbers are called *complex numbers;* the rules for using complex numbers are given in most algebra textbooks. (If you want a correct analysis, we will show no mercy.)

The general technique for filter analysis involves recognizing that the presence of reactive components induces phase shifts (see next section) so that the voltage peaks for the various circuit elements occur at different times. To take account of this problem, we use the X_C and X_L notation where now

$$X_C = \frac{1}{i\omega C}$$

$$X_L = i\omega L$$

and $2\pi f = \omega$. To handle these complex numbers, we use exponential notation to write $Z = A + iB$ as

$$Z = M e^{i\theta}$$

where

$$M = \sqrt{A^2 + B^2}$$

$$\theta = \arctan \frac{B}{A}$$

The impedance of a low-pass resistor–capacitor network (Figure 1-13) is

$$Z = R + X_C = R + \frac{1}{i\,\omega\,C} = R - \frac{1}{\omega\,C}$$

or

$$Z = \left[\sqrt{R^2 + \left(\frac{1}{\omega\,C}\right)^2}\,\right] e^{i\,\arctan\,(-1/\omega CR)}$$

To apply this to a low-pass filter, we write the input voltage, V, in exponential notation as Ve^{i0}. In this notation, since cos 0 = 1, and sin 0 = 0

$$Ve^{i0} = V(\cos 0 + i \sin 0) = V$$

The current, I, is

$$I = \frac{V}{Z} = \left[\frac{V}{\sqrt{R^2 + \left(\frac{1}{\omega\,C}\right)^2}}\right] e^{i\theta}$$

If X_C is written as

$$X_C = \frac{1}{\omega\,C}\,e^{i(-\pi/2)}$$

then the input-output voltage ratio is

$$\frac{V_C}{V} = \left[\frac{1}{\sqrt{(R\,\omega\,C)^2 + 1}}\right] e^{i\,[-\pi/2\,+\,\arctan\,(1/\omega R C)]}$$

This is the correct input-output ratio for a low-pass filter. Note that as $2\,\pi\,f = \omega$ increases, the ratio V_C/V goes to zero. As $2\,\pi\,f = \omega$ goes to zero (DC), the voltage ratio goes to one.

The high-pass *RC* filter formula is obtained the same way. The input-output voltage ratio is

$$\frac{V_R}{V} = \frac{R\,\omega\,C}{\sqrt{(R\,\omega\,C)^2 + 1}}\,e^{i\theta}$$

where $\theta = \arctan - 1/(\omega RC)$. Here, as ω goes to infinity, the ratio V_R/V goes to one; as ω goes to zero, the voltage ratio goes to zero. This is why it's a high-pass filter.

As a final example, consider a resistor, capacitor, and inductor in series. Now,

$$Z = R + i\left(\omega L - \frac{1}{\omega C}\right) = M e^{i\theta_1}$$

$$M = \sqrt{R^2 + \left(\omega L - \frac{1}{\omega C}\right)^2}$$

and

$$\theta_1 = \arctan \frac{\left(\omega L - \frac{1}{\omega C}\right)}{R}$$

If the applied voltage is Ve^{i0}, the current is $I = V/Z$. The voltage across the various elements of the circuit is I times the impedance of that element. For the resistor,

$$V_R = \left(\frac{VR}{M}\right) e^{-i\theta_1}$$

For the capacitor,

$$V_C = \left(\frac{V}{M \omega C}\right) e^{i[-(\pi/2) - \theta_1]}$$

And for the inductor,

$$V_L = \left(\frac{V \omega L}{M}\right) e^{i[(\pi/2) - \theta_1]}$$

At resonance, $\omega_1^2 = 1/(LC)$ or $f_1 = \dfrac{1}{2\pi\sqrt{LC}}$, $M = R$, and $\theta_1 = 0$. In this case, then, the expressions above take on a simple form:

$$V_R = V$$

$$V_C = \frac{-V}{R \omega C} e^{-\frac{i\pi}{2}}$$

$$V_L = \frac{V \omega L}{R} e^{\frac{i\pi}{2}}$$

In this condition, the voltage V is *entirely* across the resistor because $V_L + V_C = 0$ when $\omega_1^2 = 1/(LC)$. This is the condition for resonance, and at resonance the resistance of the string of elements is just R. It is interesting to note that at resonance, the voltages V_L and V_C are by no means zero. If you measure them with a good AC voltmeter or an oscilloscope, you will find that they can be quite large.

The *quality* of the circuit as a filter, also called its *Q factor,* is defined by

$$Q = \frac{\omega_1 L}{R} = \frac{1}{\omega_1 R C}$$

The greater the voltage across L and C at resonance, the better the "quality" of the device. Don't forget that the voltages V_C and V_L are, at resonance, equal but opposite in sign, so their sum is zero.

The parallel *RLC* circuit shown in the next section in Figure 1-16 has an impedance Z given by

$$\frac{1}{Z} = \frac{1}{R} + \frac{1}{i \omega L} + i \omega C$$

or

$$Z = \frac{i \omega L R}{R (1 + -\omega^2 L C) + i \omega L}$$

At resonance, Z equals R as before. When ω doesn't equal $1/\sqrt{LC}$, the value of Z will decrease. This makes a parallel circuit useful for picking a particular signal out of a mixture of signals at other frequencies. Measuring the voltage across this *RLC* network will show that the voltage is greatest when Z has its largest value (assuming that the current is constant). To pick out a frequency $f_1 = \omega_1$, we set $\omega_1 = 1/\sqrt{LC}$; by doing so, the voltage across the network will be greatest for signals of frequency f_1. For all other signals, the value of Z will be smaller and the voltage across the network, $V = IZ$, will be smaller (still assuming that I is constant). We call it *tuning the radio* – more discussion on that point later.

PHASE SHIFT

Now let's discuss a new idea: *phase shift.* To understand this concept, think of a power supply driving a resistor, capacitor, and inductor in parallel as shown in Figure 1-16.

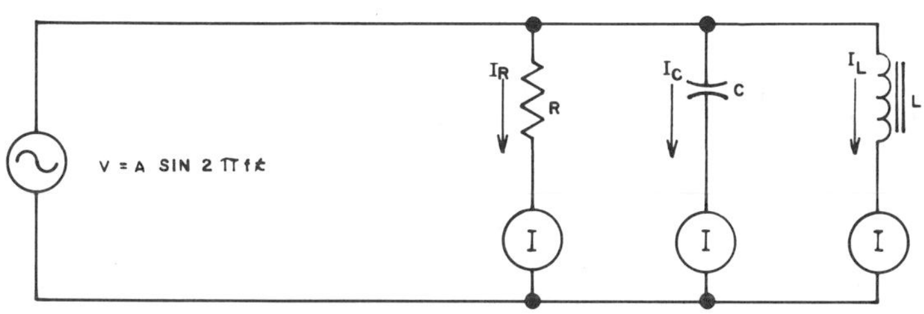

Figure 1-16. Phase-shift circuit.

The voltage is the same to each of the elements ($V = V_R = V_C = V_L$), but what about the currents I_R, I_L, and I_C? Elementary AC circuit theory tells us that

$$I_R = \frac{V}{X_R} = \frac{V_R}{X_R}$$

$$I_C = \frac{V}{X_C} = \frac{V_C}{X_C}$$

$$I_L = \frac{V}{X_L} = \frac{V_L}{X_L}$$

What we need now are some formulas relating voltage and current in capacitors and inductors. You may have seen them in elementary physics. For those of you who didn't take physics (or have forgotten it by now), we list the formulas below. Note: V is the applied voltage. It is called V_C when applied to a capacitor, V_L when applied to an inductor.

$$V_L = L\left(\frac{dI_L}{dt}\right)$$

$$I_L = \frac{1}{L}\int V_L\, dt$$

$$I_C = C\left(\frac{dV_C}{dt}\right)$$

$$V_C = \frac{1}{C}\int I_C\, dt$$

If you didn't take calculus either, don't worry. All you really need to know is *what phase shift is,* not how to derive the formulas.

If our applied voltage is $V = A \sin (2\pi ft)$, by substitution and integration and differentiation in the respective formulas we obtain

$$I_R = \left(\frac{A}{R}\right) \sin (2\pi ft)$$

$$-I_L = \left(\frac{A}{2\pi fL}\right) \cos (2\pi ft)$$

and

$$I_C = 2\pi f C A \cos (2\pi ft)$$

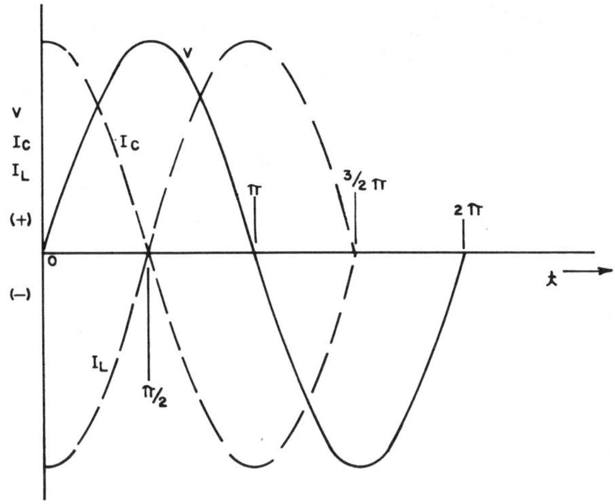

Figure 1-17. Phase-shift relationships.

Both I_L and I_C are cosine functions, but the negative sign on I_L tells us that when $t = 0$, I_C is positive and I_L is negative. This means that I_C and I_L are 180 degrees out of phase with each other and 90 degrees out of phase with V. Since I_C reaches its positive peak before V, we say it *leads* V. Conversely, I_L reaches its positive peak after V, so we say that I_L *lags behind* V. These relationships are illustrated in Figure 1-17. The current through the resistor is in phase with the applied voltage and is not shown in Figure 1-17.

Now you can impress (or bore) your friends by talking about phase shift in capacitors. To be even more esoteric, we can note that a sine wave repeats itself endlessly with a frequency f. This means that the period or time between cycles is $T = 1/f$. A point going around a circle arrives where it started after 360 degrees, so $T = 360° = 2\pi$ radians. This means that the phase shift between I_C and V_C is $T/4 = 90° = \pi/2$ radians.

INDUCTORS AND TRANSFORMERS

Inductors have interesting properties, one of which is that two inductors can produce a device called a *transformer*. What is a transformer? A transformer is really a pair of coupled inductors that we can use for all sorts of interesting projects.

If you apply a voltage to a coil of wire, you pass a current through that coil. This current generates a magnetic field, which in turn generates another voltage in the coil. This generated voltage, the back-emf, opposes the initial voltage. This is an instance of Lenz's Law, which in turn is an instance of the more general law: *Nature always fights back.*

It follows (take our word for it) that the *more rapidly* you try to change the current through an inductor, the bigger the back-emf. This is why the reactance ($X_L = 2\pi fL$) of an inductor *increases with frequency.*

A current passing through a coil generates a magnetic field. If this magnetic field intercepts another coil, a voltage will be generated in the second coil. This combination of two coils is a *transformer.* The symbol for a transformer is two coils back to back (see Figure 1-2).

In the world of transformers we refer to the coil that you apply the voltage to as the primary coil. The magnetic field generated by this primary coil interacts with the nearby secondary coil and a voltage is generated in the secondary coil. Interestingly enough, this process only works if the primary current is either AC or pulsating DC. You can apply a constant voltage DC signal to a transformer primary coil, and a current will flow in the primary coil, but no signal will appear in the secondary coil. Normally we think of a transformer as an AC device used to raise or lower a voltage (more on that point later).

While we are on the subject of transformers we should remind you that they are inductances having rather low DC resistance and an AC resistance (or reactance) given by the familiar law

$$X_L = 2\pi fL$$

We can put some numbers in this equation by thinking of a 1 henry inductor to which we apply a signal of 1 volt AC at 60 Hz. Using the above formula the current is about 2.7 mA (0.0027A). If we tried a DC signal the only thing limiting the current would be the wire resistance (about 0.0001 ohm) and something would melt rather quickly. This paper experiment might help you remember the difference between AC and DC in an inductor – now back to the transformer.

A transformer, then, is a device to change one AC voltage to a higher or lower voltage. Transformers are useful gadgets but *THEY CAN KILL YOU* (or at least provide you a shocking experience) if you are careless. The laws of the transformer are given below. N_{pri}, N_{sec} are the number of wire turns in the primary and secondary.

$$\frac{V_{sec}}{V_{pri}} = \frac{N_{sec}}{N_{pri}}$$

$$I_{pri} V_{pri} = I_{sec} V_{sec}$$

where all voltages and currents are rms. If you use a transformer to get high secondary voltage, then $(N_{sec}/N_{pri}) \gg 1$, and $(I_{sec}/I_{pri}) \ll 1$. You get out *only* what you put in.

Transformers with many taps are available, but center-tapped transformers are most often used. Their voltage system is shown in Figure 1-18.

The autotransformer (Variac) is a useful gadget. It can give you any voltage from zero to full line voltage, but you had better know how it works. (There are old EEs and careless EEs, but no old, careless EEs!) An autotransformer system is shown in Figure 1-19. You should note that when the male plug is in with orientation $A-A$ and $B-B$, the output lead (C) is at 110 volts *regardless of the dial setting*

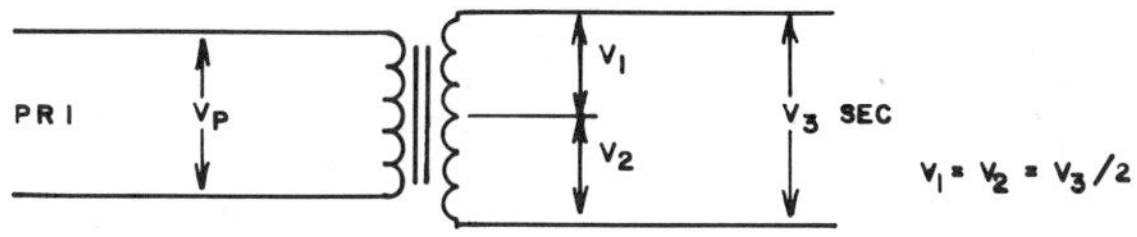

Figure 1-18. Center-tapped transformer.

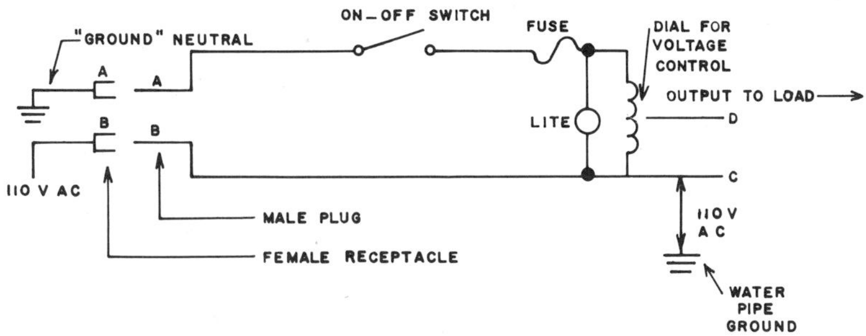

Figure 1-19. Autotransformer (Variac).

or the on/off switch. If you turn the plug over so it's $A-B$ and $A-B$, things are okay; output C is "ground" neutral, and D is whatever voltage you set on the dial. The point here is that an autotransformer *does not* isolate you from ground. A regular transformer *does* provide isolation.

In Figure 1-20 we show a typical application where a transformer is used as a voltage step-down device for a DC power supply. You might want to learn how to build a DC battery charger, and that leads us to our next gadget: the diode. (See how we can lead you on from one thing to another?)

DIODES

Diodes are devices with unidirectional current-carrying characteristics. An *ideal* diode would carry current in one direction with no voltage drop, i.e., with zero voltage across its terminals. An attempt to pass current in the opposite direction

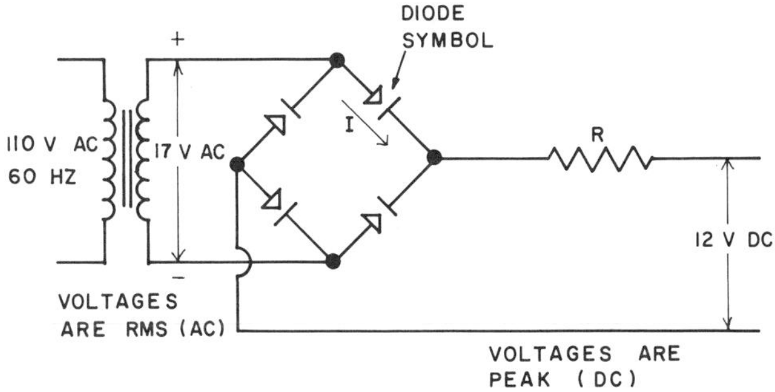

Figure 1-20. Twelve-volt DC power supply.

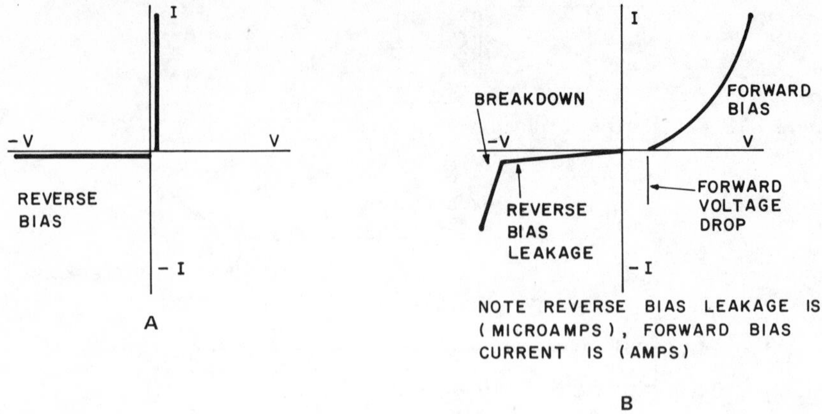

Figure 1-21. Diode characteristics. *A*. Ideal. *B*. Real.

would be almost completely futile. This ideal characteristic is shown in the *V* versus *I* curve in Figure 1-21A. The *V-I* curve of a real diode is shown in Figure 1-21B.

The current direction shown is for positive current. Electron or "negative" current is in the opposite direction. The most common use of diodes is to convert AC voltage to DC, i.e., to act as *rectifiers.* Examples of diode rectifiers are shown in Figures 1-22 and 1-23.

As you can see from the waveforms shown, the output of straight diode rectifiers is not exactly a constant voltage. It is sometimes called *pulsating DC.* With a large capacitor and a transformer, we can obtain a usable DC power supply.

Before you run to your local radio shop to buy parts for a DC power supply, we need to mention the concept of diode *peak inverse voltage* (PIV) and *current rating.* Diodes are rated simply in terms of current carrying capacity (amps) and the reverse voltage that can be safely applied (peak inverse voltage or PIV). If you want to rectify 15 volts at 10 amps, you should purchase a 20 volt PIV diode with a 12 amp current rating. (A 20% safety factor is part of *every* good electronic circuit design; we know that 20% of 15 is 3, but a 20 volt PIV diode may be easier to get than an 18 volt type.)

Suppose you want to put your new knowledge to work and build a 12 volt,

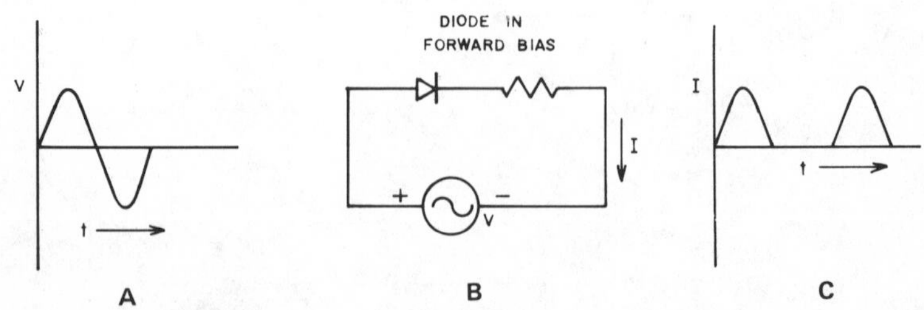

Figure 1-22. Diode operation. *A*. Input signal. *B*. Diode circuit. *C*. Current waveform.

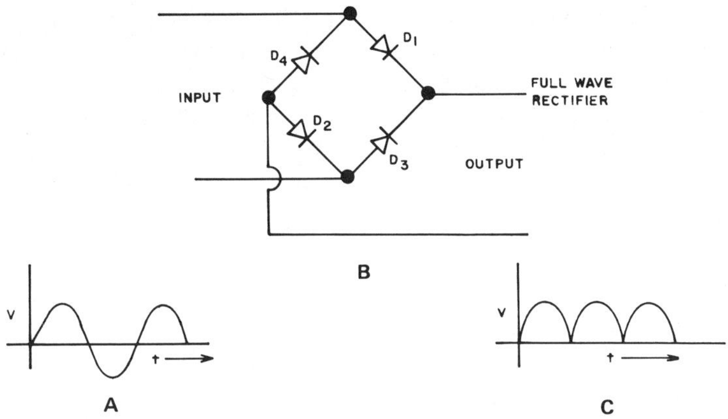

Figure 1-23. Diode rectifier bridge circuit. *A*. Input waveform. *B*. Full wave rectifier. *C*. Output waveform.

1 amp battery charger. We look back at Figure 1-20 and choose a 110 volt input, 12 volt output (1 amp) transformer. An Allied 6K94HF (1980 catalog) has a rating of 12.6 volts at 1.2 amps for about $7.70. Remember these are rms voltages: 12.6 volts rms is 17.8 volts peak.

Now we pick a diode, or, if we want to use a bridge rectifier, four diodes. From Allied's catalog, VS148 diodes are rated at 70 volts PIV at 2 amps (for $1.93 each, why look further?). If we put four of them into a bridge, our 17.8 volts peak drops to about 16 volts due to diode resistance. This diode bridge gives us 16 volts DC peak, but suppose we want to charge the battery at 13 volts peak with a 1 amp current. We then add a resistor in series with the diode bridge. Since $V = IR$, if $V = 3$ volts and $I = 1$ amp, $R = 3$ ohms. At a current of 1 amp, that's 3 watts. Last, but not least, we add a 1.2 amp slow-blow fuse to prevent short-circuits from destroying the diodes, and the job is done. The complete circuit and the charging cycle are shown in Figure 1-24.

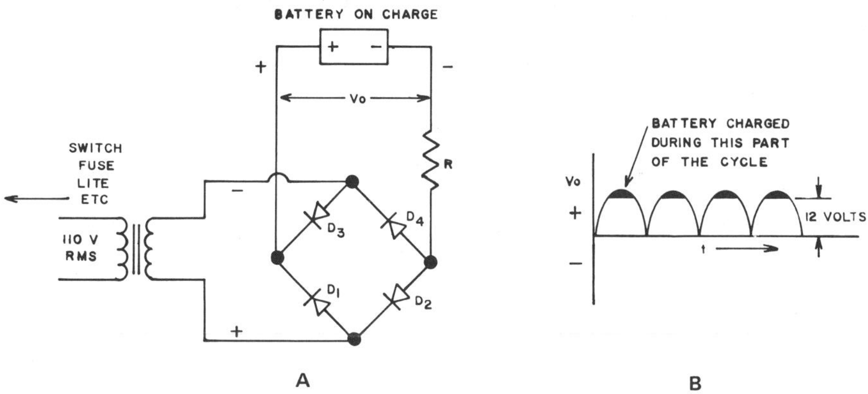

Figure 1-24. Battery charger. *A*. Circuit. *B*. Output voltage.

The circuit is completed, but now do you really know how the rectifier bridge works? As the transformer output swings (+) and (-) as shown, diodes D_1 and D_4 carry the current. Diodes D_2 and D_3 are *reverse biased;* therefore they are carrying no current. (You can pretend they aren't there). On the other half-cycle, the situation is the same except that D_2 and D_3 conduct, while D_1 and D_4 are reverse biased. The important thing to note is that in *both cases* the current flows in the *same direction in the load,* i.e., it is DC.

NEON BULBS AS PILOT LIGHTS

This topic should really come in Chapter 7, but since you will want to build circuits right away, we won't delay you. You must have seen the red glow of neon bulb pilot lights on all sorts of apparatus. They are low in cost, last a long time, and take very little power. The only problem is that they don't limit their current: if the current goes up, their resistance goes down, so the current goes up more and zap – it's all over.

To prevent this, we put a resistor in series with the neon bulb *before* we put it across the 110 volt line. At the instant the voltage comes on, the voltage drop across the resistor is zero (because the current is zero) so the full 110 volts is across the neon bulb and it lights. But then current flows and a voltage drop appears across the resistor. The current, however, is limited to a value that doesn't destroy the neon bulb. Neat, isn't it?

What size resistor is best? A good starting value is 60 kΩ, but note that the *lower* the resistor value, the *brighter* the light *and* the less the time until it burns out. You can use neon bulbs for all sorts of clever timing, switching, and counting circuits. Go to your local radio shop and buy a copy of the *General Electric Glow Lamp Manual* for about $1.00. Sometimes G.E. even gives away manuals to educational types; try writing to their Lamp Department and see what happens.

FUSES: ELECTRICITY'S SAFETY VALVE

Buying small electronic equipment fuses can be the most irritating business in the world. There are about 20 lengths, three diameters, and God knows how many voltage ratings, to say nothing of fast-blow types, slow-blow types, and so on. It can make you want to blow your brains out.

We have found it best to stick to type 3AG in standard and slow-blow styles. Buy holders for them and stock them in assorted sizes. Newark Electronics sells an assortment in a rack called an "Industrial Fuse Caddy" for about $16.95 and it is worth it. When you install fuses, use the standard type except for cases in which a heavy starting current is drawn. In that case, use slow-blow fuses.

VOLTMETERS AND AMMETERS

Use of Voltmeters

For use in the laboratory or in electric circuits, a voltmeter can be thought of as an ammeter with a series resistance and a scale pointer system that indicates the amount of current flowing. Since the ammeter resistance is low, the current flow is controlled by the series resistance. Since we know the resistor value and can measure the current flow from the scale of the ammeter, we can calculate the applied voltage. In many cases the scale will be marked directly in volts. There are some types of voltmeters whose resistance is so high that in ordinary work they do not disturb the circuit being measured. They are vacuum-tube voltmeters, electrometers, or electrostatic voltmeters and are used much like the conventional units.

Voltmeters are usually marked in ohms per volt, and *if* you know the secret code, you can figure out the total effective voltmeter resistance from the ohms-per-volt rating and the meter scale. The full-scale voltage (V_{fs}) across the meter is the voltage drop across the meter resistance R_M *plus* the drop across any resistor R_S in series with the meter. We can write this as $V_{fs} = I_{fs}\,(R_M + R_S)$. We must note that I_{fs} is a constant of the *meter coil itself* whereas V_{fs} and R_S are at our choice. The ratio

$$\frac{R_S + R_M}{V_{fs}} = \frac{1}{I_{fs}}$$

is the ohms-per-volt constant for that particular meter.

Given I_{fs}, the actual resistance of the voltmeter circuit is therefore a function of the full-scale voltage range of the meter. For example, if $1/I_{fs}$ is 20 kΩ/volt and the V_{fs} of the meter is 1 volt, then $R_S + R_M = 20$ kΩ. If V_{fs} were 100 volts, $R_M + R_S$ would equal 2×10^6 or 2 MΩ. In each case, the sum $R_M + R_S$ is equal to the ohms-per-volt rating times the full-scale voltage reading. Note that in both cases, the meter current was the same: 50 μA. This follows from the fact that 1 volt divided by 20 kΩ is the same as 100 volts divided by 2 MΩ. We repeat, the ohms-per-volt rating is the reciprocal of I_{fs} in amps, which agrees with the dimensions of Ohm's law. It may be a screwy way to define things, but that's the way it is. Note that you never do find out what R_M is, but in most cases R_S is many times larger than R_M, so it doesn't hurt to assume that $R_S + R_M \approx R_S$.

Realizing that a voltmeter is an "indicating resistance," we can now study the effect of connecting a voltmeter into a circuit to measure the voltage across another resistor. Consider the simple circuit shown in Figure 1-25, which consists of two 20 kΩ resistors connected across a 100 volt power supply or battery. We want to

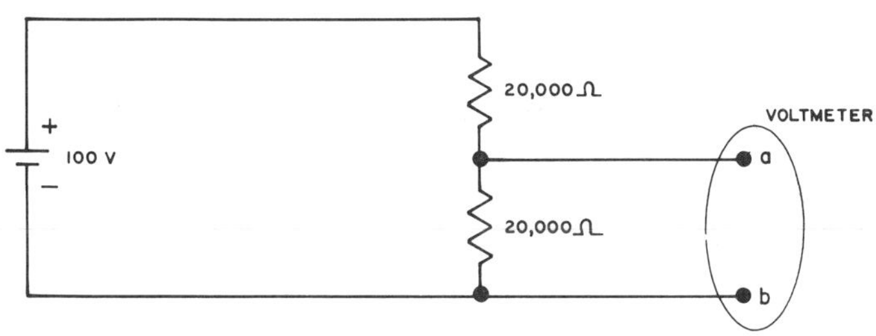

Figure 1-25. Simple series circuit containing a voltmeter: Error induced by voltmeter.

measure the voltage between terminals *a–b*. It is obvious from the voltage division formula developed earlier (p. 8) that it is 50 volts.

Now let's check this by measuring it with a voltmeter. Assume that two voltmeters are available. Voltmeter A has a full-scale reading of 100 volts and a rating of 100 ohms per volt. Voltmeter B has a full-scale reading of 100 volts and a rating of 1000 ohms per volt. Voltmeter A has a resistance of 10 kΩ. When connected across terminals *a–b*, the complete circuit is as shown in Figure 1-26A. (The voltmeter is depicted as a resistance within the circle in the diagram.) The voltage appearing across terminals *a–b* under these conditions can be calculated from the voltage division formula to be

$$V_{ab} = 100 \times \frac{\dfrac{(10\ \text{k}\Omega)(20\ \text{k}\Omega)}{10\ \text{k}\Omega + 20\ \text{k}\Omega}}{20\ \text{k}\Omega + \dfrac{(10\ \text{k}\Omega)(20\ \text{k}\Omega)}{10\ \text{k}\Omega + 20\ \text{k}\Omega}} = 25 \text{ volts}$$

Thus, the voltmeter resistance effectively reduces the measured voltage to *one-half* its previous value.

Now let's use voltmeter B. Its resistance is 100 kΩ, and the circuit with the voltmeter connected is shown in Figure 1-26B. Use of the voltage division formula shows that under these conditions, $V_{ab} = 45.5$ volts.

It is obvious that voltmeter B has not disturbed the circuit as much as voltmeter A, but it has still caused an error of 9.1%.

If a voltmeter of infinite resistance were used, V_{ab} would be measured as 50 volts. A voltmeter with a rating of 20 kΩ per volt and a full-scale reading of 100 volts would have a resistance of 2 MΩ, and its effect when connected across terminals *a–b* in Figure 1-26A would be negligible. On the other hand, if the circuit shown in Figure 1-26A were made up of two 1 MΩ resistors across the power supply, even the 20 kΩ-per-volt voltmeter would disturb the circuit. From this it is seen that *each case must be considered individually to determine the error caused by the voltmeter.*

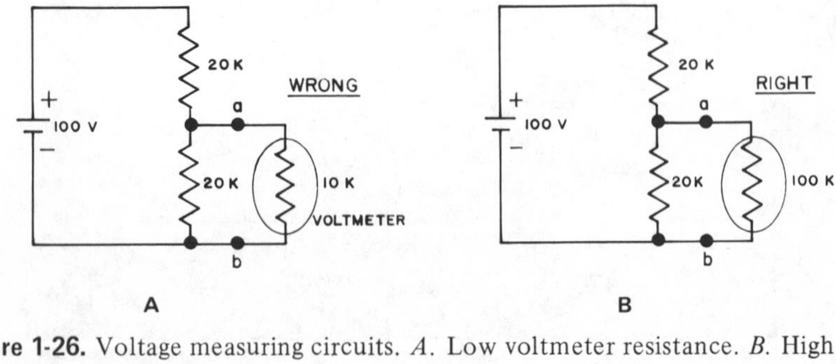

Figure 1-26. Voltage measuring circuits. *A.* Low voltmeter resistance. *B.* High voltmeter resistance.

The *ideal* voltmeter would be one with infinite resistance. Certain types of voltmeters approach this ideal, but they have other disadvantages. The best rule to follow is to use a voltmeter with the highest ohms-per-volt rating available, and then use it intelligently, while appreciating that the voltmeter has a finite internal resistance. Of course, voltmeters with a high ohms-per-volt rating may be more expensive than ones with a low ohms-per-volt rating because of the additional sensitivity required of the meter movement.

VTVM (vacuum-tube) or FET (field effect transistor) voltmeters have resistances of 6 to 20 MΩ (depending on the manufacturer) even on low full-scale ranges. They have the disadvantages, however, of being slightly more expensive and requiring the use of a 115 volt AC power line or batteries. (Recently, the Heath Company has offered a battery operated FET-VOM; it is a winner.)

Use of Ammeters

Ammeters are connected in series with a circuit to measure the current through that circuit. Figure 1-27 shows the simplest possible circuit containing an ammeter.

Again, the ammeter should be thought of as an indicating resistance, this time indicating the current flowing through it. If the ammeter were not present in the circuit of Figure 1-27, the current would be, by Ohm's law, 0.5 amp. When the ammeter is connected into the circuit, the current drops to 0.375 amp. The ammeter has disturbed the circuit appreciably. If the ammeter had a resistance of 0.01 ohm instead of 1 ohm, the current in the circuit – including the ammeter – would be 0.484 amp, so the circuit would not be disturbed so much by the insertion of the ammeter. From this, it can be concluded that an *ideal* ammeter is one with zero resistance (compare this with the infinite resistance that was the case for the ideal voltmeter).

In actual practice the ideal ammeter is approached quite closely, much more closely than is the case for voltmeters. The following list shows the resistance of ammeters of various ratings as advertised by a leading manufacturer:

Full-Scale Rating	*Resistance*
50 amps	0.001 ohm
10 amps	0.005 ohm
1 amp	0.050 ohm
100 milliamps	0.5 ohm
1 milliamp	70.0 ohms
0.1 milliamp	1625.0 ohms

These ratings will vary from one manufacturer to another, of course.

Sometimes it is difficult or impossible to connect an ammeter in the circuit where the current is to be measured. Several companies manufacture a clip-on type of milliammeter, for AC only, which is very convenient to use. Every wire carrying an electric current is surrounded by an electric field. The clip-on milliammeter

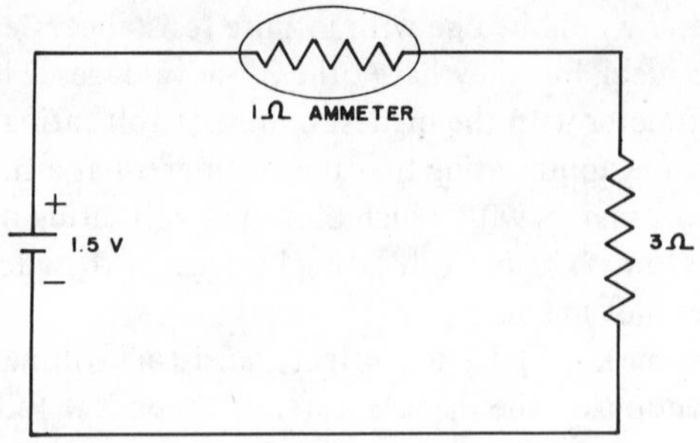

Figure 1-27. Simple circuit containing an ammeter.

senses this field and interprets it electronically in terms of the current flowing in the wire. Clip-on meters in the ampere range are sold by several manufacturers (you can find them in the Allied Electronics Catalog).

Ammeters and Shunts

Any single-range DC ammeter can be made into a multi-range ammeter by a system of "shunts," which are resistors connected in *parallel* across the ammeter. The resistance of these shunts can be calculated very simply from the current division formula. Notice that with ammeters, we put the added resistors in *parallel* with the meter. If you will look back at page 35 where we discussed voltmeters, you will see that in that case we put the resistors in *series* with the meter.

Consider an ammeter with a full-scale reading of I_M amps and a resistance of R_M ohms. We want to change the full-scale reading of this ammeter to I amps by connecting a shunt with a resistance of R_1 across its terminals. The circuit is shown in Figure 1-28. The terminals a–b are now the terminals of the new ammeter with a full-scale reading of I amps. The full-scale current through the original ammeter will still be I_M, and the excess current $I - I_M$ must be shunted around the original ammeter by R_1. By the current division formula (see page 14),

$$I_1 = I\left(\frac{R_M}{R_1 + R_M}\right) \tag{1-7}$$

and by Kirchoff's node law,

$$I = I_1 + I_M$$

or

$$I_1 = I - I_M \tag{1-8}$$

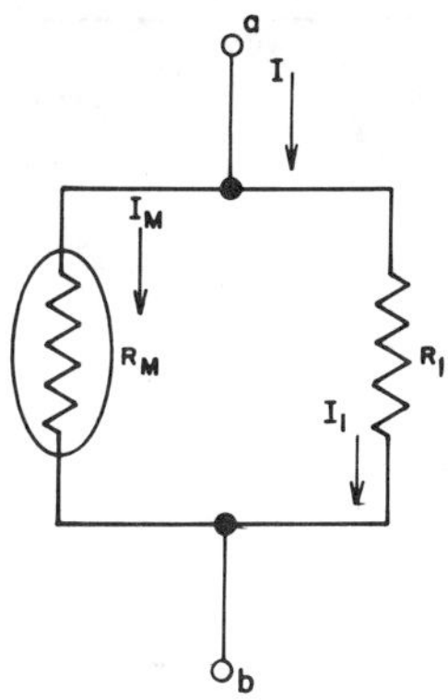

Figure 1-28. Ammeter with shunt.

Substitution of equation (1-8) into (1-7) yields

$$I - I_M = I\left(\frac{R_M}{R_1 + R_M}\right)$$

Solving for the value of the shunting resistor R_1 yields

$$R_1 = \frac{R_M I_M}{I - I_M} \tag{1-9}$$

As an example, assume that an ammeter with a full-scale reading of 1 mA and a resistance of 70 ohms is to be made into an ammeter with a full-scale reading of 100 mA. We are to calculate the value of R_1 required. Using equation (1-9),

$$R_1 = \frac{70 \times 0.001}{0.1 - 0.001} = \frac{0.07}{0.099} = 0.707 \text{ ohm}$$

In this new ammeter, 1 mA will still be flowing through the meter movement at full scale, and 99 mA will be flowing through the shunt R_1. The total current flowing into the combination will be 100 mA.

Multi-range ammeters can be homemade by using a new shunt for each new range required. Such multi-range ammeters are readily available on the market, and any single-range ammeter can be readily converted to a multi-range ammeter.

Voltmeters and Multipliers

Any single-range DC voltmeter can be made into a multi-range voltmeter by a system of resistors connected in *series* with the voltmeter. These series resistors, or

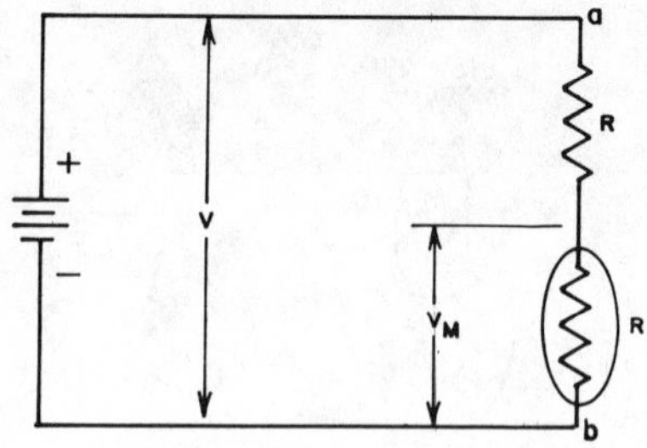

Figure 1-29. Voltmeter with a series resistor.

multipliers, as they are usually called, can be calculated very simply from the voltage division formula.

Consider a DC voltmeter with a full-scale reading of V_M and a resistance of R_M ohms. We want to increase the range of this voltmeter by means of a series resistance so that it will have a full-scale reading of V volts. The complete circuit is shown in Figure 1-29. In effect, R and R_M must divide the voltage V so that there are only V_M volts across R_M.

The required value for R can now be calculated from the voltage division formula as follows:

$$V_M = V\left(\frac{R_M}{R + R_M}\right)$$

Solving this equation for R gives

$$R = \frac{R_M\,(V - V_M)}{V_M}$$

As an example, assume that a voltmeter with a full-scale reading of 100 volts and a resistance of 100 kΩ (1 kΩ ohm per volt) is available, but a voltage of 300 volts is to be measured. A series resistor, R, can be used, and its value is calculated as

$$R = \frac{100{,}000\,(300 - 100)}{100} = 200{,}000 \text{ ohms}$$

The terminals a–b are now the terminals of the new voltmeter with a full-scale reading of 300 volts. Multi-range voltmeters can be made by using a new series resistor for each new range required. Multi-range voltmeters are readily available on the market, and you can buy one from the Heath Company.

The problem with using voltmeters for precision measurements is that they draw current from the circuit being measured. The need for a device that does not draw current brings us to the potentiometer, which will be discussed in the next section.

POTENTIOMETERS: PRINCIPLE OF OPERATION AND SOME APPLICATIONS

The potentiometer is a device for measuring an unknown emf (or voltage) without drawing *any* current from the source being measured. With thermocouple devices, you *must* use a device of this type or the measurement will be useless. In a sense, the potentiometer can be considered as an infinite resistance voltmeter.

Its principle of operation is illustrated in Figure 1-30. The element enclosed in the box with the three terminals *a, b,* and *c* is called a *potentiometer* because the potential or voltage between terminals *b* and *c* can be varied from zero to some maximum value that is set by V_1. Assuming that *emf* is the unknown potential, we adjust the variable resistor until the ammeter, A, reads zero. (You can also use super-sensitive ammeters, called *galvanometers,* for this measurement.) The point is that when the ammeter reads zero, *no* current is being drawn from the unknown emf. In this case, we know that the voltage from point *b* to ground is the same as that from *d* to ground. At this point, we can close the switch shown in Figure 1-30 and measure the voltage from point *b* to ground. The fact that the battery V_1 has enough output capacity so the current drawn by the voltmeter *can be neglected* makes this complex procedure worthwhile.

Before you go on to the next (and more complex) circuit, we suggest that you sit down and think this one through. We have to measure the output of an unknown source without drawing *any* current. We set up a battery and a variable resistor, making sure that the battery we use will have the necessary current capacity to drive a voltmeter. Then we close switch A and balance off some fraction of the battery voltage against the unknown voltage until the two voltages are equal. Now we only need to open switch A and close switch B to measure the voltage across the variable resistor with our voltmeter. Now we know exactly what the output of the unknown source is *without* having drawn any current from it.

In more complex instruments, such as those that you will find in laboratories,

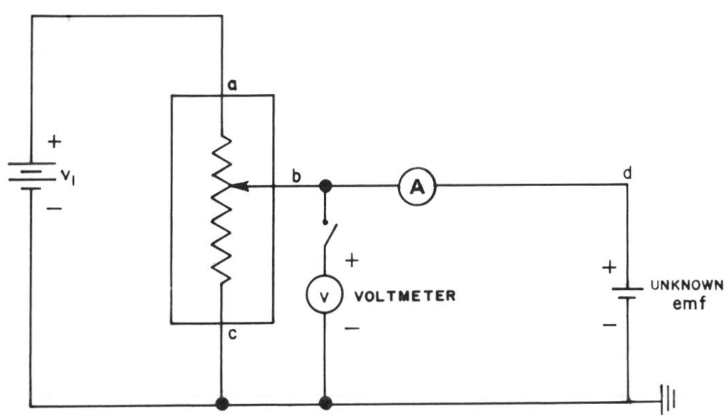

Figure 1-30. Potentiometer: Principle of operation.

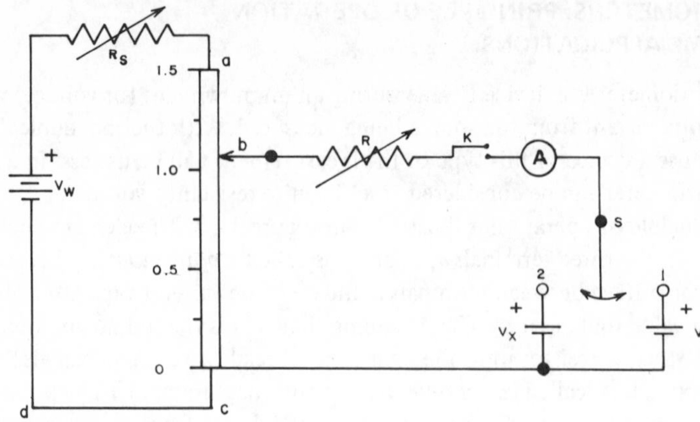

Figure 1-31. Calibrated potentiometer.

the simple voltmeter just isn't good enough. Instead, we use a *standard cell* and a *linear resistance.* A "linear resistance" is a fancy term for a long resistance wire that has been carefully constructed so that its resistance is a function of how long the wire is. (You thought all resistance wires did that; well, this one is accurate to four decimal places.) Turning now to Figure 1-31, we have a battery V_W, a linear resistor (so called because it has a linear scale along the wire), some resistors R and R_S, and a standard cell V_{std}.

In this circuit, the voltage of V_{std} is known to a high degree of accuracy. It is called a *standard cell* or Weston cell and has a voltage of approximately 1.0184 volts (its actual value is usually indicated by a tag attached to the cell). This voltage is extremely stable, but it is accurate only if *no* current is being drawn from the cell. Care should be taken to draw the least possible current from the standard cell during the calibration or standardization operation explained later.

The potentiometer in Figure 1-31 is set up for determining the value of the unknown potential V_x by comparing it with the emf of the standard cell. The first step in the use of the potentiometer is to *standardize* it, that is, to adjust the system in such a way that the mechanical position of the pointer on the slide wire reads directly in volts. This is done by putting the switch S in position 1, setting R (the protection resistor) at its maximum resistance value (usually 10 kΩ), and setting the slider to a reading on the slide wire corresponding to the emf of the standard cell. The key switch K is tapped, and R_S is adjusted so that the ammeter reads zero. Now the protective resistance of R is decreased, and the setting of R_S is refined until the ammeter reads zero with R at zero resistance and the tap key K closed. Note that the use of a series protective resistor protects not only the ammeter by limiting the current, but also protects the standard cell. If a protective resistor in parallel with the ammeter had been used, this would have protected the ammeter by shunting current around it, but it would not have protected the standard cell.

The potentiometer is now standardized, i.e., the physical position of the slide wire is a direct reading of the voltage appearing across terminals *b–c*. Now, provided (1) the working battery V_W maintains its voltage, (2) the value of R_S does not change because of excessive current, and (3) the slide wire has been accurately marked by the manufacturer, the slider can be moved to any other place on the slide wire and the voltage between terminals *b–c* can be read directly from the slide wire. Here we assume, of course, that for each measurement of V_x you go through the sequence of tapping the switch (K) and adjusting the slide wire until the ammeter reads zero.

The next step is to put the switch S in position 2, return the protective resistor R to its maximum value, tap the key, and adjust the slider again until the ammeter reads zero with R at its minimum value and the tap key closed. The value of the unknown voltage V_x is now read directly from the slide wire.

Note that the maximum value of V_x that can be measured is determined by the slide wire. This is usually 1.6 volts. Actually, a slide wire potentiometer of this type is primarily for education because of the low precision with which the slide wire can be set and read. Laboratory potentiometers usually have fine and coarse adjustments for standardization, and the actual slide wire is only a very small portion of the total resistance between terminals *a–c*. Precision potentiometers have an additional dial enabling the operator to check standardization by merely throwing a double-pole-double-throw switch. This switch selects either the standard cell or the unknown emf, and standardization can be checked just before and just after obtaining a balance with the unknown emf. Such an arrangement is shown in Figure 1-31.

In addition to the feature enabling the standardization to be readily checked, precision potentiometers have an 11-turn slide wire for the fine emf dial. With this, a more accurate balance can be achieved and the position of the slider can be more accurately read.

The potentiometer is in effect a precision voltmeter. It is a device for comparing an unknown emf with the emf of a standard cell that may be known to five significant digits. However, by applying Ohm's law and using precision resistors, the potentiometer can also be used to measure current. Furthermore, by using a precision ammeter and again relying on Ohm's law, the potentiometer can be used to measure resistance.

THE SELF-BALANCING RECORDER

An even more interesting and useful application of the above ideas occurs in the laboratory or commercial recorder. To see how the system works we look back at Figure 1-30 and think about having a motor drive (called a *servo system*) to move the contacter on the resistor. The movement would be controlled by a signal generated by the voltmeter; when the two voltages are balanced, the movement stops. If we have a pen attached to the moving resistor contact arm and a piece of chart paper underneath, we can call this a *recorder*. The whole system is shown in Figure 1-32.

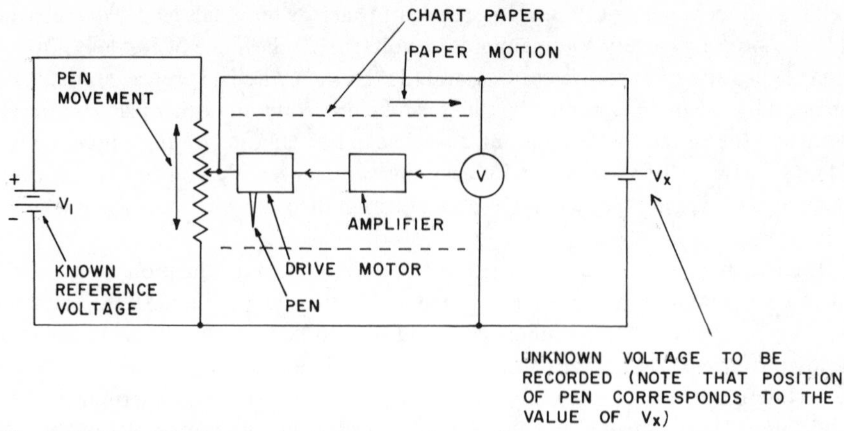

Figure 1-32. Self-balancing potentiometric recorder.

THE WHEATSTONE BRIDGE AND ITS APPLICATION TO RESISTANCE MEASUREMENT

The basic principle of the Wheatstone bridge that makes it so valuable for measurements is that it matches the amplitude of two voltages *without regard to their absolute values.* The ammeter, which is used to detect only the *difference* between the two voltages, can be a very sensitive instrument allowing the resistance values to be matched quite closely. In effect, the circuit of the Wheatstone bridge is that of two voltage dividers connected across a battery with an ammeter that indicates the *difference* between the two divided voltages. When the ammeter indicates zero, the two voltages are equal and the bridge is said to be balanced.

To really appreciate the advantages of the Wheatstone bridge we will think of a simple experiment, measuring the resistance of an unknown resistance with a known voltage source and an ammeter as shown in Figure 1-33.

This is the system used in the familiar VOM and we shall assume that the battery voltage is 1.5 volts and the resistance is about 1 kΩ. The current is about 1.5 mA, and we shall assume that the ammeter has a range of 0–10 mA with an accuracy of 1% (this is a precision meter). The error in the meter is then 1% of 10 mA or 0.1 mA. Therefore, the error in measuring the resistance is also 1% or 10 ohms. In other words the best measurement we can make of this resistance is ± 10 ohms. We might get a more accurate meter, but that would be very expensive and there is a limit to how far we can go in that direction. What we need is a way of balancing most of the resistance against a known resistance and then measuring the remainder. If the remainder is 10 ohms we can measure it to ± 1% or 0.1 ohm, which is a lot better than the ± 10 ohm accuracy we had before. We do this with a Wheatstone bridge, which should indicate to you what a remarkable gadget it really is. The Wheatstone bridge principle has many applications, and you would do well to understand it.

The Wheatstone bridge circuit is shown in Figure 1-34. When the ammeter reads zero, points *b* and *c* are at the same potential. The voltage drop across the resistance

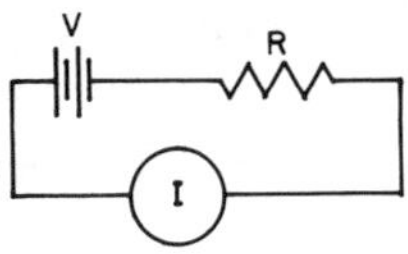

Figure 1-33. Measuring resistance with a voltage source and an ammeter.

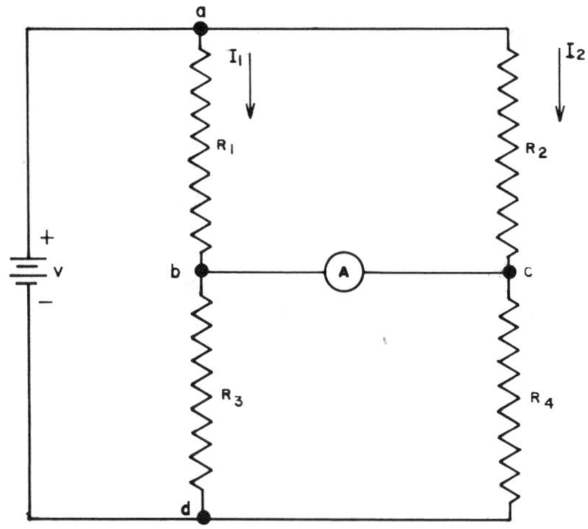

Figure 1-34. Wheatstone bridge.

R_1 then must be equal to the voltage drop across R_2, and the voltage drop across R_3 must be equal to the voltage drop across R_4. It follows that the current through R_3 must be equal to the current through R_1, which is I_1 (there is no current through the ammeter, so there is no other place for the current to go except through R_3). Likewise, the current through R_4 must be equal to the current through R_2, which is I_2. Applying Ohm's law to the two circuits we adjust R_1 until:

$$R_1 I_1 = R_2 I_2 \tag{1-10}$$

$$R_3 I_1 = R_4 I_2 \tag{1-11}$$

Dividing equation (1-10) by equation (1-11) and cancelling out the current yields

$$\frac{R_1}{R_3} = \frac{R_2}{R_4}$$

This is the basic equation of balance for a bridge circuit. If the bridge is being used to measure an unknown resistance, say R_2, then

$$R_2 = R_4 R_1 / R_3 \tag{1-12}$$

It is interesting to note that according to equation (1-12), when measuring an unknown resistance, R_2, it is *not* necessary to know the absolute value of R_1 and R_3 but only the *ratio* of R_1 to R_3. For this reason, in commercially available Wheatstone bridges, the ratio of R_1 to R_3 usually appears as a ratio arm or multiplier whose value can be selected by a knob and pointer. These ratios are usually 0.001, 0.01, 0.1, 1, 10, 100, and 1000. R_4 usually appears as a four-knob decade resistor.

In general, voltages from 1.5 volts to 6 volts are used as the driving voltage. Voltages in excess of 6 volts might damage the precision resistors built into the Wheatstone bridge, thus ruining its accuracy.

One more comment before we leave the Wheatstone bridge. The principle involved is at least 5000 years old and is by no means limited to electricity. In fact, what we might call the *mechanical Wheatstone bridge* was invented by the first great traders of the Mediterranean, the Phoenicians. If you think about the problems of accurately weighing something on a typical spring balance (like that in a grocery store) you can see it is comparable to the circuit of Figure 1-33. Its accuracy is limited. In contrast is the beam balance, where the unknown weight is balanced against a known series of weights. The ultimate accuracy may be in the range of 0.25 mg when the weight itself is $\geqslant$ 25 grams (an accuracy of 10^{-3}%). Clever people those Phoenicians.

SIGNAL GENERATORS

The signal generator is a variable frequency source of AC voltage. It is used in testing circuits, for instance, in measuring the frequency response of a filter or amplifier. Most signal generators provide either a sine or a square wave output. The available output frequencies range from 10 Hz to about 0.6 MHz. For frequencies below 10 Hz, ramp generators are used; above 0.6 MHz, we often speak of pulse generators, which are just signal generators by another name.

You should be aware that *most* audio frequency signal generators do *not* have a low output impedance. It is often 600 ohms; for this reason, short-circuiting the output *usually* does no harm.* In addition they may have three-wire cords and three-output terminals allowing either floating or grounded operation. In this instance, "ground" refers to the *earth ground* provided by the power company. If the supply has a two-wire cord, ground is really only "ground" neutral. While we are on this topic, it wouldn't hurt you to review the material on pp. 6–8 so you can be sure that you understand the difference between neutral, ground, and the abomination of "ground" neutral. The reason this is so important is because of the many different ways that signal generators have their outputs arranged. We can't describe all of them, but we will give a few examples to show you how they go.

*The 600 ohm output impedance protects the signal generator by limiting the current if you accidentally short-circuit the output. The only time the 600 ohms can cause you trouble is when you try to drive a low impedance load. For example, suppose you set the signal source for an output voltage of 1 volt and then hooked it up to a 1000 ohm load. What would the voltage across the load be? Before you say "1 volt," draw the circuit and convince yourself that it will really be 0.625 volt.

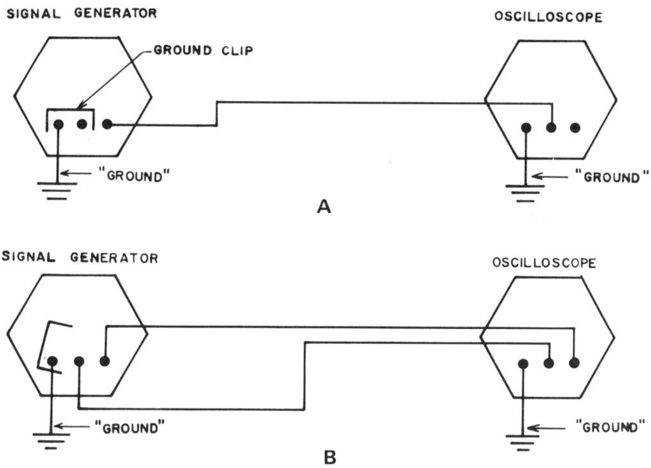

Figure 1-35. Signal generator connections. *A*. Grounded system. *B*. Floating system.

A signal generator that has only two outputs may have one of them connected to the case, or both outputs may be floating with respect to the case. If it has a three-wire cord, the case will be connected to the third, or "ground," wire. If it has a two-wire cord, the case will be floating with respect to power ground. Most modern laboratory power supplies have a three-wire cord and three-output terminals. In this situation – which is shown in Figure 1-35 – the case and one terminal are connected to power ground. You can operate these sources as either grounded or floating systems, as shown in Figure 1-35.

If you were to use the grounded system (which is generally preferable) and you hooked the output of the signal generator to the grounded input of the oscilloscope, you would be short-circuiting the signal generator. Short-circuits are *bad;* that is why signal generators, at least those for student use, have a 600 ohm internal impedance to keep accidental short-circuits from damaging the equipment.

POWER SUPPLIES

Power supplies are sources of AC or DC electricity, but they differ from signal generators in one significant way: they have a low output impedance with considerable current capability, and if you short-circuit them, there may be sparks flying. The better and more expensive supplies have current-limiting circuits or fuses to *protect the supply,* but that will not protect *you* or some other delicate piece of equipment, so be careful. In any case, read the directions *first.*

There are all sorts of power supplies and output configurations, but the first thing to do is look at the cord. If it is a *two-wire cord,* the case (the box it is built in) of the supply will be floating and so will the two output jacks. Be warned that their resistance to the case may be only 10 kΩ or less but it should be something.

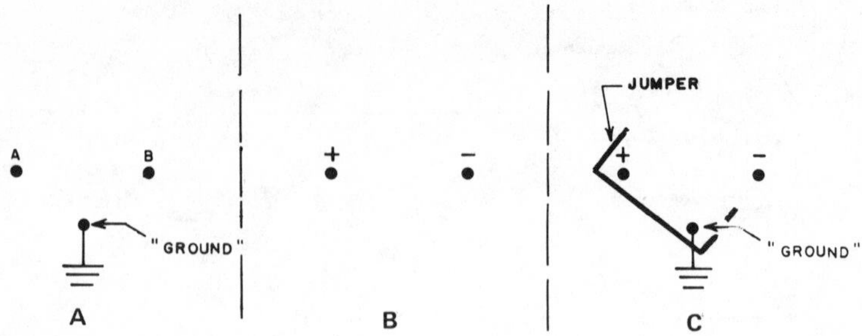

Figure 1-36. Power supply terminals.

That is the problem with two-wire instruments: the case is usually at some potential with respect to the outputs and to ground, but no one can predict what it will be. For power supplies with *three-wire cords,* the case is always at ground potential. Where there is a three-wire cord, there are usually three output jacks: two floating with respect to ground and the third at ground potential. You can use either grounded or floating operation by putting in jumper wires between the terminals shown in Figure 1-36C.

As you get more interested in electronics, you will want to use constant current or constant voltage supplies. A *constant current supply* puts out whatever voltage is necessary to keep the current to the load constant. The *constant voltage system* produces whatever current is necessary while keeping its output at a constant voltage. In Chapter 3, we will discuss how to turn a simple battery charger into a constant voltage or current source.

Other types of power supplies have dual outputs with respect to ground, e.g., ±6, 12, 15, or 24 volts. They are used to power operational amplifiers or other devices requiring two voltages. (We will have more to say about these later.)

If you have specific power supply problems or just want to learn more about them, we suggest you write to the Kepco organization, and ask for their *Power Supply Handbook* (see the Appendix). It is well written, very informative, and free!

CATHODE RAY OSCILLOSCOPE

The cathode ray oscilloscope is the most useful and versatile indicating and measuring device available in the field of instrumentation. It can be used as an AC or DC voltmeter, an ammeter, a frequency meter, and a phase difference meter. However, its greatest usefulness comes from its ability to give a visual display of voltage or current waves, at frequencies from DC to 500 MHz. These waves may be either periodic, in which case a constant image on the face of the tube, or transient, in which case a single trace of the phenomenon occurs across the tube and then disappears. In the early days of its development, it was a tool exclusively used by electrical engineers. Today, however, its usefulness, indeed its indispensability, extends far beyond the field of electrical engineering into mechanical engineering, physics, chemistry, physiology, medicine, and other fields too numerous to mention.

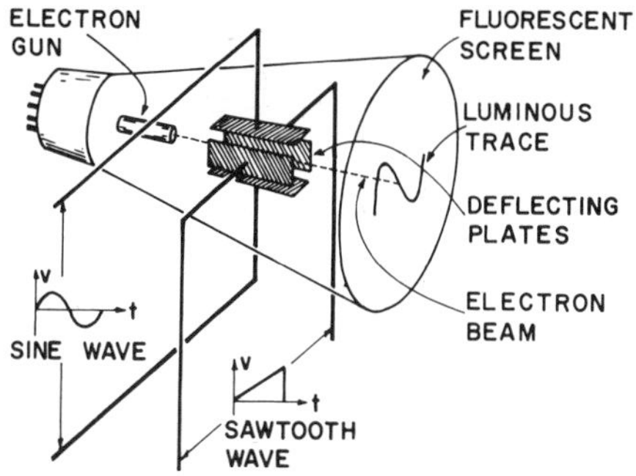

Figure 1-37. Cathode ray tube of the electrostatic deflection type.

The heart of the cathode ray oscilloscope is the *cathode ray tube,* which contains an electron source with a beam-forming arrangement and accelerating and focusing electrodes that serve to form a fine stream of high velocity electrons. The electrons are projected between two sets of plates arranged so that, when charged, one set of plates will deflect the electrons horizontally and the other set will deflect them vertically. A typical cathode ray tube of the electrostatic deflection type is shown in Figure 1-37.

A bright spot is formed when the electron beam strikes the fluorescent screen. Because of the very low mass of the electrons, their inertia is extremely low and the beam can be made to follow high frequency variations well into the 500 MHz frequency range. Because of the persistence of the screen and the persistence of vision, the pattern appears stationary on the fluorescent screen. However, it is actually a dynamic pattern formed by a single spot in rapid motion.

Many controls are necessary to make the cathode ray oscilloscope the versatile instrument that it is. Amplifiers are required to increase the deflection sensitivity, and a sweep generator with a trigger circuit is required to form the apparently stationary patterns.* A typical cathode ray oscilloscope is shown in block form in Figure 1-38. To help you understand how this system works, we will discuss it section by section. Figure 1-39 shows the systems that you will find on a typical oscilloscope.

Vertical Deflection Section

This section consists of three parts: A Y-positioning control to position the spot in the Y direction, an attenuator to take care of high-level input signals, and an amplifier to take care of low-level input signals.

The *Y-positioning control* is a DC control device that sets the undeflected position of the spot in the vertical direction. With this control, the spot can be

*Some of these terms may seem a little strange to you, but stick with us. It will gradually become less confusing as we discuss how the system works.

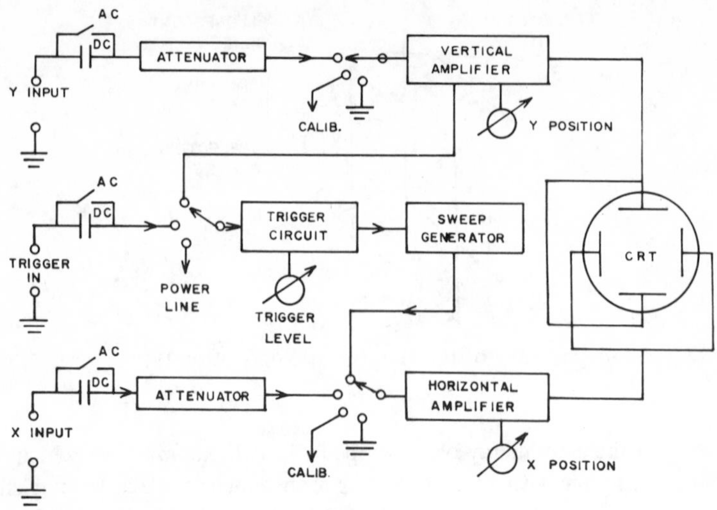

Figure 1-38. Block diagram of cathode ray tube oscilloscope.

placed at any position on the face of the tube in the vertical direction, and even off the face of the tube for special applications.

The *attenuator* reduces the sensitivity between the vertical input terminals and the vertical deflection plates so that signals of high amplitude, which would drive the spot off the face of the tube on peaks, may be viewed.

The *amplifier* increases the sensitivity between the vertical input terminals and the vertical deflection plates so that signals of low amplitude, which would not deflect the spot sufficiently, may be viewed. The frequency range of this amplifier allows signals from DC to radio frequencies to be viewed. There is usually provision for shorting a blocking capacitor at the input of this amplifier when DC signals, or signals of very low frequency or with a DC component, are to be viewed. Sometimes it is desirable to block DC from the input of the oscilloscope, sometimes it is immaterial, and at other times it is necessary not to. So the position of this switch may be important.

Some oscilloscopes are calibrated directly in Y volts per centimeter; others are not calibrated. The calibration of this control is always made with DC, so on DC inputs, the peak-to-peak volts per centimeter are indicated by this control. Provision is made in some oscilloscopes to check the Y-deflection calibration against an internal signal in the oscilloscope itself. This is usually a square wave of known amplitude. The square wave is used, instead of a sine wave, because with square waves there is no question about peak versus root-mean-square values – you just take the top and bottom of the square wave. Other uses for the square waves will be developed in later chapters. They are a great way to test an amplifier for its response to AC and DC signals.

So far all we have talked about is the vertical deflection circuitry in the oscilloscope. If we stopped here you would be left (in the figurative sense) with a trace that went up and down but did not travel from left to right. This travel from left to

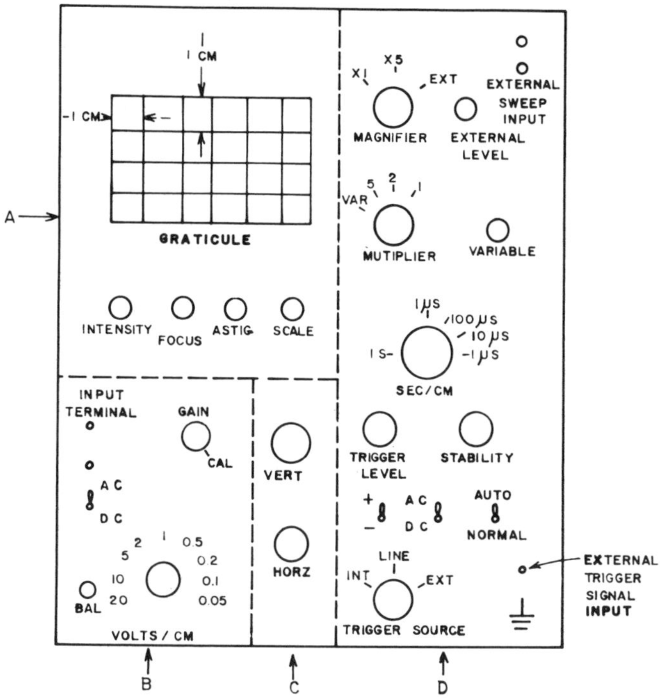

Figure 1-39. Typical oscilloscope controls. *A*. Cathode ray beam controls. *B*. Vertical amplifier. *C*. Positioning. *D*. Time base.

right is controlled by the horizontal deflection section. It applies the necessary voltage to sweep the trace across the screen at some prechosen velocity usually rated in centimeters per second or per millisecond. It also provides a means of "stretching out" the trace to permit the experimenter to view only the most "interesting" part.

Horizontal Deflection Section

The horizontal deflection section consists of the same three parts as the vertical deflection section: an X-positioning control, an attenuator, and an amplifier. In some oscilloscopes the horizontal amplifier is less sensitive than the vertical amplifier because in most applications, the X-deflection is usually brought about by the sweep or trigger voltage present inside the oscilloscope, and sufficient voltage is present so that a sensitive amplifier is not needed. On the other hand, the signal voltage under investigation is usually applied to the vertical amplifier, and this may be of very low amplitude.

Sweep or Trigger Section

The sweep or trigger voltage consists of a saw-tooth voltage wave that can be applied to the horizontal deflection plates through the horizontal amplifier. This horizontal sweep provides a linear time base in the X direction. This voltage moves the spot from left to right at some pre-chosen and constant speed, and then snaps it back to its original position on the left in a very short time. In most oscilloscopes, this "return trace" is automatically blanked out. The repetition rate of this sweep is adjustable by a control marked either *SWEEP RANGE* or *TIME/CM*. The setting of this control determines the amount of spread on the time axis or the time base of the pattern, i.e., the seconds per centimeter along the X axis. In some oscilloscopes this control is marked in sweeps per second instead of centimeters per second. It is usually obvious which is implied.

A very important part of the sweep section is the synchronization of the sweep with the signal applied to the Y input. If this sweep is not started at exactly the right time for each sweep, a stationary pattern will not appear. There are three ways of synchronizing or triggering this sweep: one is from the signal itself, another is from the 60 cycle per second (Hz) power line, and the third is from some external signal. Provision is made on the front of the oscilloscope for selecting one of these methods by a rotary switch labeled *INTERNAL, LINE,* or *EXTERNAL.* The level of this synchronizing signal can be adjusted by a control labeled *SYNC AMPLITUDE, TRIGGER LEVEL,* or some other self-explanatory expression.

Other features on most oscilloscopes include (1) being able to select a single sweep for nonrecurrent or transient phenomena, (2) an automatic, free-running or recurrent sweep for periodic waves, or (3) a driven sweep for nonperiodic waves (but which is often useful for viewing recurrent waves). In this third mode of operation, the input signal triggers the sweep so no pattern is present on the screen in the absence of an input signal at the Y terminals.

Intensity and Focus Controls

These two controls are for controlling the intensity of the electron beam and for focusing it down to a small diameter on the fluorescent screen. The least intensity needed (depending on room illumination) should always be used. A low-level beam gives the sharpest focus and prevents burning of the screen. Never leave a small intense spot fixed on the screen for more than a few moments, because this may leave a permanent black spot on the screen.

Display of Waveforms: Amplitude versus Time

When a recurrent waveform is applied to the vertical plates and a saw-tooth waveform is applied to the horizontal plates, a stationary pattern is observed on the screen if the two waveforms are properly synchronized. This type of pattern can be understood by a very simple experiment: move a pencil back and forth on a piece of paper in such a way as to repeatedly trace a straight line about one inch long. Now pull the paper underneath the pencil perpendicularly to the line being traced, and a crude sine wave is traced out by the pencil. The pencil point represents the spot on the screen tracing out a vertical line in the absence of a sweep voltage. The moving of the paper is analogous to sweeping the spot across the screen at the same time it is moving up and down. Synchronization of the vertical movement of the pencil and the horizontal motion of the paper produces a sine wave.

Oscilloscope Operation

How do we begin to use the oscilloscope? This stops many students cold. Let's look at a typical oscilloscope as shown in Figure 1-39. We have grouped the controls into four sections: *A, B, C,* and *D*. We will discuss other capabilities on page 56.

A. CATHODE RAY BEAM CONTROLS

1. *ON/OFF:* Turns on the AC power. It may be a separate switch or incorporated with another beam control.

2. *INTENSITY:* Controls the brightness of the beam trace or spot. This control should be adjusted for an easily observable trace. An excessively intense beam may burn the cathode ray tube phosphor. Almost all oscilloscopes feature *retrace blanking.* This means that after the beam is swept from left to right, the beam intensity is automatically reduced below the level of visibility until the beam is returned to the left side. On some oscilloscopes a spot will be seen when the beam is not being swept from left to right. This spot is usually at a reduced intensity. The intensity control may be adjusted until the spot disappears. When the beam is swept, the intensity will automatically increase to a visible level. On some oscilloscopes the beam spot is completely blanked and will not appear for any setting of the intensity control. These oscilloscopes often have a *FREE RUN* switch setting in the time base controls that always produces a trace.

3. *SCALE ILLUMINATION:* Controls the graticule brightness.

4. *ASTIGMATISM:* Controls spot shape.

5. *FOCUS:* Controls spot size. The astigmatism and focus control should be used together to obtain a sharp trace.

B. VERTICAL AMPLIFIER

1. *INPUT TERMINALS:* The terminals to which the signal to be viewed is attached. Many oscilloscopes have only two terminals, one of which is taken through the power line ground to earth. Other terminals may be in the form of a coaxial* connector. Some oscilloscopes have three input terminals. The signal may be attached to the ungrounded pair if floating or ungrounded operation is desired. Still others may have a switch that grounds one of a pair of input terminals.

(**Warning:** Many other types of electronic equipment have a grounded terminal. *Always attach all grounded terminals to a common circuit point;* otherwise, a short-circuit will exist via the power line.)

2. *VOLTS/DIVISION:* This control will adjust the height of the waveform

**Coaxial:* A fancy word for a shielded connector consisting of a central wire (that carries the signal) fixed by insulation inside a grounded shield.

present. There is usually an electric switch with several positions labeled in volts per inch or centimeter.

3. *GAIN CONTROL:* This is a continuously variable control that may be put in the position marked *calibrated.* When the gain control is in this position, the *VOLTS/DIV* control setting is calibrated to a known value. The *GAIN CONTROL* provides a continuous gain adjustment between the calibrated positions of the *VOLTS/DIV* switch.

4. *AC–DC:* A switch that determines whether only AC beam deflection or AC and DC deflection will occur. This is illustrated in Figure 1-40. If the trace or spot is previously adjusted to a zero reference line, the waveforms shown will produce displays similar to those shown in Figure 1-40.

5. *BALANCE:* This control pertains to operation with the AC–DC switch in the DC position. It should be adjusted to minimize the vertical movement of the trace or spot when the *GAIN* or *VOLTS/DIV* controls are moved between their extreme positions. Only one may be affected by the balance control. This, as well as other details, may be determined by consulting the equipment manual or by trial and error (if all else fails, read the directions).

C. POSITIONING

1. *VERTICAL:* Moves the spot or trace up or down. Internally, this control is a part of the vertical amplifier, but this is often grouped separately from the other vertical amplifier controls.

1. *HORIZONTAL:* Adjusts the spot or trace left or right.

D. TIME BASE

Time base controls are the most numerous and difficult to understand. Lack of appreciation of their functions can lead to hours of frustration.

1. *TIME/DIV:* This control is a selector switch with positions marked in microseconds, milliseconds, or seconds per inch or centimeter. Various settings of this control will adjust the sweep speed to various calibrated values, thus controlling the spread of a waveform.

2. *MULTIPLIER:* This control is a selector switch with positions often marked *variable, 1, 2, . . . 5.* In the variable position, a separate control will adjust the sweep

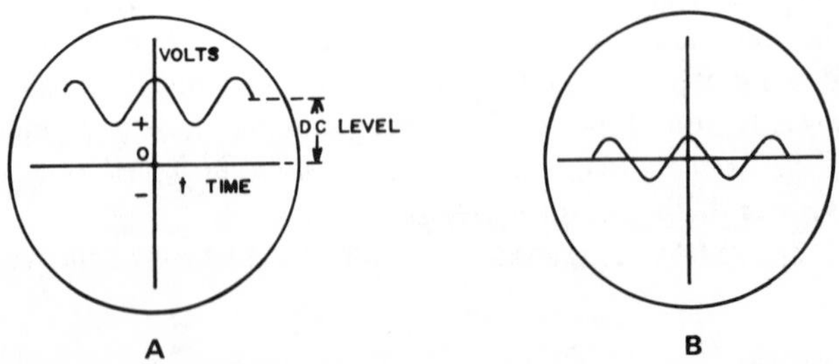

Figure 1-40. Oscilloscope traces. *A.* DC operation. *B.* AC operation.

time continously between the calibrated times per division of the *TIME/DIV* switch. In the *1, 2,* or *5* position, the sweep is still calibrated, but the time per division indicated by the *TIME/DIV* switch should be multiplied by 1, 2, or 5 respectively. Some oscilloscopes have no multiplier, but have many more positions on the *TIME/DIV* switch.

3. *VARIABLE:* Continuously adjusts the sweep time when the multiplier is in the variable position.

4. *MAGNIFIER:* Usually a two-position switch which is marked ×*1*, ×*5*, or *Off* and *On*-×*5*. In the ×*5* position the sweep speed is increased by 5; thus, the time per division value must be divided by 5. Magnification actually makes the trace five times longer. As a result you can only see about one-fifth of the trace at a time, and you must adjust the *HORIZONTAL POSITION* control to view the entire trace.

5. *EXTERNAL LEVEL* or *ATTENUATION:* Attenuates the sweep signal to obtain the desired sweep length.

6. *INTERNAL/EXTERNAL:* A switch which, when in the external position, allows the beam to be horizontally swept by an external signal applied to the *EXTERNAL SWEEP TERMINALS.*

7. *TRIGGER SOURCE:* Selects either internal, line, or external triggering. The trigger voltage is an AC, DC, or AC + DC signal that starts the horizontal oscillator for each sweep. In the internal position, the sweep triggers on the signal applied to the vertical amplifier. This is the most frequently used mode. For line triggering, a 60 cycle per second signal from the power line is used. This is convenient when investigating signals synchronized to the power line. In the external position, the sweep may be triggered by an external signal attached to the *EXTERNAL TRIGGER TERMINALS.*

8. *TRIGGER LEVEL:* This control adjusts the instantaneous voltage level at which the horizontal sweep is triggered. Thus, using internal triggering, you can have the trace begin at any point on the waveform to be viewed.

9. *STABILITY:* Allows the stability of the sweep oscillator to be adjusted, which has the effect of controlling the minimum amplitude trigger signal necessary for triggering. Adjusting this control prevents triggering by low level signals. Some oscilloscopes do not have a front panel stability control.

10. *±MODE SWITCH:* Determines that the trace is to be triggered on the positive or negative going portion of the trigger signal.

11. *AC/DC MODE SWITCH:* In the AC position, sweep triggering occurs at some point that is determined by the setting of the *TRIGGER LEVEL* and ± *CONTROL* on the AC portion of the trigger signal. In the DC position, triggering occurs anytime the instantaneous trigger voltage exceeds the value determined by the abovementioned switches.

12. *AUTO/NORMAL MODE SWITCH:* In the *auto* position, triggering will occur automatically when the stability control is set properly. The trigger level control will be disabled. In the *normal* position, the abovementioned controls are operative. The *auto* mode is less than perfect, and manual control is necessary for complex waveforms.

E. OTHER TYPES

There are as many varieties of oscilloscopes as there are manufacturers and models. Many of them have additional capabilities that are not discussed above. Several are listed here with short explanations.

1. *MULTIPLE TRACE:* Has the capacity of displaying two or more waveforms simultaneously. Most of these alternate the trace between input signals. A few have dual-beam cathode ray tubes that operate simultaneously.

2. *X–Y INPUT:* Has an additional amplifier – identical to the vertical amplifier – that can be used for calibrated, external, horizontal sweeping.

3. *STORAGE:* The cathode ray tube has an electron flood source that allows the trace of a nonrepetitive or very slow signal to be displayed for as long as several hours.

4. *SAMPLING:* Sampling is a technique by which displays of very high frequency, repetitive signals are obtained. Most oscilloscopes have a high frequency limit that is determined by the bandwidth (roll-off) of the vertical amplifier. If the bandwidth is pushed up, a further limitation encountered is the electron transit time from the deflection plates to the screen. As a result, direct trace viewing beyond a few thousand megahertz has been impossible to obtain, which resulted in the development of the sampling scope. This instrument samples the fast signal at various times and displays the result as a sequence of spots that show what the original waveform *was* like. This system can only be used if a repetitive signal is available.

Oscilloscope Applications

FREQUENCY MEASUREMENTS

When using the oscilloscope as a frequency meter, sine waves are applied to each set of plates. When the two sine waves are of the *same* frequency, an elliptical pattern appears on the screen. When the vertical frequency is *twice* that of the horizontal frequency, a butterfly pattern, or a figure eight on its side, appears on the screen. This pattern has two loops or peaks at the top indicating that the vertical frequency is twice that of the horizontal frequency. If the vertical frequency is *three* times that of the horizontal frequency, there will be three loops or peaks at the top, and so on. These patterns are called *Lissajous figures.* Frequencies may be compared in this way up to a ratio of about 10 to 1. For higher ratios, it becomes difficult to count the peaks unless the two frequencies are extremely stable and very fine frequency adjustment of one of them can be made.

Another method of comparing two frequencies is to connect both of them, either in series or in parallel, to the vertical deflection terminals and set the sawtooth sweep frequency at a very low rate. When the two frequencies are very close together, the beat* notes can be observed on the screen. Beats with higher harmonics of one of the frequencies can also be observed as stationary patterns.

*The term *beat* comes from music. When two signals of almost the same frequency are added, certain "sum" and "difference" frequencies appear. The low-frequency difference signals are called *beats.*

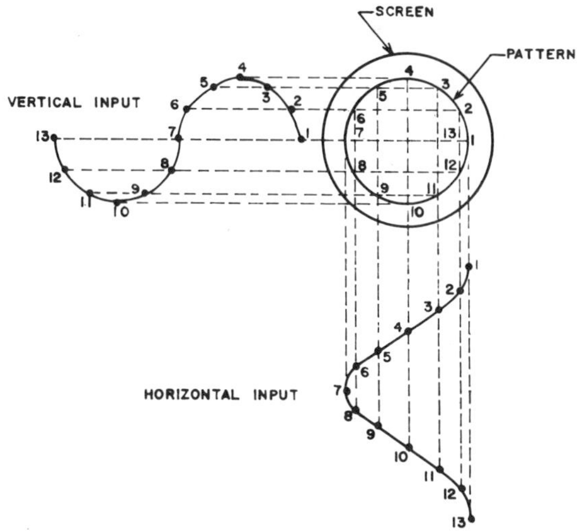

Figure 1-41. Circular Lissajous figure from two sine waves 90 degrees out of phase. (Note that when the phase difference is zero, the circle shrinks to a single line at a 45 degree angle.)

Lissajous figures can be understood from the following explanation. Figure 1-41 shows a circular pattern obtained from two sinusoidal waves 90 degrees out of phase: a sine wave applied to the vertical input and a cosine wave applied to the horizontal input. Equal intervals of time along each wave are marked with numbers from 1 to 13, and each of these numbers is joined by a straight line to the place on the screen where the spot will be at that particular instant in time. The circle swept out by the spot is shown. Figure 1-42 shows a figure eight or butterfly pattern swept out by the spot when a sine wave of frequency f_1 is applied to the vertical plates and a cosine wave of frequency f_2 is applied to the horizontal plates. Again, equal intervals of time are marked, and the position of the spot is shown at each instant in time. If the frequency ratio were 5 to 1 instead of 2 to 1, five peaks would appear.

If the frequency is not a simple ratio such as 5 to 1 or 7 to 1, but, for example, is 9 to 4 or 8 to 3, then the frequency applied to the horizontal plates is to the frequency applied to the vertical plates as the number of loops across the top is to the number of loops along a vertical side.

For a perfectly circular pattern, the two signals must be of the same amplitude.

PHASE MEASUREMENTS

In the previous explanation of the circular Lissajous figure shown in Figure 1-41, the statement was made that when the signals are 90 degrees out of phase, the pattern is a circle. If the signals were in phase, the pattern on the face of the oscilloscope would be a straight line with a slope of 1. This can be shown by making a diagram similar to Figure 1-41, with the two applied voltages equal in amplitude and in phase. As the phase relation between the two signals departs from zero degrees, the

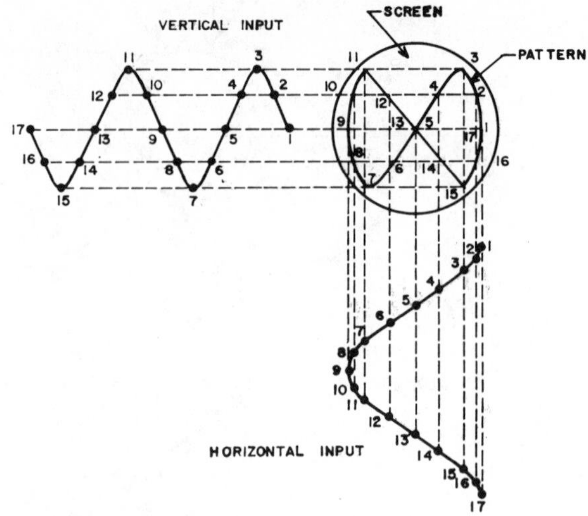

Figure 1-42. Butterfly pattern from two sine waves: the frequency of the vertical input is twice that of the horizontal input.

straight line "opens up" into an ellipse slanted to the right, and becomes a circle when the phase difference is 90 degrees. When the phase difference exceeds 90 degrees, the circle becomes an ellipse slanted to the left, and this ellipse becomes a straight line with a slope of −1 when the phase difference is 180 degrees.

The phase difference can be calculated from the relationship

$$\theta = \arcsin \frac{Y_{\text{intercept}}}{Y_{\text{max}}}$$

in which

θ = phase angle between the two signals

$Y_{\text{intercept}}$ = axis point where the ellipse crosses the vertical

Y_{max} = highest vertical point on the ellipse

When making the above calculation, the two signals must have equal amplitudes. In other words, the ellipse must be contained within a square on the grid on the face of the oscilloscope.

This is a good spot to stop, do an experiment, and learn something about oscilloscopes plus phase shift (Figures 1-16 and 1-17). In Figure 1-17 you will note that the current through capacitor I_c is exactly $I_c = 2\pi f A C \cos 2\pi ft$ or 90 degrees out of phase with the applied voltage. To make use of this fact, we set up the circuit shown in Figure 1-43.

In Figure 1-43 a signal generator drives a resistor and a resistor-capacitor circuit.

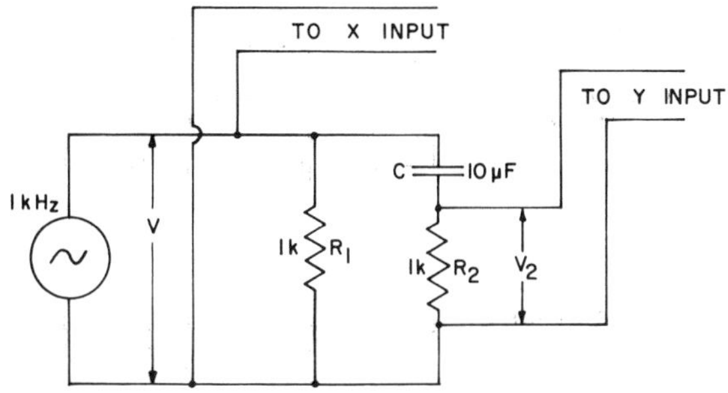

Figure 1-43. Phase shift demonstration circuit.

The current through the resistor is out of phase with V so the voltage across R_2 is out of phase with V too. If you put V into the X axis of the scope and use the V_2 signal to drive the Y axis or sweep (read the manual if necessary), you will be looking at two signals that are exactly 90 degrees out of phase. The Lissajous signal should be a circle.

Try the effect of changing the frequency of V. (Should it affect the Lissajous figure?) Try changing C and R_2 to larger or smaller values. For very small values of C, i.e., 10 pF and $R = 10^6$ ohms, you may find the phase shift is less than 90 degrees. (Why?) (Answer: Think of what happens as C becomes a short-circuit so that we have zero phase shift. When the phase shift is zero, V and V_2 are "in phase," and the trace on the oscilloscope will be a straight line.)

The Self-Balancing Potentiometric Recorder

The operating principles of this system were discussed earlier in the text associated with Figure 1-32. Here we will go through some of the how-to-use-it details. We might begin by noting that the word *potentiometric* implies that at balance the unit does not draw current from the signal source. Actually, the "infinite" input impedance does not really exist, but units like the Heath Company IR-18M come pretty close (i.e., above 5 MΩ at balance and 500,000 ohms when off balance). Another point of interest is that the "reference voltage" is obtained from a Zener diode rather than a mercury cell (the mercury cells have a nasty habit of going dead just when you need them).

These instruments, though somewhat expensive, $229.95 (1980 dollars), are very useful and no lab is really complete without one. They are essentially voltage sensitive, but current signals are easily converted to voltage for recording. Typical sensitivities are zero to 10 mV with ability to cover any voltage up to zero to 250 mV. Beyond that you have to use external voltage dividers (you do remember them, we hope). To see how recorders work, we refer to Figure 1-44 showing a typical unit. Note that it takes standard ink pens, soft-tip pen, or even a soft pencil; this avoids having to buy special pens at "special cost."

Looking at the controls on top, from left to right, the three rocker switches

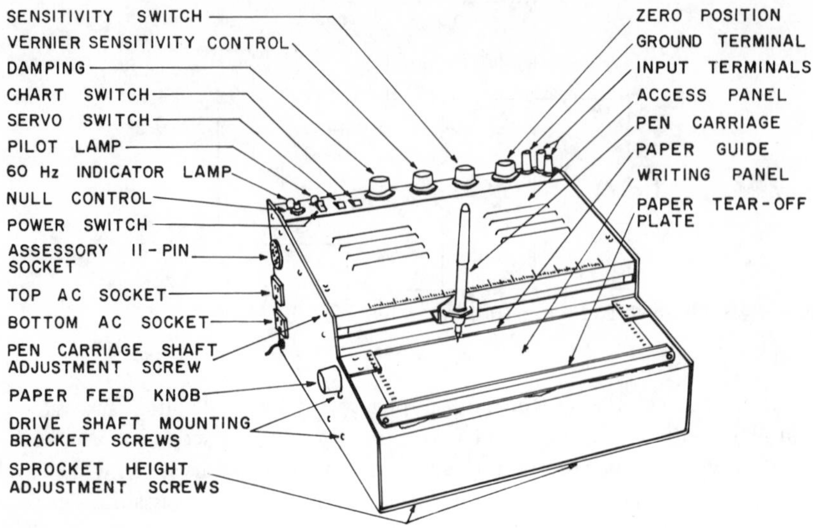

Figure 1-44. Typical pen recorder.

provide (1) power-on for warm-up and standby conditions; (2) servo-on to allow adjustment for zero and sensitivity; (3) chart-on so that you can actually record the data.

The damping control allows the operator to control the speed of response and oscillation of the pen-servo system. The best setting is one where the pen is just at the threshold of oscillation. The sensitivity control consists of two units – a decade switch for basic range control (i.e., zero to 10 mV, zero to 100 mV, etc.) and a vernier pot which allows intermediate adjustments.

The zero position control allows the operator to set the pen at zero on the chart or at some other position if necessary. This is convenient when both (+) and (-) signals are to be recorded.

There are three input terminals marked (+), (-), and ground. In most cases the (-) terminal is grounded, but for "floating operation" (see the material on page 47 for a discussion of "floating") the (-) terminal is not to be grounded. Usually the "high" side of the signal source (the ungrounded wire) is connected to the (+) terminal. In some cases this wire may have to be shielded (with the shield connected to ground) to avoid AC pickup.

You should practice hooking up something like a signal source (a sine or square wave generator will be great) to the recorder and playing (yes, we said "play") with the damping control, the gain, etc. How else can you learn?

MODES OF RECORDING

In general, recorders fall into two large classifications: (1) direct writing and (2) photographic. The charts that these types make fall into three general classifications: strip chart (Z-fold), circular chart, and X–Y charts.

Direct-Writing Recorders

There are five recording methods in general use in direct-writing oscillographs. These are (1) ink pen on hard-surface paper, (2) heated stylus on temperature-sensitive paper, (3) sharp stylus tracing on carbon-coated paper, (4) metal stylus on electro-sensitive paper, and (5) the chopper bar–typewriter ribbon method on hard surface paper. All of these have advantages, disadvantages, and limitations which will now be considered in turn.

The *ink pen* method makes a very neat record under laboratory conditions, but it is definitely limited in usefulness if the recorder is in a vehicle subject to acceleration. The ink will either flow too freely or not freely enough if the acceleration is in line with the direction of the pen. Also, this method is inconvenient if the recorder is to be used intermittently since the ink might dry in the pen and clog. However, under continuous use, very neat records can be obtained and the width of the line is not so dependent on the speed at which the recording pen moves as it is in certain other methods. Trace widths of from 0.015 to 0.005 inch are available. The pen is of the trussed capillary type with one end of the capillary tube riding below the surface of the ink in an inkwell and the other end riding lightly on the chart paper. Provision is usually made to prevent spillage of the ink and for priming the system to start the flow of ink. Once started, however, the inking system will work over long periods of time without attention.

Some recorders, such as those made by the Heath Company, use low-cost tip-wick pens. This makes it easy to change the colors of the ink. It is also much easier to put in a new pen than to clean a capillary inkwell system.

The *heated stylus on temperature-sensitive paper* method is very convenient for either continuous or intermittent use. The stylus consists of a piece of nichrome ribbon or wire that is arranged to ride on the surface of a temperature-sensitive paper as it is pulled over a relatively sharp edge. The general arrangement is shown in Figure 1-45. The temperature of the nichrome ribbon is adjustable and should be set for the best trace. The main disadvantage to this system is that the density and width of the trace depend upon the speed with which the moving arm is deflected.

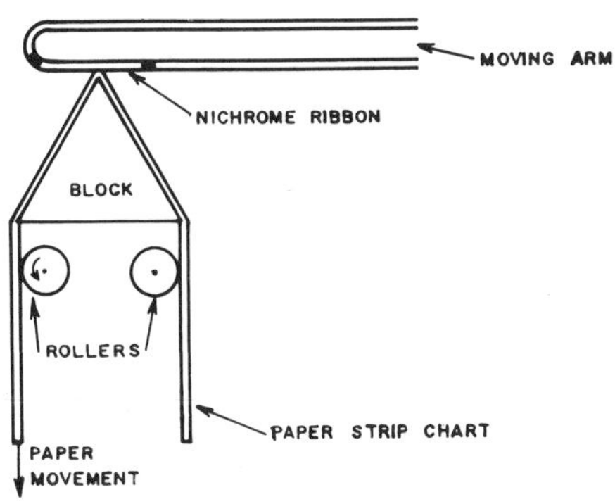

Figure 1-45. Heated stylus on temperature-sensitive paper.

The baseline that is drawn when the arm is not moving is usually much heavier than the line drawn when the arm is moving. However, this system is very convenient for intermittent service because it is ready immediately for service after a short or long period of inactivity. This is not to imply that this system is unsuitable for continuous use; it is very good for continuous use. The temperature of the stylus is adjustable while a recording is being made. This makes it possible to adjust the trace for the best resolution of any part of the trace of a periodic function.

The method of a *sharp stylus tracing on carbon-coated paper* is essentially a method of scribing or dotting a trace on blackened or pressure-sensitive paper. This method is also suitable for either constant or intermittent service. The continuous-trace type is subject to various errors, and it is thus limited to relatively high current recorders. The dotted-trace type makes a dotted line instead of a continuous line. This is accomplished by intermittently clamping the stylus arm and automatically tapping on the stylus to make a dot on the pressure-sensitive paper. Between dots, the stylus arm is free to swing to a new location, where it is again clamped and tapped to make another dot.

The *metal stylus on electrosensitive paper* is a method in which the circuit between the moving stylus above the paper and the metal platen below the paper is completed by the paper itself, which is electroconductive. This action essentially "burns" a trace into the paper. This method is quite sensitive to the "burning current": excessive current causes sparking and smoke, and too little current causes an intermittent and weak trace.

The *chopper bar and typewriter* method consists of an oscillating hammer that records a dotted line on a piece of paper in much the same way that a typewriter records letters on a piece of paper. Standard typewriter ribbon is used, and is fed through guides across the strip chart just above a metal writing table. A stylus, activated by the basic current-sensitive element and intermittently tapped, imparts blows on the ribbon, thus making a record on the chart below the ribbon. The records appear to be continuous at most standard chart speeds. For fast chart speeds or rapid deflections of the stylus, the record will be lighter and will consist of a series of dots. This system is very economical, but it is limited to relatively low current, slow speed applications.

More expensive recorders often use paper that is folded (Z-fold) rather than stored on a roll. The only advantage is that the charts tend to lie flat instead of curling up. At this writing, 1980, Z-fold paper is about four times as expensive as regular rolled chart paper.

Photographic Recorders

Photographic methods are in general of two types: one uses a standard camera mounted to photograph the screen of a cathode ray oscillograph; the other employs an internal light source and an optical system that focuses the filament of this light source on a photographic strip chart. The position of this image depends upon the position of a small plane mirror attached to a galvanometer, which is a very sensitive ammeter. One great advantage to this latter method is its multi-channel possibilities: only one light source and optical system is required for many channels, and the light beams can cross each other without interference to give overlapping tracings on high amplitude peaks. However, the photographic paper must be loaded, developed, and fixed in a photographic dark room, and the records are not available until after the photographic processing has been completed. With mechanical methods of recording, on the other hand, the records are available immediately.

Multi-Element Oscillographs

These instruments are of the photographic type and are available with up to 24 simultaneous recording channels. On most units the paper width is approximately 10 inches, and this paper must be developed by usual photographic techniques. High sensitivity ammeters or galvanometers are usually used with a single-turn coil or even a coil consisting of a single filament of flattened metal with a mirror cemented to it. The frequency response of these galvanometers extends up to 10,000 cycles per second (10 kHz), and these units are extensively employed in strain gauge and geophysical exploration work. A single light source is used, and an image of its filament is focused on the photographic paper. There is usually a provision for viewing the trace by means of a many-sided rotating mirror. Only periodic phenomena can be viewed by this rotating mirror system, but transient phenomena as well as periodic phenomena can be recorded.

The principal advantage of multi-element recording oscillographs is the large number of channels that can be viewed simultaneously. This feature is very important in geophysical exploration work in which the delay between the various reflected waves is of significance.

Multi-element recording oscillographs are widely used in the flight testing of aircraft, and special 50-channel units have been developed for this use. Each galvanometer in this unit records the imbalance of a strain-gauge bridge.

X–Y Recorders

All of the recording devices described thus far record voltage or current as a function of time. Very often it is desirable to record current as a function of some other parameter. An obvious example is in power supply testing when it is essential to know the output voltage as a function of the current supplied to a load. A device for doing this is called an *X–Y recorder,* and, in the above example, the X quantity would be the current to the load and the Y quantity would be the voltage across the load. These instruments usually employ an ink pen and graph paper 8½ by 11 inches or 11 by 16½ inches in size. Some can be converted to Y–t (that is, Y as a function of time) recorders by driving the pen mechanism horizontally with a synchronous motor.

These recorders are actually instruments intermediate between manual data plotting and complex computers. In this respect, the X–Y recorder is a combined data taker and plotter that is very slow compared with oscilloscopes but very fast compared with manual methods. It plots a complex curve in less than 1 minute, and provides a permanent record. Other independent or Y variables may be used to draw a family of curves on a single sheet. Human errors due to interpolation and plotting are eliminated, as is the age-old question of how many points determine a curve – a continuous curve is plotted.

X–Y recorders are usually of the self-balancing potentiometer type with a flat recording surface. Sensitivities of one millivolt per inch for either the X deflection or the Y deflection and writing speeds up to 40 inches per second are available.

Broad fields of application for the X–Y recorder include the recording of strain as a function of stress, pressure as a function of temperature, voltage as a function of current or vice versa, and torque as a function of speed.

Cathode Ray Oscillograph

The cathode ray oscilloscope was described in detail earlier. It is being mentioned again here because of its possible use as a recorder. Recordings are made by photographing the face of the cathode ray tube. This is very often done by mounting a 35 mm roll-film still camera on the front of the oscilloscope, using a light-tight tube. Many oscilloscopes carry a bezel around the tube opening for this use. Another method is to use a direct-developing Polaroid Land camera. These are available commercially for oscillographic use. Film speeds up to 10,000 ASA are available for these units, and the actual picture is available a few seconds after the exposure has been made.

POSTSCRIPT

At this point we have decided to stop delaying you and launch you into the next chapter, which is all about (you guessed it) op-amps. This doesn't mean there aren't other topics to discuss regarding circuits and instruments – see any EE text for proof of that. But we want you to get some experience with op-amps before you get bored. You will be making mistakes at first, but once you get interested you will find it easier to pick up what you need regarding circuit theory and instrumentation.

There isn't any book, *Circuit Theory Without Pain* (not yet anyway, but give us time). However, the books listed below will be interesting to you when you get to Chapter 3 in our book:

Diefenderfer, A. J. *Basic Techniques in Electronic Instrumentation.* Philadelphia: W. B. Saunders, 1972.

Smith, R. J. *Electronics: Circuits and Devices* New York: John Wiley, 1973.

TEST YOURSELF PROBLEMS AND SUGGESTED EXPERIMENTS

Test Yourself Problems

1. You have a 90 volt battery, but you really need two voltages of 50 and 40 volts for an experiment. Draw the circuit and show how you would provide these voltages. Note that total current taken from the battery cannot exceed 100 mA. Answer: The total resistance cannot be less than 900 ohms. The individual resistors have values of 400 and 500 ohms.
2. You have a 0 to 10 mA meter with an internal resistance of 0.01 ohm and you want to be able to read currents from zero to 10 amps. Draw the circuit and calculate the necessary shunt resistor. Answer: Shunt resistance 10^{-5} ohms.
3. A 2 volt signal source is in series with a resistance of some 10^4 ohms. You want to measure this voltage with an error of no more than 1%. How much current can your voltmeter draw from the circuit? What will its internal resistance be? Answer: Maximum current is 2×10^{-6} amps. Ohmmeter resistance must be 990 kΩ.
4. A transformer has a turns ratio (primary to secondary) of 1 to 100. If the input voltage and current are 10 volts at 2 amps, what are the output current and voltage? Answer: 1000 volts at 20 mA. What is the wattage input? Answer: 20 watts.
5. In your own words, discuss the concept of isolation via a transformer and explain why a Variac does not provide isolation.
6. A given meter has an ohms-per-volt rating of 50,000. How much current is required to deflect the meter "full scale"? Answer: 20 microamps. If the same meter has a full scale voltage of 100 volts what is the internal resistance? Answer: 5 M ohms.
7. Discuss, in your own words, how a potentiometer can measure a voltage without drawing any current from the circuit.
8. Discuss, in your own words, the similarities and differences between power supplies and signal generators. What is meant by the phrase "signal generators do not have a low output impedance?" Why would you have units of this type in an introductory student laboratory? What might happen to a low output impedance power supply if you short-circuited it?
9. Suppose you are in charge of a construction project where men are working while standing on wet ground. The engineer in charge asks, "Is it worth paying the extra money to run in three-wire grounded power versus the two-wire system we have now?" Note that a "yes" answer is not enough. He wants to know why.
10. We noted in the text that decibels are a convenient way of handling large numbers. You might convince yourself of this by taking the numbers 10, 100, and 1000 and calculating the decibel values. Even 10^6 is only 120 dB.
11. Design a simple high-pass filter that will stop at least 90% of the signals at 10 Hz

and pass at least 90% of the signals at 1000 Hz. Answer: We can solve for RC from either of two equations

$$0.1 = \frac{20\pi\, RC}{20\pi\, RC + 1} \quad \text{yields} \quad RC = 1.77 \times 10^{-3}$$

$$0.9 = \frac{2000\pi\, RC}{2000\pi\, RC + 1} \quad \text{yields} \quad RC = 1.43 \times 10^{-3}$$

If we chose $1.77(10)^{-3}$, the first equation is satisfied and the second yields an even better result.

$$\frac{2000\pi\,(1.77)10^{-3}}{2000\pi\;[(1.77)10^{-3} + 1]} = 0.916$$

If we use $RC = 1.43(10)^{-3}$, the first equation yields

$$\frac{20\pi\,(1.43) \cdot 10^{-3}}{20\pi\;[(1.43)10^{-3} + 1]} = 0.082$$

more than satisfying the 0.1 requirement. If $RC = 1.77(10)^{-3}$, we can choose $R = 1000$ and $C = 1.77(10)^{-6}$F.

Suggested Experiments

Rules of the Lab:

1. *Always check device dial settings before turning on.*
2. *Always turn off all devices before making circuit changes.*

A. VOLTAGE DIVIDER

Using a VOM, two 10 kΩ resistors, and a DC power supply:

1. Draw a voltage divider circuit diagram.
2. Write down circuit equation.
3. Solve equation for the resistor values you are using.
4. Check solution with experimental values.

B. OSCILLOSCOPE

Draw a schematic of the following circuit, then construct it.

1. Connect a diode in *series* with a 10 k ohm resistor. Now connect that pair to the function generator set for about 3 volt peak to peak output. The *cathode* of the diode should be connected to the positive terminal of the function generator. Place the scope probe and scope ground across the 10 k ohm resistor.

2. Draw the schematic.
3. Which is your choice?

	Scope	*Function generator*
Ground	_______	_______
Earth ground	_______	_______
Chassis ground	_______	_______
Neutral	_______	_______
Floating ground	_______	_______

4. Indicate, on the schematic, how the scope is connected.

C. TIME BASE SELECTOR

Set the oscilloscope trigger mode to INTERNAL AUTOMATIC. Set the function selector, on the function generator, to SINE WAVE. Adjust the frequency on the sine wave to 1 kHz.

1. If the frequency is about 1 kHz, where should the time selector be to see the waveform conveniently on the screen? (**Remember**: Time Base = seconds/cm and 1 kHz = 1000 cycles/sec. Or if you wanted to see 1 sine wave cycle per centimeter, 1 sine wave cycle/cm = [1000 cycles/sec] [10^{-3} sec/cm]
   Input frequency Time base
2. You should be ready to turn everything ON now. (Remember Rule #1!)
3. At about what setting, on the time base selector, does only 1 cycle fit on the oscilloscope screen?

D. TRIGGERING

There are three modes of triggering: Internal Normal Mode; Internal Automatic Mode; and External Mode.

1. Normal Mode: Set the trigger mode to the "normal" mode. Set the slope of the input waveform, on which triggering is to occur, to POSITIVE (may not be available on your scope). Now adjust the TRIGGER LEVEL until a stable waveform is obtained.

   The scope is now triggering on a (1) ____________ (positively or negatively) sloping portion of the input waveform. (**Note**: The scope triggers on the voltage that first appears at the left hand edge of the screen.) The voltage at the triggering point is (2) ____________ (positive or negative).

   Now switch the direction of the diode. (Remember Rule #2!) Adjust the "trigger level" until a stable waveform is obtained. The scope is now triggering on a (1) ____________ (positively or negatively) sloping portion of the input waveform. The voltage at the triggering point is (2) ____________ (positive or negative).
2. External Mode: Repeat the above exercise for normal mode except use the external mode selection and external input.

The answers are: (1) __________ (positively or negatively)
(2) __________ (positive or negative).

Where was your scope external input connected in the test circuit?

Now connect the scope external input to a different spot on the test circuit (either at the function generator positive terminal or between the diode and resistor). Vary the TRIGGER LEVEL adjustment and note any changes. What is the advantage of external mode over other modes?

E. SIMPLE DC POWER SUPPLY

Construct the following circuit:

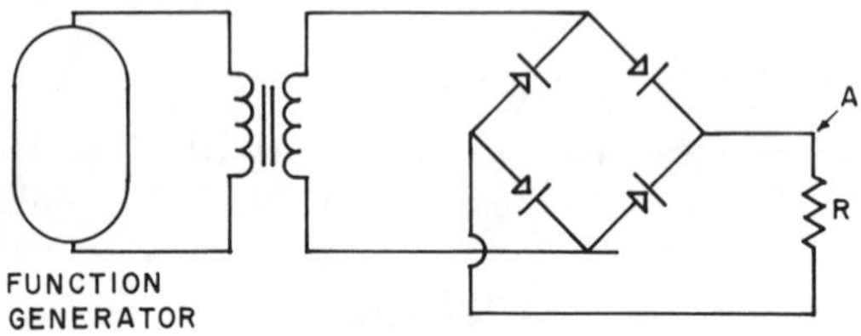

Figure 1-46.

R = 10 kΩ

Label the polarities of the transformer output and at point A. Draw a sketch of the output of the transformer (sine wave input).

Draw a sketch of the waveform at point A.

Now place a .05μ F capacitor in *series* with *R*. (Remember Rule #1!)

Draw a sketch of the waveform at point A.

Repeat with the capacitor in *parallel* with the resistor *R*.

Insert a 100 kΩ resistor in place of the 10 kΩ and sketch point A.

What's happening and why??

F. TRANSFORMERS

Observe how many inputs and outputs there are; then draw a schematic labeling primary and secondary lines. Using your function generator as an input to the transformer, determine the values of *N* for each output.

$$\left(\textbf{Remember: } \frac{V(\text{secondary})}{V(\text{primary})} = N \right)$$

2 The Op-Amp and How to Make It Work for You

BASIC TYPES OF COMMERCIAL OP-AMPS

In Figure 2-1, you are looking at a typical op-amp. An op-amp is a black box or a small can with anywhere from 6 to 14 pins or lead wires.* To find out which pin is which, you must first realize that op-amp manufacturers are *not* consistent about how they mark their products. We will start with a variation of the old Burr-Brown system† because I think it is easier to learn. At the same time, we will use the newer system that most manufacturers seem to be using. Manufacturers of op-amps give out instructions with their products, so you can easily identify the proper terminals and correlate our instructions with whatever op-amp you happen to have.

There are four basic types of op-amp packages: cans, flat packs, dual inlines, and

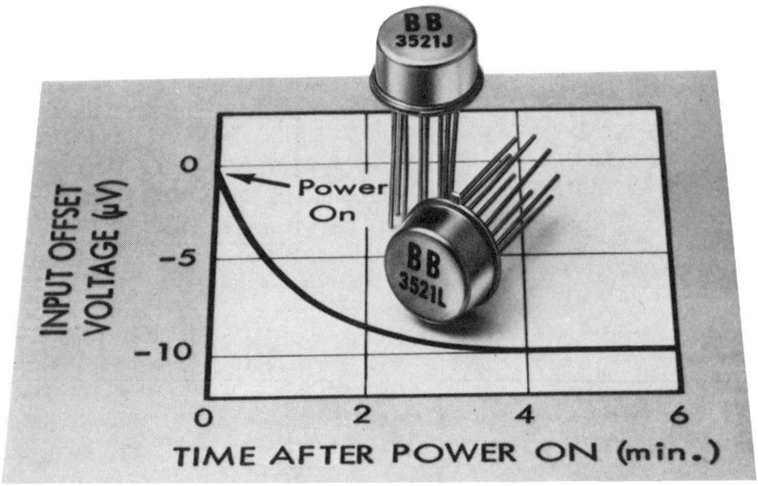

Figure 2-1. A typical op-amp: the Burr-Brown model 3521L. (Courtesy of Burr-Brown Research Corporation, Tucson, AZ)

*I assume that you can find or borrow some op-amps in whatever organization you are connected with. If this is not the case, you may as well turn to the section on "How To Buy Op-Amps" (pp. 95–96).

†Burr-Brown Research Corporation, P.O. Box 11400, Tucson, AZ 85734.

"First tell me what you can afford, and we'll have a good laugh and go on from there."

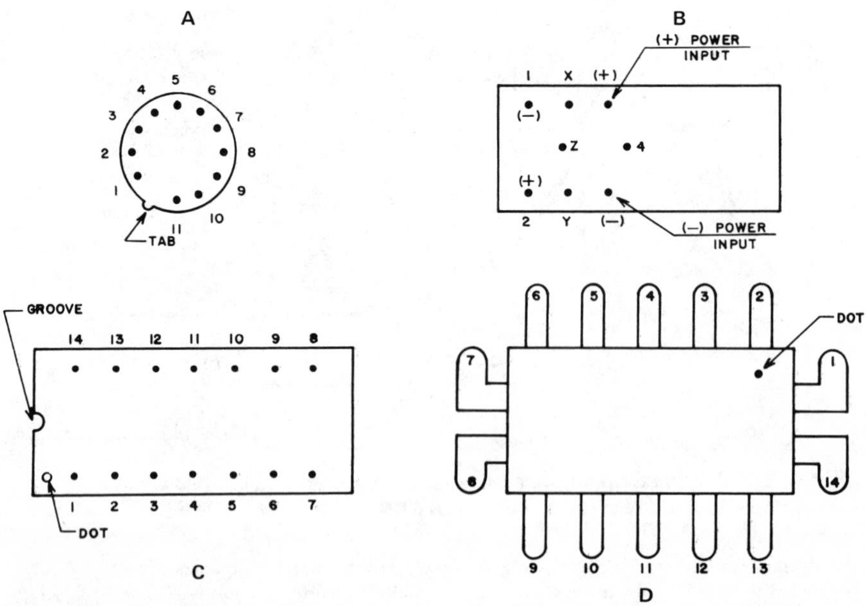

Figure 2-2. Op-amp packages. *A.* Op-amp can, bottom view. *B.* Op-amp module, bottom view. *C.* Dual in line package, top view. *D.* Flat pack.

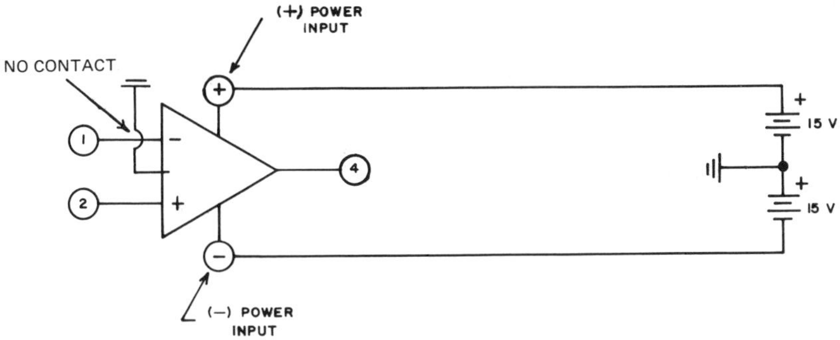

Figure 2-3. Power supply connections for an op-amp.

modules. Figure 2-2 shows the four types and how to identify the terminals. You *should not* – repeat, *should not* – be frightened by the large number of pins. The system is simple once you know the secret code. A few examples will make this clear.

Consider the Burr-Brown 4008/40 module op-amp. The module is shown in Figure 2-2B. The first thing you do is hook up the power supply. Hook up some batteries as shown in Figure 2-3 (batteries are much cheaper; power supply circuits will be discussed later). Numbers 1, 2, and 4 are Burr-Brown's old system. The plus and minus signs on pins 1 and 2 are the more universal notation that we mentioned before. Note that the center tap to ground *could* be run to earth through a water pipe, but for now "ground" can be any one point on your circuit breadboard. It doesn't *have* to be connected to anything; it is just your ground point. *Every* connection marked "ground" goes to this ground point.

If you are using a dual inline op-amp like the Motorola MC1437P (Figure 2-2C), the power inputs are –15 volts at pin 7 and +15 at pin 14. In future discussions *we will assume* that the power supply is connected to the op-amp. Don't be afraid: count the pins and hook it up.*

For the moment we will ignore the pins marked *X, Y,* and *Z* on the Burr-Brown module (Figure 2-2B), and the tabs 1, 3, 4, 9, 10, 11, 12, and 13 on the Motorola package (Figure 2-2C). We will use only three terminals for now: two inputs and one output. The inputs we will call (-) or (1) and (+) or (2); the output is obviously (4). On your op-amp, they may correspond to other numbers or letters. A look at your op-amp instruction book or catalog will tell you which pins are inputs and outputs. The *inverting* input is marked (-) or (1), and the *noninverting* input is (+) or

*When we wrote the first edition of this book, most op-amps required + and –15 volts from the power supply, and in many circuits you will see that a –15 and +15 supply has been drawn in. Before you say, "I can't find that much voltage in my application," you should be aware that in early 1979 two manufacturers announced op-amps that operated on –1.1 and +1.1 volt power supplies. Such is progress: for details see G. Mhatre, Low-voltage op-amps due from two vendors. *Electronic Engineering Times,* Feb. 5, 1979, P. 24.

(2). The other pins and the meaning of inverting and noninverting will be explained later.

A WORD OF CAUTION: we will assume that you have read and worked through Chapter 1 *or* that you know how to use VOMs, diodes, resistors, inductors, capacitors, signal generators, and oscilloscopes. If you don't, *please* go back to Chapter 1. You can't build bricks without straw. If you are at least vaguely familiar with the above instruments and devices, great! Proceed with vigor, and if you get stuck on a device or an instrument, go back to Chapter 1 and read up on it.

SOME ELEMENTARY OP-AMP CIRCUITS

Having hooked up the power supply, the next thing to do is learn how to draw an op-amp circuit and how to hook one up. Figure 2-4A shows a symbolic op-amp in which the inputs are designated by (-) and (+) or (1) and (2). The output is numbered (4). We have *assumed* that a power supply has been attached to the proper terminals and that the op-amp is ready to go. The *(-) or (1) input* means that *if a voltage that is positive with respect to the (+) or (2) input were applied, the output would go negative.* **Please note that the (+), (-) notation does *not* mean (+) voltages go in one terminal and (-) at another; it means that the (-) terminal *inverts* and the (+) one does not.**

Other and still worse forms of notation are shown in Figure 2-4B and C. In Figure 2-4C the *only* input terminal shown is the (-) or number (1) input. You are supposed to *know* that number (2), the (+) terminal, is connected to ground (which isn't shown either). In this book we will use the notation of Figure 2-4A, but be on your guard when venturing into strange territory.

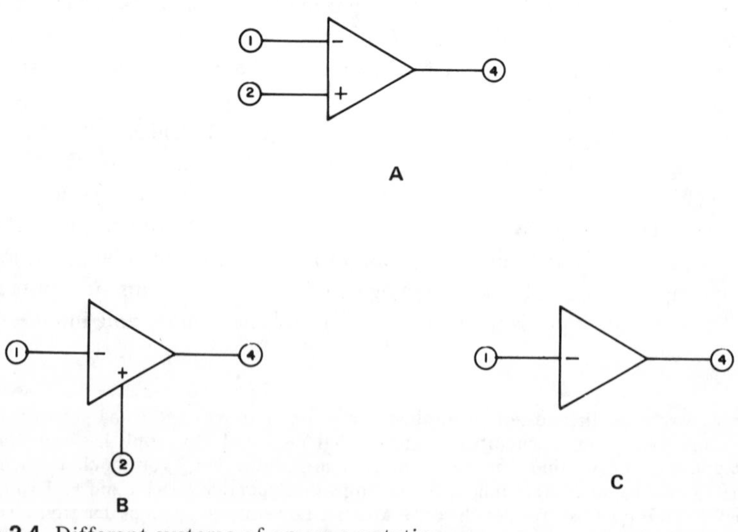

Figure 2-4. Different systems of op-amp notation.

Open Loop Amplifier Circuit

The first and simplest op-amp circuit is what we call an *open loop amplifier,* which is shown in Figure 2-5. The *ratio* of the output to imput voltage, V_0/V_a, is called the *open loop gain.** The gain of a good op-amp is large, sometimes as much as 10^5. This is great, but if our trusty op-amp has a maximum output voltage of, say, 10 volts, and if $V_0/V_a = 10^5$, an input of $V_a = 10^{-4}$ volt will drive the op-amp to saturation.† Stray voltages of 10^{-4} volt are all too common, and open loop circuits are used *only* for switches that are ON or OFF.

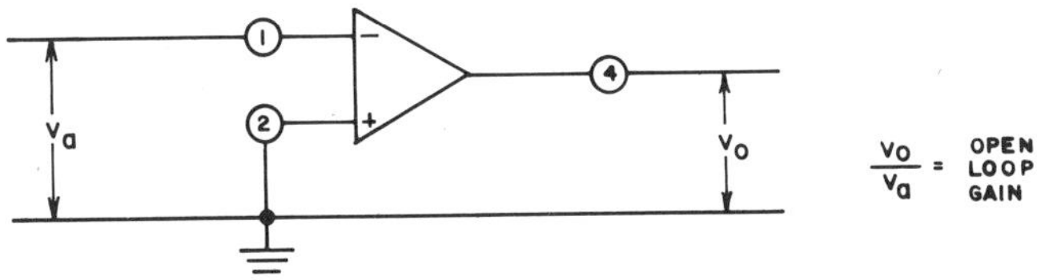

Figure 2-5. Open loop amplifier circuit.

Inverting Amplifier Circuit

We would like to have a *linear amplifier,* so that *if the input is a sine wave, the output looks like a bigger sine wave.* A linear amplifier is shown in the circuit of Figure 2-6, which can be considered our first practical op-amp circut, i.e., one you might really use. Notice that a resistor R_0 has been connected between the output and the input; this is called a *feedback resistor* because it feeds back some of the output signal to the input. This improves the amplifier in several different ways, as we shall see below. For the moment, the important point is the *new* relation between V_0 and V_1, to wit, $V_0/V_1 = -R_0/R_1$. Now the gain of the op-amp is *controlled* by the external resistors. More importantly, if R_0 and R_1 are *variable* resistors, we have an amplifier that can be controlled quite easily. We shall refer to the ratio V_0/V_1 as the

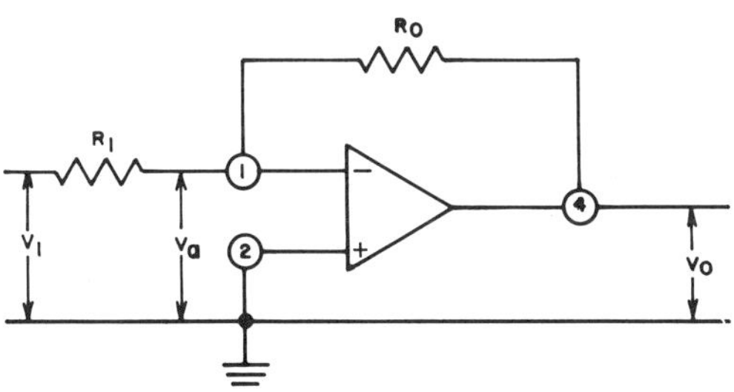

Figure 2-6. Inverting amplifier circuit.

*This definition of *gain* as output over input is a most important one and we urge you to remember it. It will be used extensively throughout Chapters 2, 3, and 4.

†Saturation means that the op-amp output goes to its maximum value (about 12 volts either plus or minus) and stays there. This isn't necessarily bad for the op-amp but it isn't very useful, so we try to avoid it if possible.

closed loop gain because we have closed the feedback loop by connecting R_0 between the output and the inverting input. Note that the closed loop gain is a function *only* of R_1 and R_0; it is *not* dependent upon the characteristics of the op-amp. That is the big difference between open loop gain and closed loop gain.

You should set this circuit up and play with it – yes, I said *play!* If you can't have fun with electronics, why bother learning about it? So have fun and gain confidence; op-amps are almost indestructible. Use a signal generator to produce V_1 (1 volt peak to peak is about right). If R_0/R_1 is about 5, the output will be about 5 volts peak to peak. Notice that this circuit inverts the sign or any signal you apply to it.

Try varying V_1 from 10 Hz to 10 kHz while holding the peak-to-peak amplitude constant and, at the same time, watch the amplitude of V_0 with an oscilloscope or a voltmeter (remember most meters read rms voltage, not peak-to-peak voltage). You will observe that V_0/V_1 drops off as the input frequency goes up above 10 Hz: this is called *amplifier roll-off.* This decrease in gain with frequency is a characteristic of all amplifiers. (You have to know the habits of an animal in order to catch and tame it.)

Gain versus input frequency data is presented in what EEs call a *Bode plot.* A typical Bode plot is shown in Figure 2-7. Notice that in Figure 2-7, gain is expressed in *decibels* (dB). We defined the decibel in Chapter 1, and we use it here as a convenient way to handle large numbers. The dB (gain in decibels) = $20\log_{10}(V_0/V_1)$. It is convenient to remember that a gain of 10^5 is 100 dB, 10^4 = 80 dB, 10^3 = 60 dB, 10^2 = 40 dB, 10 = 20 dB, and 1 is 0 dB. The decibel scale is confusing at first, but it has its virtues.

Looking back at Figure 2-7, we see that for this op-amp the gain is constant (and very large) up to 10 Hz. The gain drops off linearly, on a logarithmic plot, until at 1 MHz (10^6 cycles per second) there is *no* gain. It is important to note the frequency at which the gain begins to fall off (often called f_o) and the frequency at which the gain falls to zero dB or *unity gain* (called f_t). The slope is 20 dB per decade or 6 dB per octave. This means that if we double the frequency of the input signal (from, say, 100 Hz to 200 Hz), the gain drops from 80 dB to 74 dB, or a drop

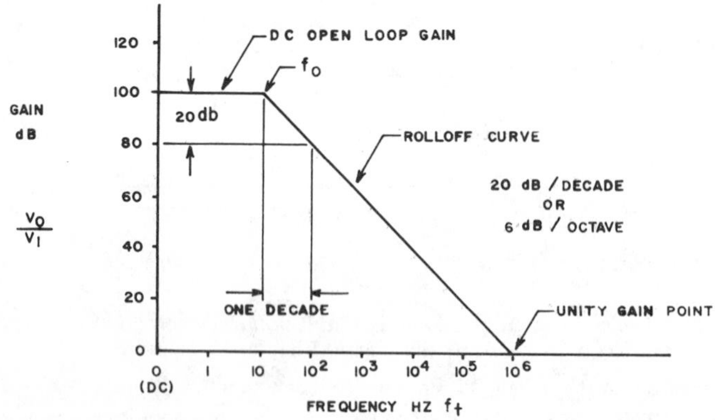

Figure 2-7. A typical Bode plot.

of 6 dB. If we go from 100 Hz to 1000 Hz, the gain drops to 60 dB, a loss of 20 dB. This notation may seem complex (and it is), but it is also useful in evaluating amplifiers.

To see what roll-off* means in an actual circuit, let's consider a circuit operating at a gain of 20 dB ($V_0/V_1 = 10$). The gain stays constant up to 0.1 MHz, so the circuit can be expected to operate without problems from DC to 0.1 MHz. If a gain of 1000 is needed, the gain begins to fall off at 1 kHz. This is called the *gain* × *bandwidth product,* and, roughly speaking, *the product of the gain and bandwidth is a constant.* High gain means low bandwidth (remember that concept).

Op-amp manufacturers try to get the highest gain-bandwidth product, but of course every improvement adds its factor of increased cost. The fight for increased gain-bandwidth has led to another problem, namely, a change in the slope of the roll-off curve with frequency. We will come back to this problem later; for now, just don't expect both high gain and high frequency response at the same time.

Noninverting Amplifier Circuit

Having put together an op-amp circuit, seen it work, and observed its gain-bandwidth behavior, we are ready for our next circuit: the *noninverting amplifier* (Figure 2-8). This circuit has several interesting features: first, the gain is given by $V_0/V_2 = (R_0 + R_1)/R_1$; second, we note that now the output and input are of the *same sign* – the amplifier does *not* invert the input signal. The gain of this amplifier can be controlled by varying either R_1 or R_0, but our gain equation indicates that varying R_0 is more effective (R_1 also occurs in the denominator). For best sensitivity, R_0 should be greater than R_1.

Once again, *you should set up this op-amp circuit, play with it, and test its response to signals of different frequencies.* The gain-bandwidth effect shown previously with the inverting op-amp circuit is easily observed with the noninverting circuit. You may wonder about the resistor R_3: this is something you *might* have to put in if you use certain types of signal generators. The reasons for this will be discussed in more detail in Chapter 6; for now, consider R_3 as something you put in *if*

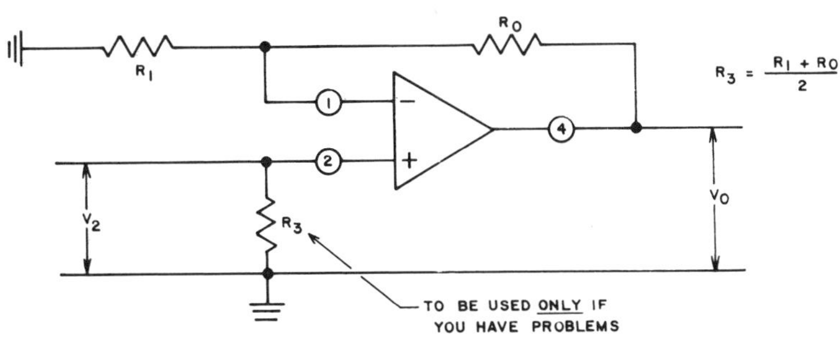

Figure 2-8. Noninverting amplifier circuit.

*Recall – or review – the discussion of roll-off in connection with filters in Chapter 1 (p. 22).

the op-amp fails to work or saturates. Of course, with R_3 in place, the input impedance is no longer infinite but R_3. (Life is like that!)

As long as we are talking about problems, we must prepare you for that old devil: *oscillation.* Op-amp circuits go into oscillation sometimes, and then the output, V_0, looks like a messed-up sine wave no matter what you do to the input. Don't worry about it. We will discuss this problem and its complete cure later. For now, put a capacitor of about 100 pF (1 picofarad is 10^{-12} F) across R_0 and the problem should go away (if 100 pF doesn't work use 10 μF). This will work on either the inverting or noninverting amplifier; it will cut your high-frequency gain but the DC gain will not change.

Returning to the op-amp system, we must introduce some new concepts if you are to use the capability built into every op-amp by the manufacturer. One of these concepts is the notion of *impedance.* When we speak of op-amp impedance, we are talking about *how much* electrical resistance we find at the input terminals. This resistance in turn defines how much current the op-amp will take from a voltage source. For example, if a hypothetical op-amp has an imput impedance (resistance) of 10,000 ohms, how much current will it draw from a 1 volt source? The answer is provided by Ohm's law; since $V = IR$, $I = 1/10^4 = 0.0001$ amp. A 1 MΩ resistance will draw only 1 μA, and so on. This is important to us because some voltage sources have *very little* current capability. A thermocouple, for example, might require a 5 kΩ minimum load impedance; a pH meter will yield a poor or negligible reading when attached to a voltmeter with less than 1 MΩ of impedance.

The concept of impedance is important enough – and confusing enough – to warrant some further discussion. Suppose that our signal source is a 10 volt battery in series with a 1 MΩ resistor (this might well be a certain type of pH meter), as shown in Figure 2-9. A VOM and a VTVM* are used to measure the source voltage. The VOM has an internal resistance of, say, 20 kΩ, and the VTVM has an internal resistance of 10 MΩ. The VOM will draw a current I of

$$I = \frac{10}{10^6 + (2 \times 10^4)} = 9.8 \times 10^{-6} \text{ amps}$$

The VTVM will draw

$$I = \frac{10}{11 \times 10^6} = 9.1 \times 10^{-7} \text{ amp}$$

The voltage drop across the VOM will be

$$(2 \times 10^4)(9.8 \times 10^{-6}) = 0.196 \text{ volt}$$

*Look up VOMs and VTVMs in Chapter 1 if you have forgotten them.

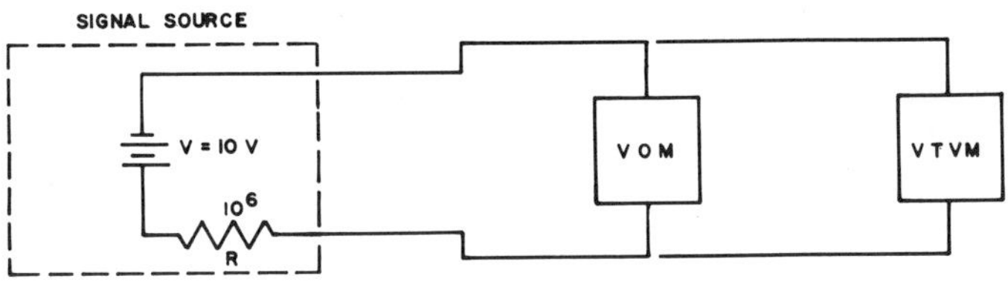

Figure 2-9. Measuring the voltage of a high impedance source.

The remaining 9.8 volts is *across* the internal resistance of the signal source. Our VOM reading is in error by a factor of

$$\frac{10 - 0.196}{10} \times 100 = 98\%$$

With the VTVM, the voltage drop is $(9.1 \times 10^{-7})(10^{7}) = 9.1$ volts; the remaining voltage is dropped inside the signal source. The error is now

$$\frac{10 - 9.1}{10} \times 100 = 9\%$$

This extreme example demonstrates the importance of impedance control and the matching of source output and detector input resistances. The term *matching* is used here to mean that a device with a high output impedance should be monitored by a detector with a still higher input impedance; in this sense, they are matched.

The next question might be: What is the input impedance of an op-amp? Here we must introduce a bit of philosophy. You probably didn't know EEs were philosophers; indeed they are and they work very hard at it (of course, most of them don't realize that some of these ideas are over 2000 years old). Plato introduced the idea of the "ideal" in the sense that there existed, apart from the various imperfect particular instances that we see in the world, an ideal chair, table, man, and so on. The properties of this ideal form could be defined and understood better than those of individual instances. It followed that the more our "real" chair, table, etc., became like the ideal, the more perfect it would be. The ideal need not "really" exist except in one's mind, but the definition of the general properties of the ideal can be useful in our acquiring knowledge of the real world. In electrical engineering we use the term *models* instead of ideals, but Plato would feel quite at home with an up-to-date book on the theory of modeling in electrical engineering. ("The more things change, the more they remain the same.") We will define the ideal op-amp and use the concept to analyze circuits; then in Chapter 6 we will see how real op-amps and circuits differ from the ideal. We can come far closer to the ideal op-amp than any man has ever come to Plato's ideal man, which might or might not be of some significance. The solution of that question is left as an exercise for the reader.

IDEAL OP-AMPS

The ideal op-amp has infinite input impedance and draws zero current from a voltage source. The ideal op-amp has *infinite gain, zero offset voltage, zero phase shift, and zero output impedance.* To see what these strange terms mean, let's return to our first useful circuit, the inverting op-amp.

In Figure 2-10, op-amp inputs (1) and (2) are summing points.* Point (2) is at ground potential and may be considered to have zero voltage.

The input voltage V_1 is applied as shown in Figure 2-10. If the op-amp has zero input current, any input current through R_1 *must* flow out through R_0. The ideal op-amp has infinite impedance, so *no* current flows into the op-amp at terminal (1).

Let's see what we can deduce from these exciting facts. First, the current I_1 is *entirely* controlled by V_1 and R_1, so $I_1 = V_1/R_1$. No current flows into the op-amp, so $I_1 = I_0$. Second, since $I_0 = I_1$, then V_0 must be large enough to push a current I_0 through R_0. Hence, $V_0 = I_0 R_0 = I_1 R_0 = V_1 R_0/R_1$. This leads us to the relationship between V_0 (the output voltage) and V_1 (the input voltage):

$$\frac{V_0}{V_1} = -\frac{R_0}{R_1}$$

The minus sign, which appeared without warning, comes from the fact that this is an *inverting* amplifier: if V_1 is (+), then V_0 is (-). The thing to remember is $V_0/V_1 = -R_0/R_1$. This is an important relationship and you will be seeing it again and again.

To make sure that you really appreciate these great ideas, we will go through

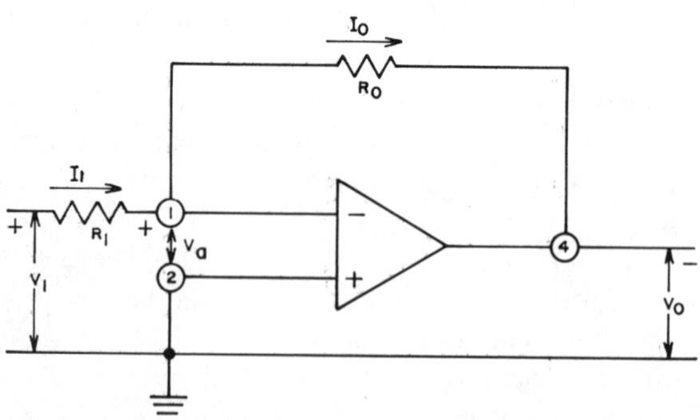

Figure 2-10. Inverting amplifier circuit showing current relationships.

*A *summing point* is any point where we apply the current conservation law: *The sum of all currents in and out of a summing point must be zero.* In this case, at point (1) we have $I_1 - I_0 = 0$. A *summing point* is sometimes called a *nodal point*.

this analysis again in a slightly different way. Applying our current conservation rule to junction 1, we write:

$$I_0 = I_1$$

Assume for the moment that the op-amp input (1) is at some voltage, V_a (we don't really need to know how much). The current through resistor R_1 is, according to Ohm's law,

$$I_1 = \frac{V}{R_1} = \frac{V_1 - V_a}{R_1}$$

Similarly, the current in resistor R_0 is

$$I_0 = \frac{V_a - V_0}{R_0}$$

Since $I_1 = I_0$,

$$\frac{V_1 - V_a}{R_1} = \frac{V_a - V_0}{R_0}$$

If $V_a = 0$, we would have $V_1/R_1 = -V_0/R_0$, or $V_0/V_1 = -R_0/R_1$, the *gain law* for this circuit. The next problem is to convince ourselves that V_a – the voltage at point (1) with respect to ground – is zero. Recall from our discussion of an open loop amplifier (Figure 2-5) that the gain, which we will now call A, was defined as V_0/V_a. It follows that $V_0/A = V_a$. If A, however, is infinite, then $V_a = 0$. And one of the properties of our ideal op-amp is that it has infinite gain. This was what we needed to prove that $V_0/V_1 = -R_0/R_1$.

One question that frequently bothers students is "why" the voltage V_a is zero when a feedback resistor has been connected in the circuit. The answer is simple when you think about it. If V_a is (+) with respect to ground, the voltage fed back from the output is negative – it drives V_a to zero. In fact, the feedback voltage is just large enough to keep $V_a = 0$; the bigger V_a, the bigger the feedback voltage and vice versa.

At this point you probably feel that you have been had, and in a sense you have. However, the trick was played for a good cause; you should be convinced that if A (the gain) is large and there is feedback, the *voltage between points (1) and (2) is vanishingly small.* This is important since it allows us to assume that for our ideal op-amp, points (1) and (2) are at the *same voltage.* There is an important and subtle point here: the op-amp input impedance is is infinite, but if *any* feedback exists from (4) to (1), then inputs (1) and (2) are at the *same* voltage. This means that the effective resistance between points (1) and (2) is zero, and the op-amp circuit impedance is controlled by R_1 *only.*

If we connect a 1 volt battery as V_1 and if R_1 = 10 MΩ, then I_1 = 0.1 μA. This current does not really flow into (1) and out of (2); rather, it flows in through R_1 and out through R_0, but the effect of the circuit is as though (1) and (2) were connected together.

This inverting op-amp circuit allows us to match the output impedance of any device by choosing R_1 properly. However, we *must avoid* making the common beginner's mistake that is shown in Figure 2-11. Here the source has an impedance of 100 kΩ, and the student matches it with a resistance of R_1 = 100 kΩ. The current is I = 1 volt/200 kΩ = 5 μA, and the voltage drop across the internal resistance is IR = 0.5 volt. So the best op-amp in the world still would give us a 50% error. Why? The answer is we *drew* current from the voltage source.

To get away from this problem we have to use another op-amp circuit, which will be presented in due course. At this point, we urge you to remember that the input impedance of the op-amp is *infinite.* However, when we look (in the electrical sense) into the inverting amplifier circuit, we see an impedance of R_1 ohms, not infinity. The reason is simple: point (1), the (-) terminal, is at ground potential because V_a = 0. This means that the input impedance is R_1, the external resistance in the input circuit. It follows that the current take is controlled by both R_1 and the source resistance in the voltage source.

This is a point that causes much confusion for beginners in the "art of op-amps." The op-amp has certain properties of its own, i.e., infinite input impedance. However, when we hang resistors, capacitors, etc., across the op-amp, its characteristics change. An analogy might be the familiar auto engine. It has certain torque/speed characteristics, but we can put various transmissions and differentials in the system that provide a variety of properties. An example might be the 6 cylinder engine: its behavior in a Chevie truck and a Datsun 280 Z are very different.

There is another point that we must emphasize: if there is feedback from the output to the (-) input, the voltage V_a between the (-) and (+) inputs is zero. This does *not* mean that the (-) and (+) inputs are short-circuited inside the black box; rather, it is a property of the feedback circuit we have built.

Remember that the *effective* input impedance of this circuit is R_1, and that we can match the output of any device by simply adjusting the value of R_1. Those of

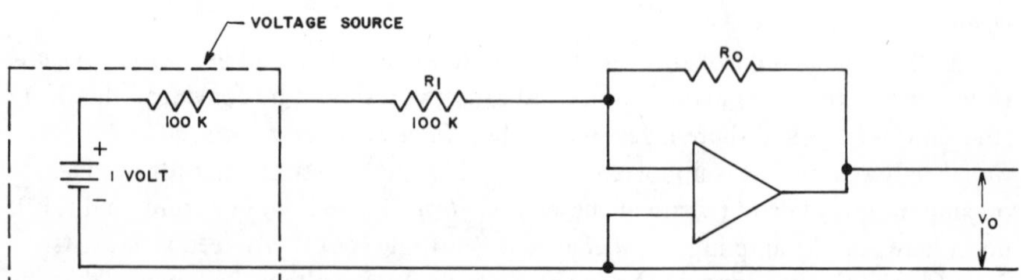

Figure 2-11. How *not* to measure the output voltage of a high impedance source.

you who have tried to match a high-fidelity amplifier with a high-output impedance to a loudspeaker with an 8 ohm coil will understand how useful this property of op-amps can be. Of course, a simple op-amp won't drive an 8 ohm speaker by itself, but we will show you how to handle that problem later in Chapter 3.

While we are on the question of impedance matching, we should discuss two *separate and distinct* uses of that word.

The first application is *measuring a voltage.* In this case we want a very high input impedance so that essentially *all* the source voltage will be across the measuring device. We saw, in the example of measuring a signal source with a VTVM (Figure 2-9), that 10 MΩ was not really enough when the source impedance was 1 MΩ.

The second case of interest is when we wish to transfer *maximum power to the load.* Every electrical engineering text goes through a derivation that proves that maximum power transfer occurs when the load impedance is *equal* to that of the source. This is illustrated in Figure 2-12. In Figure 2-12,

$$I = \frac{V_1}{R_1 + R_2}$$

and

$$V_2 = I R_2 = \frac{V_1 R_2}{R_1 + R_2}$$

Therefore, if $R_1 = R_2$, then $V_2 = V_1/2$; that is, only *half* the source voltage appears across the load. This is a good matching system for the *best delivery of power* to the load. Once again, however, we repeat that it would be a *very bad* system if you wanted to measure the source voltage V_1.

Another point that sometimes causes confusion is the *current* that an op-amp delivers to a load. With normal inverting or noninverting circuits, we control V_0 by means of resistors, i.e.,

$$\frac{V_0}{V_1} = -\frac{R_0}{R_1}$$

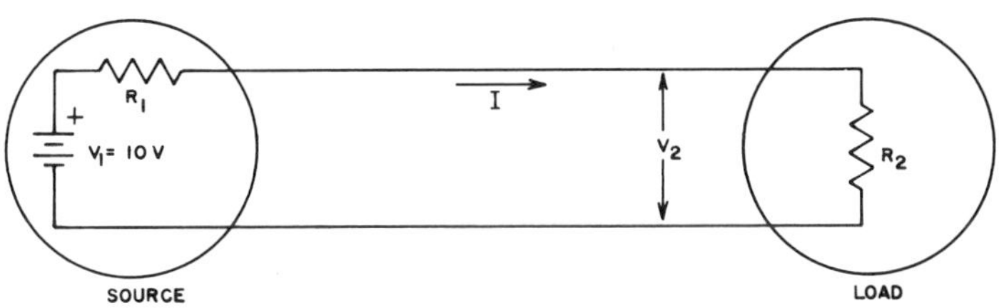

Figure 2-12 Power transfer relation between source and load.

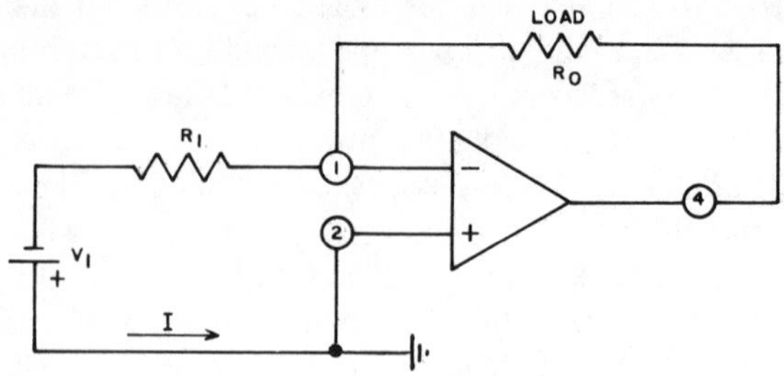

Figure 2-13. Constant current circuit.

or

$$\frac{V_0}{V_2} = \frac{R_0 + R_1}{R_1}$$

The current through the load is set by V_0 and the *load resistance*, R_L. If $V_0 = 8$ volts and $R_L = 10\ \text{k}\Omega$, then

$$I_L = \frac{8}{10^4} = 0.8\ \text{mA}$$

This is entirely *independent* of the input current I_1; it *only* depends upon V_1, R_1, and R_0. In this case the op-amp acts as a *constant voltage generator.*

However, op-amps can also act as *constant current generators* or current sources. A typical constant current circuit is shown in Figure 2-13.* Here V_1 delivers a constant current through R_1 that *must* flow through R_0. Now if R_0 varies, the current through R_0 *remains constant* because V_1 and R_1 are constant. So if R_0 is our load, the current in the load is constant *regardless* of how the load resistance changes with time. The op-amp has become a *constant current source.*

Flushed with these victories, we can resume analyzing the noninverting op-amp circuit of Figure 2-8 in the light of what we have learned from studying ideal op-amps. Redrawing Figure 2-8 as Figure 2-14, we put in current arrows and the hypothetical (but useful) voltage V_a between op-amp inputs (1) and (2).

Since *no* current enters the ideal op-amp at point (1), we know that $I_0 = I_1$. From Ohm's law,

$$I_1 = \frac{V_2 + V_a}{R_1}$$

*A "neat" application of this circuit is shown in Figure 3-6, where you will discover how to make a cheap ammeter act like one that costs a lot more.

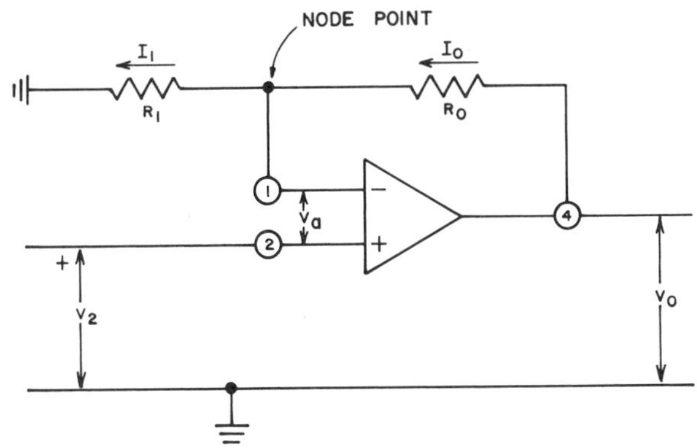

Figure 2-14. Noninverting amplifier circuit showing current relationships.

and

$$I_0 = \frac{V_0 - (V_2 + V_a)}{R_0}$$

hence,

$$\frac{V_2 + V_a}{R_1} = \frac{V_0 - (V_2 + V_a)}{R_0}$$

If we now assume that $V_a = 0$ (which we can do since this is an ideal op-amp) and use some algebraic tricks, we can show that

$$\frac{V_0}{V_2} = \frac{(R_0 + R_1)}{R_1}$$

This was the gain law that was given earlier for this circuit. This is a *very* important equation. We suggest you not merely learn it, but actually understand where it comes from.

Now we might ask, what is the input impedance of *this* circuit? The answer is infinity: *no current* flows into the ideal op-amp at point (2). For commercial op-amps, however, the actual input impedance ranges from perhaps 300 kΩ for cheap ($2) devices to 10^7 MΩ for expensive op-amps. In actual practice there are a few tricks to make the op-amp act as though the input impedance were almost infinite. We will tell you about those tricks in Chapter 6.

By the way, if you put this op-amp circuit on the voltage source of Figure 2-11, the current drain would be *zero*. The voltage drop across the internal resistance would be zero, and the op-amp would measure the *correct* 1 volt output voltage.

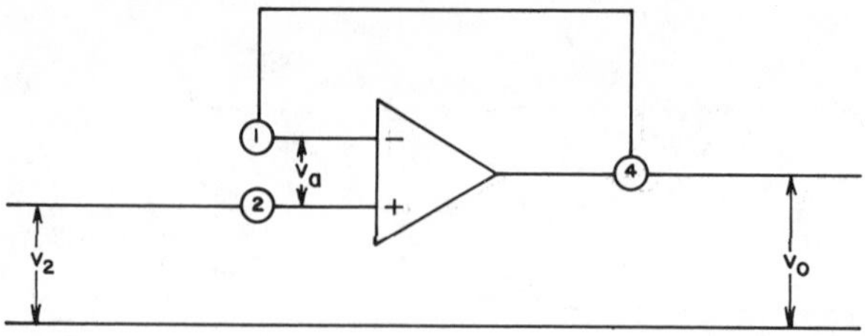

Figure 2-15. Voltage follower circuit.

Another important op-amp circuit is the *voltage follower*. A typical voltage follower circuit, shown in Figure 2-15, is often used to match high-output-impedance devices to low-input-impedance recorders. The voltage follower does *not* produce any gain, but it is extremely stable and effective as an impedance matching circuit.

The analysis of the voltage follower circuit is quite simple: since $V_2 + V_a = V_0$ and $V_a = 0$ (being an ideal op-amp), then $V_0 = V_2$. This is a *unity gain* circuit with *infinite input impedance* and *zero output impedance*. Circuits of this type are frequently used for "signal conditioning" if signals are to be sent over long wires.

The concept of signal conditioning is worth some added discussion. Suppose you are in a copper refining plant where a pH meter is measuring the acidity of a leaching solution. The man in the control center wants to know what the pH is on a continuous basis, but somehow you suspect that a conventional pH meter circuit will not pass current through the two miles of wire between the leach tank and the control center. What to do? You put in an op-amp circuit as shown in Figure 2-15. The input impedance is very high (infinite for an ideal op-amp) so the current taken from the pH meter is very low. Great! The output impedance of the op-amp is very low so it easily drives the necessary current through the two miles of wire – presto, the problem is solved. Industrial operations make use of many, many op-amps.

Here again we emphasize that you should be building op-amp circuits while reading this book. Just reading will never get you to the point of being confident in the use of electronics. ("Be ye doers of the word and not hearers only.")

SOME IMPORTANT OP-AMP CONCEPTS

Common Mode Versus Differential Mode

At this point we must clarify a sometimes confusing definition relating to the input impedance and characteristics of real op-amps: namely, *common mode* versus *differential mode.* The meaning of these terms is illustrated in Figure 2-16. In this diagram there is a resistance R_{DM} between inputs (1) and (2), and a resistance R_{CM} from (1) and (2) to ground. You should note that all these resistors are *inside* the op-amp; you *cannot* change them. The manufacturer will give a *common mode impedance* in the specification sheet. This is the resistance R_{CM}. If we connect input points (1) and (2) together and apply a voltage V, the current to ground will be $I = 2(V/R_{CM})$ because the R_{CM} resistors are in parallel.

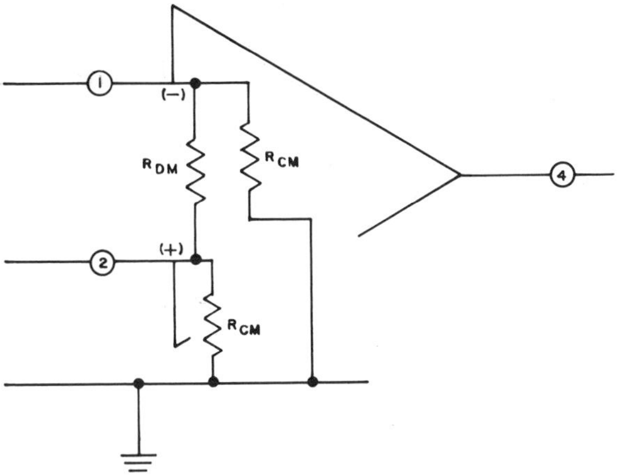

Figure 2-16. Op-amp input impedances.

If we apply a voltage difference *between* inputs (1) and (2), this is called *differential input.* The impedance in differential input, R_{DM}, is usually smaller than R_{CM} for common mode input. If it is 5 MΩ, the current drain due to a 1 volt difference between (1) and (2) is 0.2 μA.

An important characteristic of op-amps is called the *common mode rejection* (CMR). This simply means that if the voltages applied to inputs (1) and (2) are equal in both sign and magnitude, the output of the ideal op-amp is *zero.* This is easy to see if we note that terminal (1) is the inverting input, whereas (2) is the noninverting input. Suppose V_1 is a positive signal; it thus produces a negative output at (4). However, if $V_2 = V_1$, then V_2 produces a positive output at (4), and the net output is zero. This is illustrated in Figure 2-17. Note in Figure 2-17 the ground connection between V_1 and V_2 is not needed in a real circuit. We use it here as an analytical convenience.

In Figure 2-17, $I_1 = I_0$ and $I_2 = I_3$, because no current can flow into the op-amp. Now V_a is zero so

$$I_1 = \frac{V_1 - V_3}{R_1}$$

and

$$I_0 = \frac{V_3 - V_0}{R_0}$$

but, since $I_1 = I_0$,

$$V_0 = V_3\left(1 + \frac{R_0}{R_1}\right) - V_1\left(\frac{R_0}{R_1}\right)$$

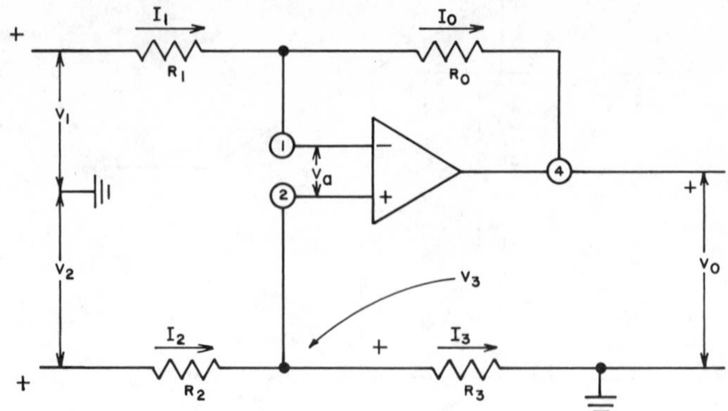

Figure 2-17. Common mode input.

(Remember that V_a is zero.)

In the lower part of the circuit, $I_2 - I_3 = 0$, and

$$I_2 = \frac{V_2}{R_2 + R_3}$$

So since $V_3 = I_3 R_3$ and $I_2 = I_3$,

$$V_3 = V_2\left(\frac{R_3}{R_2 + R_3}\right)$$

This can be substituted into the above equation for V_0 to yield

$$V_0 = V_2\left(\frac{R_3}{R_1}\right)\left(\frac{R_1 + R_0}{R_2 + R_3}\right) - V_1\left(\frac{R_0}{R_1}\right)$$

Now if $R_2 = R_1$ and $R_3 = R_0$, this reduces to

$$V_0 = \left(\frac{R_0}{R_1}\right)(V_2 - V_1)$$

and therefore if $V_1 = V_2$, *the output is zero.*

The common mode rejection capability of an op-amp is rated in terms of a standardized test. Assume that we apply a 1 volt signal to both the (+) and (-) inputs *after* connecting them together. The output voltage is some value B volts, because the op-amp common mode rejection is not perfect. Now we apply a voltage of C

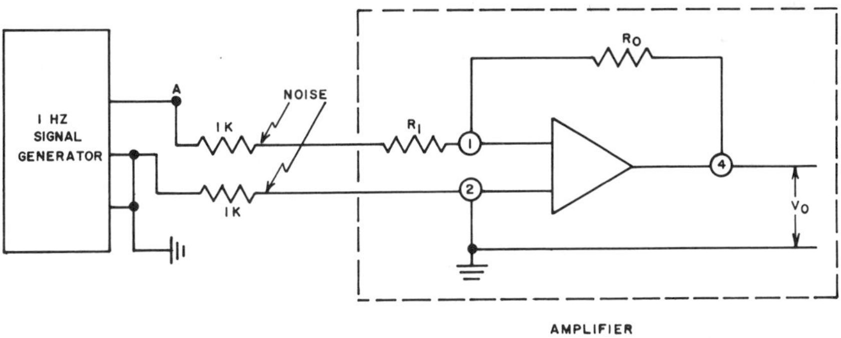

Figure 2-18. Difference circuit.

volts *between* the (+) and (-) inputs until the output again reaches *B* volts; the common mode rejection of the op-amp is the ratio of $1/C$. Since C is usually about 10^{-5} the common mode rejection ratio is anywhere from 80 to 100 dB. There will be all sorts of applications of this concept in Chapter 3 so we urge you to learn it now!

To demonstrate how the common mode rejection of an op-amp is used, we will do an example. This is a problem you might well encounter in actual practice. In Figure 2-18 we show a signal source that we want to run through wires to an op-amp. Let's assume that the signal is a 1 Hz square wave from a grounded signal source. At one part of the cycle, point *A* is (+); at some time later, point *A* is (-). This is our signal. To this signal, we add 60 Hz noise from lights and motors. Thus, if the signal is 1 volt peak to peak and the noise is 0.5 volt peak to peak, the result is a mess. If we use an op-amp amplifier in the usual mode (Figure 2-18), it amplifies everything, both *signal and noise.*

However, if we use a differential input mode as shown in Figure 2-17, the 60 Hz noise disappears: it is rejected as being common to both inputs. The 1 Hz square wave comes through just great. To make sure you appreciate this achievement, we have designed a laboratory demonstration, which is shown in Figure 2-19.

In Figure 2-19, signal source *X* is the "good guy" signal going to the op-amp in the differential mode. Note that *X* is ungrounded – this is an important point. Of course, this requires a signal source with a *floating output* (recall that from Chapter 1, Figure 1-35). Signal source *Y* is the bad guy, the *noise.* Notice that it is delivered to *both* leads at the same time. Now look at the op-amp output and notice that the *X* signals are *amplified* and the *Y* signals are *rejected.* This demonstrates the ability of the differential op-amp circuit to reject common mode signals.

To see what happens without differential input operation, hook up the circuit shown in Figure 2-20. It is a grounded input or *difference* (*not* differential) circuit. Once again, *X* is the good signal and *Y* the bad signal. However, now both *X* and *Y* are amplified and that means trouble. So you should remember the use of differential input and the meaning of common mode rejection. Common mode rejection is a *most important* part of industrial instrumentation technology.

As an example of the common mode rejection we might consider the applica-

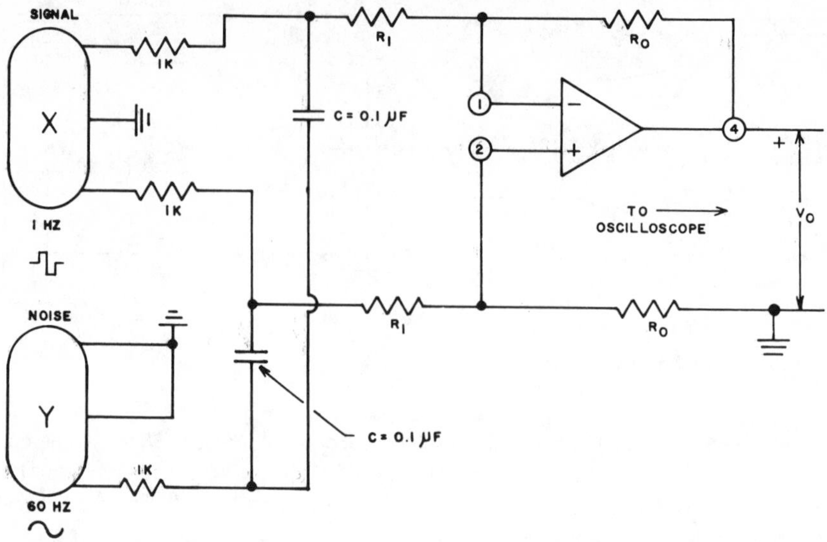

Figure 2-19. Experimental differential input system. (Note that the lower resistor R_0 is needed to balance the R_0 feedback resistor.)

tion to thermocouples. In Figure 2-21 we show a typical copper-constantan thermocouple pair (constantan is an alloy having specific thermocouple properties). The voltage developed is a function of the difference of temperature $T_h - T_c$ and a constant K as $V = K(T_h - T_c)$. In an "electrical" sense we have a voltage source

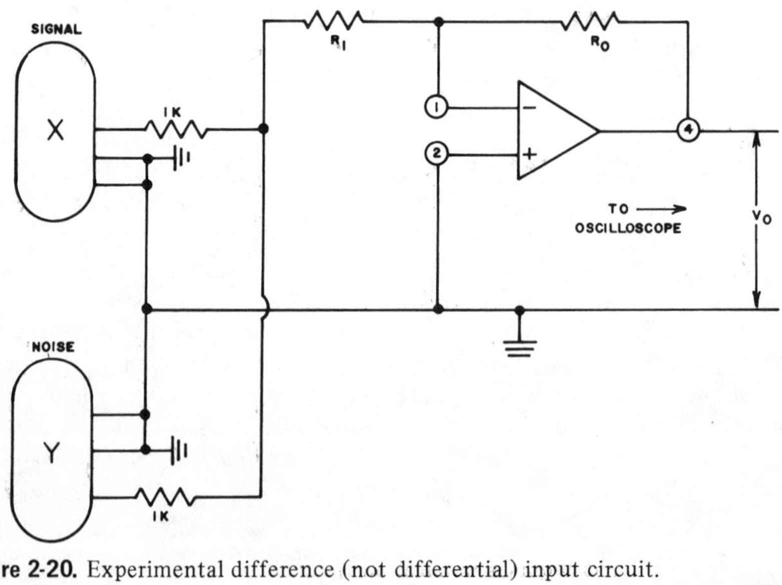

Figure 2-20. Experimental difference (not differential) input circuit.

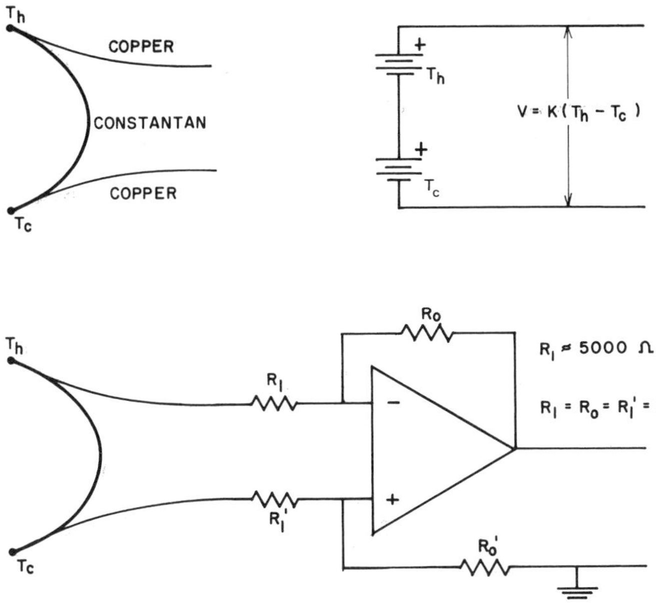

Figure 2-21. Thermocouple connected to an op-amp in the differential mode.

with a "high" 5000 ohm internal resistance. (Thermocouples are *not* current devices.)

In Figure 2-21 we also show the thermocouple connected to a differential op-amp circuit. (A circuit of this type is used in Arizona for the control of air conditioning and evaporative cooler systems. The evaporative cooler works fine as long as the air is dry; when the summer monsoons come, you have to switch to the much more expensive air conditioning. The thermocouple circuit is used to sense the relative humidity by having one junction "dry" and the other inside a "wet" sock. At low relative humidity, the temperature difference is large and the signal turns on the evaporative cooler and turns off the air conditioner. When the air is moist, the temperature difference decreases and the system turns the cooler OFF and the air conditioner ON. The unit sells commercially for $65 and you can build one for $10.)

Notice in Figure 2-21 that resistors R_1 and R_0 are about 5000 ohms and R_1' and R_0' are also 5000 ohms. This is done so that the resistance to ground will be the same for both the T_h and T_c leads.

In most circuits the signal source "knows" where ground is (for example, Figure 2-18), but with thermocouples it is poor practice to ground either of the junctions. So you insure proper grounding as shown in Figure 2-21. Note that T_h "sees" ground at the (-) input via a total of 10,000 ohms plus the load resistance, which is usually rather low. To make everything balance we have to insure that the "cold" junction T_c has the same resistance to ground; this is done by making R_1' and R_0' equal to 5000 ohms each. The reason for all this fuss is that there will always be some current drawn by any real op-amp and that current has to come from the power supply via the "ground" connection. Another trick that might be used if we don't have the CMR circuit shown in Figure 2-21 involves connecting 2 MΩ resistors from each of the thermocouple connections T_h and T_c to ground. The 2MΩ resistors will not draw enough current to upset the op-amp or the thermocouple reading.

Returning for the moment to the topic of difference versus differential circuits we might note that there is a confusing problem of nomenclature. Correctly speaking, the circuit of Figure 2-17 should be called a *common mode noise rejection* or *differential input circuit.* A circuit like those shown in Figure 2-18 or 2-20, where the signal input is with respect to *ground,* ought to be called a *difference circuit.* Unfortunately, there is no consistency in this matter, either in the industry or in the literature. You just have to look at the circuit and figure it out for yourself. We have tried to be consistent with our notation in this book, but we *do* have to show you how things are done in the real world.

The Instrumentation Amplifier

The previous section should have convinced you of the advantages of differential input for common mode noise rejection. In fact the advantages of this circuit are so great that a whole spectrum of specialized op-amps with very high CMR have sprung up. They are called *instrumentation amplifiers,* and a word or two on their properties might be in order.

We begin by noting that with a regular op-amp the gain is under your control, you simply change the feedback resistors. However, the design of the unit is such that it is hard to get high gain and good CMR at the same time. The situation is even worse if the two input resistors (Figure 2-21) are not identical in value. (This can happen in industrial or medical applications and is obviously a problem.) Another difficulty with the usual op-amp circuit, with CMR, is living with relatively low input impedance and obtaining a good CMR at the same time.

The solution is – you guessed it – an instrumentation amplifier. It is a high-impedance, high-gain device with excellent CMR. One thing you have to give up is control over the gain. It is usually fixed or adjustable only within narrow limits. This is really no problem; you can always follow it with a standard op-amp amplifier.

To appreciate how this is done, we look at Figure 2-22, which shows the input as two op-amps in the buffer mode. The input impedance is high, and this makes for good CMR. The buffers are followed by a regular differential input op-amp that provides gain. All of these goodies are built into a single package to spare you the problem of having to hook up and balance the circuits.

For further details on instrumentation amplifiers we refer you to our friends at Burr-Brown Research, or the book by J. Graeme, *Applications of Operational Amplifiers – Third Generation Techniques.* New York: McGraw-Hill, 1973. One comment on the use of these devices with thermocouples or signal sources that have

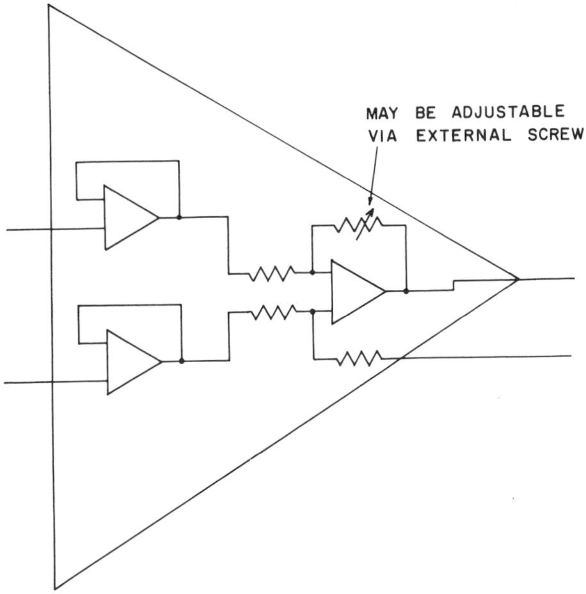

Figure 2-22. The instrumentation amplifier.

no ground connection – you do have to provide some information about ground to the op-amp input, and we show a typical setup in Figure 2-23. The 1 MΩ resistors will reduce the input impedance of the circuit from the value set by the instrumentation op-amp (10^{11} Ω in some cases) and if that is a problem you might try 5 MΩ instead. If that doesn't work you will just have to look at another book (such as Graeme's).

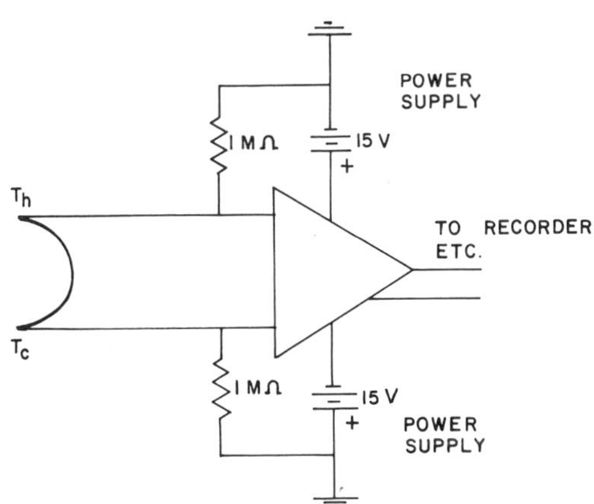

Figure 2-23. The instrumentation amplifier connected to a thermocouple pair.

Other Useful Concepts

To complete our discussion of op-amp characteristics, we must define a few other terms like *response time, slew rate, settling time, phase shift,* and *feedback factor.* These are not very important in our immediate application of op-amps, but you might be buying op-amps some day and you will want to know what the words mean.

RESPONSE TIME

To define *response time,* we assume that a small voltage (about 50 mV) has been applied to the op-amp input. This signal input causes a change in the output voltage, but this change does *not* occur in zero time. The actual response follows a curve like that shown in Figure 2-24. The output voltage time curve is given by the equation

$$V(t) = V_0 \, (1 - e^{-t/k})$$

Here V_0 is the final "correct value," k is a constant, t is time in seconds, and e = 2.718 (the base of natural logarithms). Looking at our equation, we see that when $t = k$ the output voltage is

$$V(t) = V_0 \, (1 - 1/2.72) \approx 0.63 \, V_0$$

or about two-thirds of its final value. This 63% point (or time) is often called the *eith time* because the difference between the actual and final voltages has been reduced by the factor 1/e. The eith time gives us a measure of how fast things are going. The important point is that some manufacturers refer to the eith time as the *response time.*

Another term that you will see is the *90% time.* This is the time for $V(t)$ to equal 0.90 V_0. Attaining this value requires a time equal, not to k, but to $t = 2.3\,k$. So a device that has an advertised eith response time of 1 second is no better or worse than one having a 90% time of 2.3 seconds.

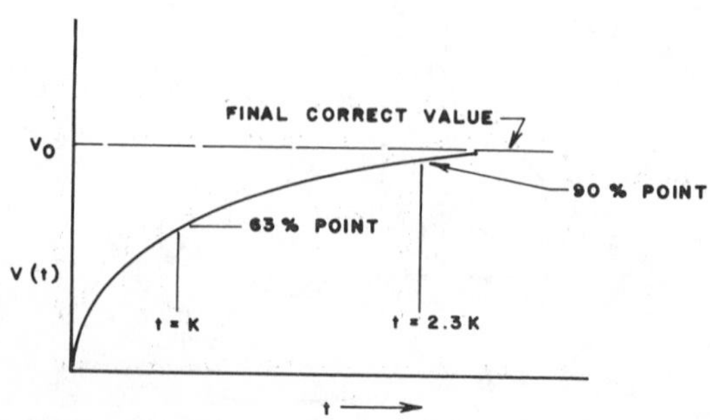

Figure 2-24. Definition of op-amp response time.

SLEW RATE AND SETTLING TIME

To define *slew rate* and *settling time,* we assume that an infinite step voltage has been applied to the op-amp input. An *infinite step* means that it goes from, say, 1 volt to 2 volts in zero time; it does *not* mean that the voltage becomes infinite. If the voltage stays at 2 volts, we call it a *1 volt step function.* The rate of change of the output due to this step function input is called the *slew rate.* In general, the higher the slew rate (in volts per microsecond), the better. However, every rose has its thorns, and a high slew rate means that the output may overshoot the correct value and have to settle down to the proper level. This takes time: the *settling time.* (Be on guard: various manufacturers may have their own definition for this.) In Figure 2-25, we illustrate our definition of these terms. Here, by *final value* we mean a value within some tolerance that suits our purpose at the time.

PHASE SHIFT

The concept of *phase shift,* which was introduced in Chapter 1 (see Figure 1-16), requires a little mathematics. A little mathematics won't hurt you; at least we haven't heard of anyone dying from it yet. You can even skip this section without missing much.

Assume that we have a black box whose input is a signal $V_1 = A \cos 2\pi ft$, where f is the frequency in hertz. The output is also a cosine wave (we hope), and takes the form of $V_0 = B \cos(2\pi ft + \phi)$. Now B/A is the gain, but the important term here is ϕ, the *phase shift.* Notice that V_1 is zero when $2\pi ft = \pi/2$, whereas V_0 is zero when $(2\pi ft + \phi) = \pi/2$. So if ϕ is not zero, V_1 and V_0 are *out of phase:* there is phase shift in the black box.

If our black box is an op-amp, we know that when input (1) is (+), output (4) is (-). This is a phase shift of 180 degrees. However, it is *customary* to ignore that and to refer to an op-amp that has only a 180 degree phase shift as having "zero" phase shift. If the 180 degree phase shift (which is built into the op-amp) *changes*

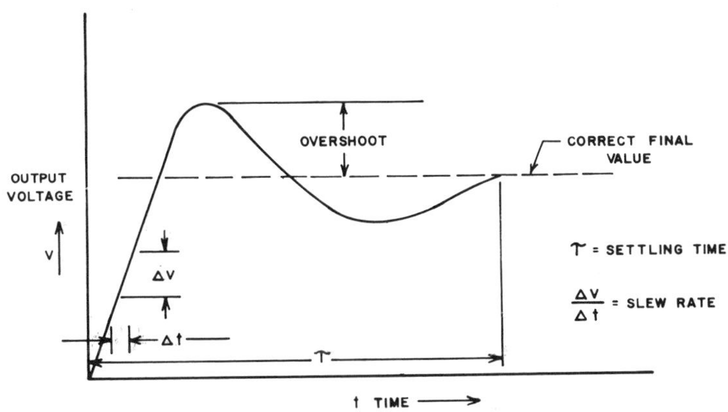

Figure 2-25. Response of an op-amp to a step input.

with frequency or signal amplitude, *then* EEs refer to it as "phase shift." This is just one of the ways we confuse the uninitiated.

Why does phase shift cause problems? To begin with, a good op-amp has zero phase shift at low frequencies, but the phase shift increases as higher input frequencies are used. If the input signal has a number of frequencies, some of them may be shifted more than others because the phase shift is frequency-dependent. This can cause distortion, so the output signal doesn't look like the input. If the phase shift is large enough to produce positive feedback, there may be oscillation. For now, though, don't fret about it; we will discuss compensation techniques for this problem in Chapter 6.

FEEDBACK FACTOR

Feedback factor is a term we haven't used yet in discussing op-amps but you often see it in the literature, so it might be well for us at least to define it for you. The general feedback equation (for all types of amplifiers) is

$$A_F = \frac{A_O}{1 + A_O \beta}$$

Here A_F is the closed loop gain and A_O is the open loop gain. The symbol β stands for the *feedback factor,* which is a way of measuring the fraction of the output voltage V_0 that is fed back to the input.

Referring back to Figure 2-8, the current through R_0 and R_1 is

$$I = \frac{V_0}{R_1 + R_0}$$

The fraction of V_0 applied to input (1) is

$$I R_1 = \frac{V_0 R_1}{R_1 + R_0} = V_0\left(\frac{R_1}{R_1 + R_0}\right) = \beta V_0$$

so

$$\beta = \frac{R_1}{R_1 + R_0}$$

That is all there is to feedback factor. For an ideal op-amp, A_O is infinite, whereas $A_F = (R_0 + R_1)/R_1$ for the noninverting op-amp and $A_F = R_0/R_1$ for the inverting circuit. If you insert these terms in the feedback equation above, it reduces to an identity. Feedback factory analysis is used in circuit stability studies.

HOW TO BUY OP-AMPS – OR A FOOL AND HIS MONEY ARE SOON PARTED

By this time you must be thinking of buying op-amps and will have obtained catalogs from people like Burr-Brown, Teledyne, Philbrick, and Analog Devices. The problem is they have so many devices available, with such a bewildering variety of specifications, that you don't know what to buy.

The first thing to do, if you are in a university, is to write some begging letters asking for free op-amps. As one of my students put it, "There is nothing cheaper than free." If something along the line of our sample begging letter (Chapter 8) doesn't move them, you might just have to pay cash. In that sad event, we suggest that you order the cheapest op-amps that are available and use them to learn on. Yes, we know there are people who say "buy something good and you will have it for a lifetime." Those are the people who, when their darling son says he wants to learn the saxophone, go out and buy a $400 instrument. After 2 weeks of lessons, the little ingrate takes up girls instead of music and they have the saxophone sitting around for a lifetime. Go ahead and buy the cheapest op-amps, learn how to use them, and by the time you are really limited by their characteristics, we hope you will have the money for a few good units. Then your students can play with the cheap ones.

While we are on this subject, you may be wondering about all the ads you see for modules to accomplish some specific purpose. There are thermocouple modules that allow you to put on ¼ mile of extension wire, weighing modules, counting modules, and so on. The only limit is the ingenuity of the human mind and some company's idea of what will sell.

Should you buy these things instead of making them yourself? The answer is probably "yes" in the cases of circuits to multiply, divide, do square roots, or take logarithms. It would take you a long time to learn how to build things like that. Regarding other types of modules, it is a case of your own ratio of money to time. Buying modules is a fast way to get in a position to do something. In industry, time is expensive, so you tend to buy. At universities, time may well be cheap; some university administrations seem to think that faculty time is free and can therefore be wasted in unlimited paperwork. You just have to play this game by your own rules.

Another point of interest is that you can save many dollars by purchasing or leasing used instruments – voltmeters, power supplies, oscilloscopes, gas chromatographs, and so forth – from a variety of second-hand dealers. The instruments are usually in fine condition, and most dealers will guarantee their equipment. A few dealers' names and addresses are listed in the where-to-buy-it section at the back of the book. If you watch the ads in magazines like *Science* or *Industrial Research,* you will see all sorts of equipment available second hand. After a while you may feel like Fanny Brice singing "Second-hand Rose," but you can comfort yourself with the thought that "when you're poor you have to be a little smart."

To sum things up, by now you should be convinced that op-amps are the greatest things since sliced bread. If you are not convinced, throw this book away! Don't try to get your money back, we have already spent it.

In the next chapter, we will discuss the application of op-amps in a variety of circuits. In Chapter 6, we will show you how to compensate for imperfections in real op-amps. We hope that you will stay with us; if not, please recycle the paper on which this book is printed.

TEST YOURSELF PROBLEMS AND SUGGESTED EXPERIMENTS

TEST YOURSELF PROBLEMS

1. What must the internal resistance of the voltmeter V be in order to read the voltage with 1% accuracy? Answer: $V_R \geqslant 200$ kΩ.

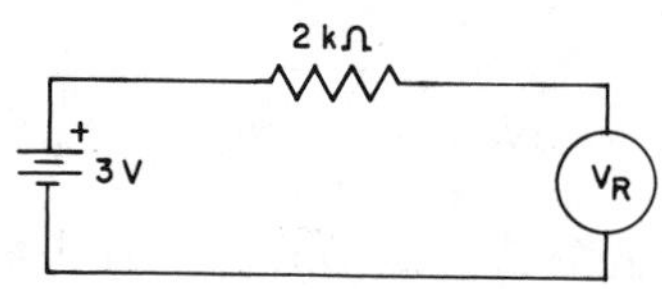

Figure 2-26.

2. What is the gain of this circuit? What is the input impedance (Z)? Answer: Gain = 2.33, Z input = 3 KΩ.

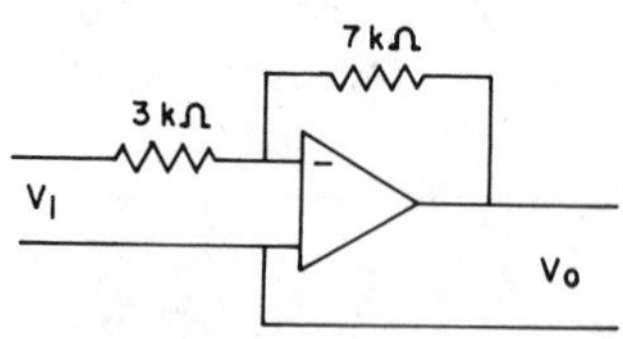

Figure 2-27.

3. Draw a common mode rejection circuit and in your own words discuss the way CMR works.
4. Draw a voltage follower circuit.
   What is its input impedance?
   What is its output impedance?
   What might it be used for?

5. Given the following

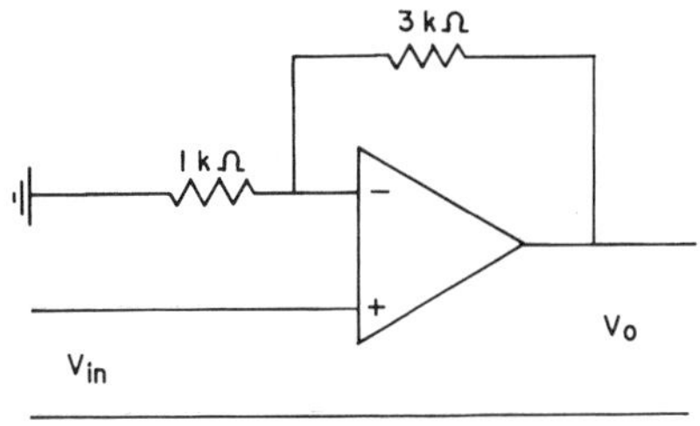

Figure 2-28.

For $V_{in} = 10 \cos \omega t$
What is V_{out}? (Answer: $V_0 = 40 \cos \omega t$)
What is the input impedance of this circuit? (Answer: ∞)

6. In your own words, what is roll-off? What does gain-bandwidth limit mean?
7. Draw an op-amp circuit to "buffer" a high-impedance signal source. What is the gain of a buffer circuit?
8. Under what conditions does a source transfer maximum power to a load?
9. Draw a simple constant current op-amp circuit. Why is the current through the load constant?
10. What do we mean by R_{om}, R_{cm}, and *common mode rejection?*
11. What is meant by the terms *slew rate, settling time,* and *overshoot?*
12. What do the terms *response time, eith* (63% point), and *90% time* mean?
13. Discuss the difference between "difference" and "differential" circuits.
14. What are the characteristics of an ideal op-amp?
15. In an ideal op-amp, with negative feedback, there is no voltage between the (+) and (−) inputs. Why?
16. What is the input impedance of an inverting op-amp circuit? (Answer: the impedance is controlled by whatever input resistor is used)
17. What is the input impedance of a noninverting op-amp circuit? (Answer: for an ideal op-amp, the impedance is infinite)
18. What is the output impedance of an op-amp? How would you use an op-amp to couple a high impedance source to a low impedance load?
19. Draw a simple constant current op-amp circuit and show why the current remains constant as the feedback resistance changes.
20. Draw a common mode rejection circuit and indicate what types of noise are rejected. What types of noise will not be rejected?

Suggested Experiments

Rules of the Lab:

1. *Always check device dial settings before turning on.*
2. *Always turn off all devices* before *making circuit changes.*

3. *If you are not familiar with the particular op-amp or device that will be used, you should have or try to find the appropriate company manual for more information.*
(Most large electronic stores – not Radio Shack – will have copies that you can look at. If that doesn't work, you will have to look in *Thomas' Register of American Manufacturers* for the address of Motorola, or whichever company is involved, and write to them for technical information.)

A. OP-AMPS

Never do this:

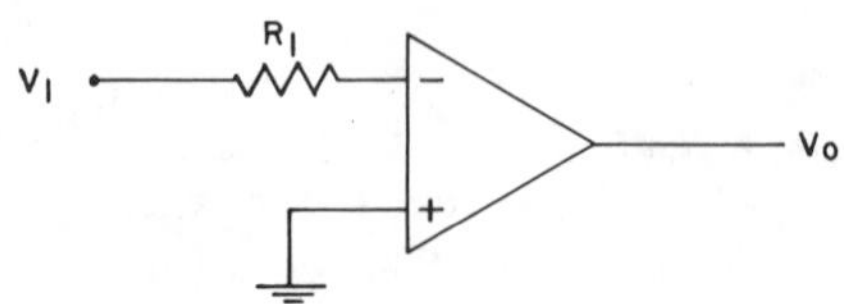

Figure 2-29.

Why?

B. INVERTING AMP

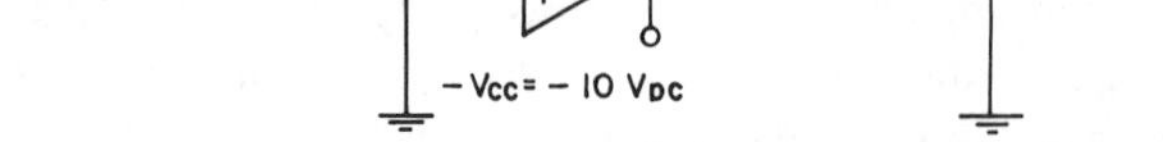

Figure 2-30.

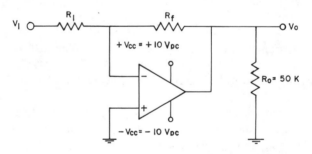

1. What is the maximum voltage that V_1 can be if: $R_1 = R_f$ __________,
$R_f = 2\,R_1$ __________, $R_f = 10\,R_1$ __________, $R_f = \frac{1}{2}R_1$ __________.
2. What is the equation for V_0 in terms of V_1?
3. How are you going to get ± 10 volts DC? __________ Diagram.
4. Why does the op-amp have offset input controls?
5. How are you going to correct any DC offset that your op-amp might have?
6. Construct the circuit of Figure 2-30 with $R_1 = 10\ \text{k}\Omega$ and R_f = a decade box.*
Draw the schematic and *label* the pin numbers on the op-amp.

*A decade box is a metal box containing an assortment of resistors, connectors, and switches that can provide a wide range of resistance values for experimental purposes.

7. Connect the function generator to V_1 and set $R_f = 2$ kΩ. What is the gain? Theoretically ______. Experimentally ______.
8. Vary the frequency of the input to the op-amp. Is there any change in the gain?
9. What is the input impedance of this circuit? Theoretically ______. Experimentally ______.
Use two 10 kΩ resistors to form this circuit; set the function generator to produce a 1 V and DC output.

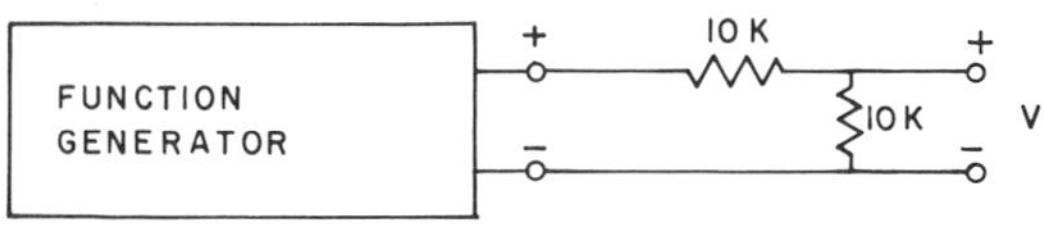

Figure 2-31.

Experimentally, what is the voltage (V)?
Now connect the function generator to the op-amp input (it is now called V_1) and measure voltage V again.

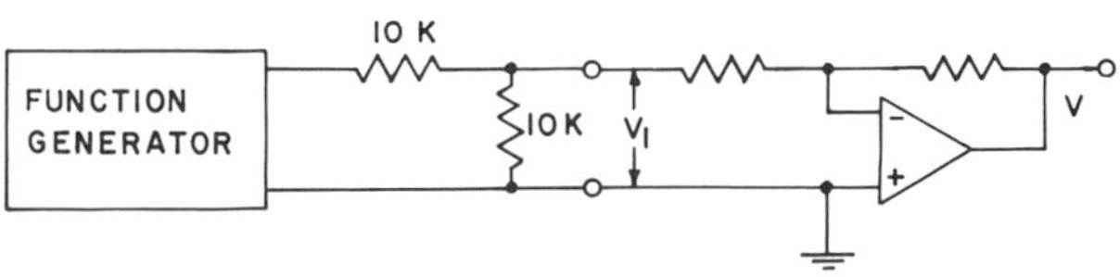

Figure 2-32.

What is it? Why?

C. NONINVERTING AMPLIFIER

If you have the following circuit:

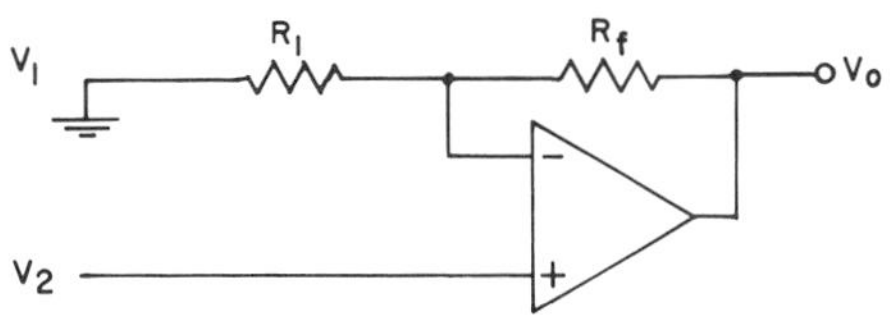

Figure 2-33.

1. What is the gain if: $R_1 = R_f$? ______. If $R_f = 2\,R_1$? ______
2. Construct the above circuit with R_1 = 10 kΩ and R_f = 10 kΩ = decade box.
3. Draw the circuit diagram, labeling the op-amp pins.
4. Connect the function generator and see if there is any variation in gain as the frequency is changed.

5. What is the input impedance of this circuit?
   Use two 10 kΩ resistors to form this circuit:

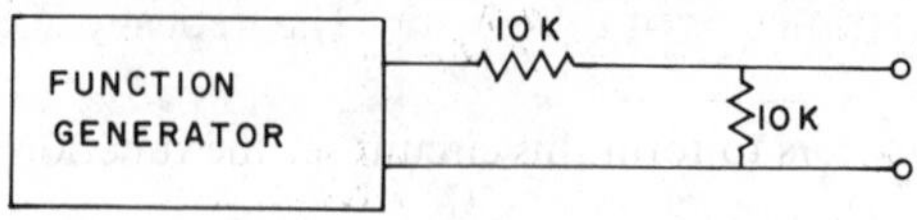

Figure 2-34.

What is the voltage V experimentally? ____________
Now connect V to the amplifier input (V_1) and measure V again.

1. What is it? ____________
2. Why? ____________
3. Can you see any advantage of this circuit over the inverting amplifier?

D. CONSTANT CURRENT SOURCE

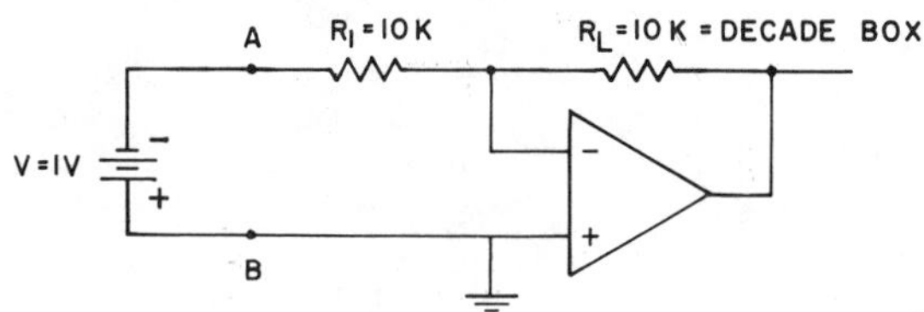

Figure 2-35.

Draw the diagram, labeling all points.

1. Is point A earth, ground, or neutral? ____________
2. Is point B earth, ground, or neutral? ____________
3. What is the current through R_L? ____________
4. Change R_L to 40 kΩ. What is the current through R_L? ____________
5. Change R_L to 20 kΩ. What is the current through R_L? ____________

E. DIFFERENCE AND DIFFERENTIAL AMPS

Using the MC 1437 *dual* op-amp, construct a unity gain difference amp and a unity gain differential amp. Use 10 kΩ resistors for the differential amp and 22 kΩ resistors for the difference amp.

1. Draw circuit diagrams for the difference amp and for the differential amp.
2. Connect the function generator and, using the oscilloscope, test both circuits (ckt) to assure yourself that they are working properly.

Theoretically:	*Differential ckt*	*Difference ckt*
What is the input impedance?	____________	____________
What is the gain?	____________	____________
What is the output impedance?	____________	____________

What is the voltage at the inverting terminal, of the difference amp, for any reasonable input to the circuit? ____________

3. Connect a long wire pair (15-20 feet) from the output of the function generator to the input of the differential amp. Stretch the excess wire out on the floor as far away from the circuit as possible.

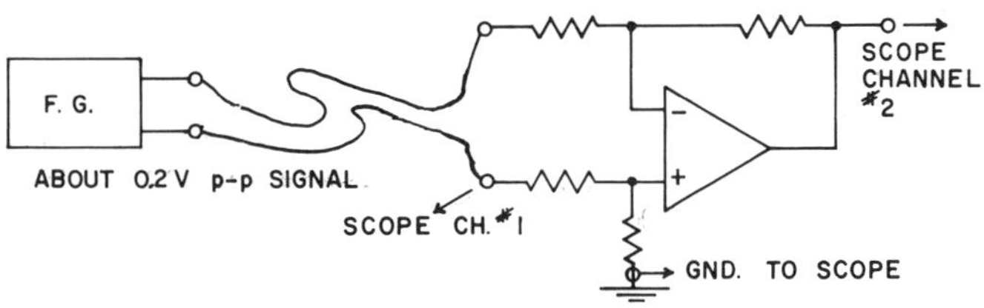

Figure 2-36.

Using the scope, observe the input and output simultaneously (*do not* put an earth ground directly on the function generator terminals).

1. Is there a significant amount of noise or is it being rejected?

Disconnect the long wire from the input of the differential amp and connect it to the difference amp as shown:

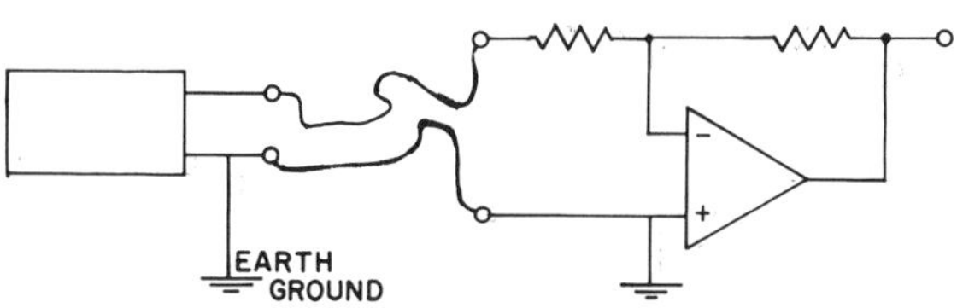

Figure 2-37.

1. Is there a significant amount of noise and is it being rejected?

Use a ½ inch drill to induce noise onto the long line (plug it into a bench across the room from yours). Keep the drill as far away from the circuit as possible but close enough to the long line to induce noise.

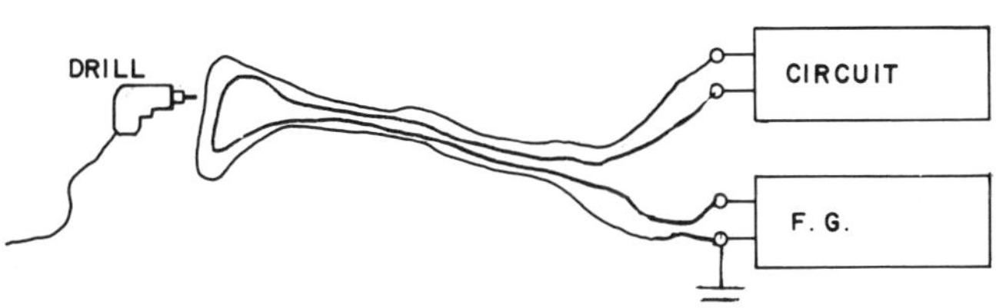

Figure 2-38.

1. Is the noise being rejected?
   Switch the long line to the differential amp and repeat the drill exercise.

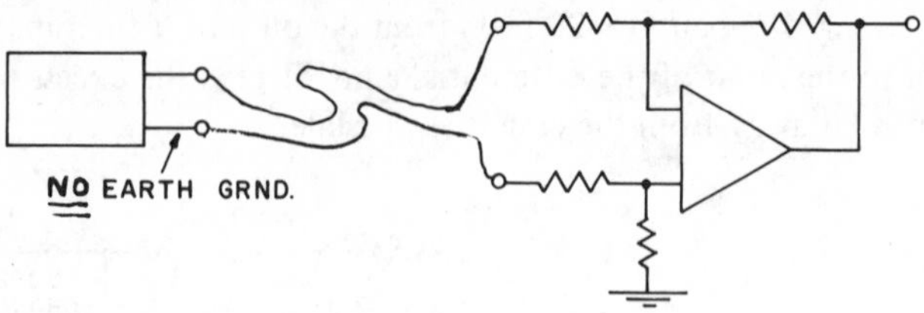

Figure 2-39.

1. Is a significant amount of noise being rejected, compared to the difference amp?
2. What is the difference between a differential and a difference amp, based on your findings?

F. ZENER DIODES

Construct the following circuit:

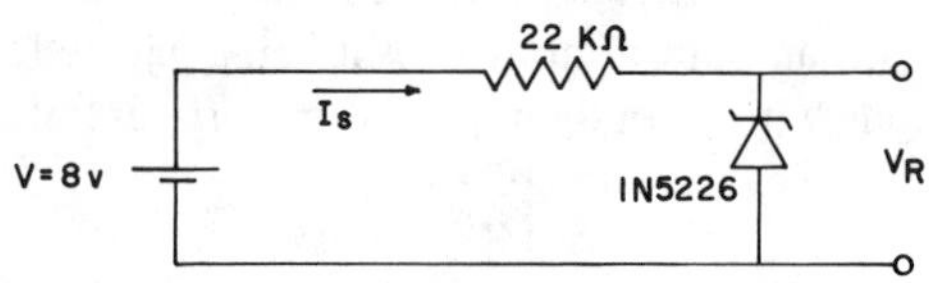

Figure 2-40.

1. What is the voltage V_R?
2. Vary V and comment on what happens to V_R (don't go above 15 volts).
3. What is the zener diode good for?

G. DARLINGTON CIRCUIT

Construct the following circuit (using 2N4265 NPN, 2N3702 PNP):

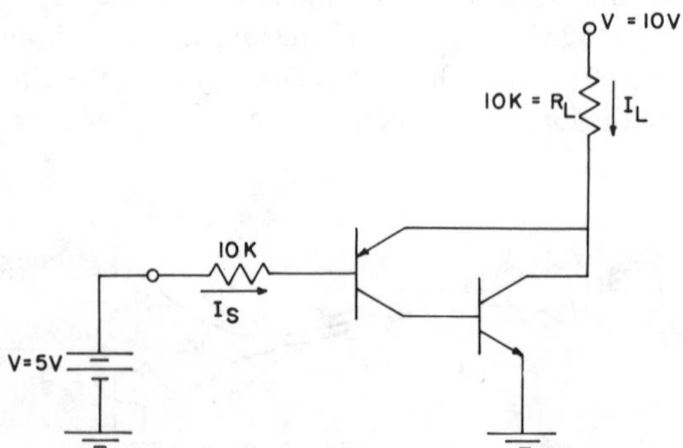

Figure 2-41.

1. What is the input current I_s?
2. What is the output load current I_L?
   Raise voltage to 8 volts and repeat measurements: I_s ______ I_L ______
   Try with the function generator as an input.
   What is the Darlington circuit good for?

H. WIRING OP-AMPS FROM SCRATCH

Using the book, page 84, construct a unity gain follower for general applications. Draw a complete schematic.

1. What does compensation do for you?
2. Check your circuit to make sure it works.

I. RISE TIME, OVERSHOOT, SETTLING TIME

Using the same circuit you made in Experiment H (unity gain follower), input a 1 kHz, 4 volt peak-to-peak, square-wave signal. Adjust your scope until you can observe the rise time.

1. Draw a picture of the input and output waveforms.
2. Approximately: What is the rise time? ____________ What is the overshoot? ____________ What is the settling time? ____________
3. Indicate how these were measured.

3 Op-Amp Applications for Fun and Profit

In this chapter, we will present a series of handy circuits for doing useful things. You should look through this section, find the circuits that you want to build, and go to it. These circuits have been used in our laboratory by students taking our "Electronics Without Pain" course, so they *should* work.

In case of trouble, check the wiring first, then try another op-amp of the same kind. If that doesn't cure the problem, you will have to put in variable resistance and capacitance boxes and try adjusting values. When all of these things fail, try another circuit or start reading Chapter 6. *Above all never give up.*

What sort of things can you expect to do with op-amps? In answering this question, our objective is to make sure that you begin by trying the things you are likely to be able to do, and vice versa. ("Nothing is more encouraging than success or more disheartening than failure.")

In pages that follow we will use the terms *engineer, geologist, chemist,* and so on. It is understood (at least by the author) that you will mentally substitute *bee keeper, physiologist,* or whatever you happen to be. Don't be afraid. You have come this far; from now on, it's downhill all the way.

An engineer who uses op-amps must recognize that this device is useful at several levels of sophistication. A "real expert" will build circuits that work at the 10^{-11} amp level with almost zero drift. A "beginner" should feel satisfied if he can accomplish this at the 10^{-5} amp level. As you get more experience, you can expect to cook up (or steal) better circuits, but in the beginning don't expect to be as good as the one who has been at it for years.

One of the problems in this area is that manufacturers' catalogs and sales sheets are written for "the expert." He may need all this complex data, but you don't (at least not for many useful circuits). The thing to remember is that your talents will limit you for the first year or two. Don't expect to work at 10^{-10} amp or 10 MHz just because the manufacturer's catalog says you can.

Table 3-1 provides some guidelines about what you can expect to be able to do. They are, to some extent, a matter of personal opinion, but we suggest that they are useful in the beginning. Keep in mind that all these parameters are interdependent: you should not try to work with current levels of 10^{-5} amp and voltages of 100 mV at the same time.

"He got the plans from some electronics catalog."

Table 3-1. Op-Amp Capabilities and You

	Beginner	Second Year	Real Expert
Accuracy	1%	0.1%	0.01%
Voltage signals	100 mV	10 mV	1 mV
Current signals	10^{-5} A	10^{-8} A	10^{-11} A
Impedance levels	10^{5} Ω	10^{7} Ω	10^{10} Ω
Highest frequency	100 kHz	1 MHz	50 MHz

Before launching into the nuts and bolts of practical op-amp circuits, we should say a few words about putting in feedback capacitors across the feedback resistor, R_0. There is a school of op-amp thought that says "put 100 pF across R_0; it can't hurt and it may help." The other school says "use no medicine unless you are sick; put in feedback capacitance only when needed." The best position is probably somewhere in between both factions (like a liberal-conservative). We don't usually show a feedback capacitor in a circuit, but if you have a high-frequency noise or oscillation problems, 100 pF of capacitance can cure a lot of ills. The *details* of how the capacitor does this will be discussed in Chapter 6.

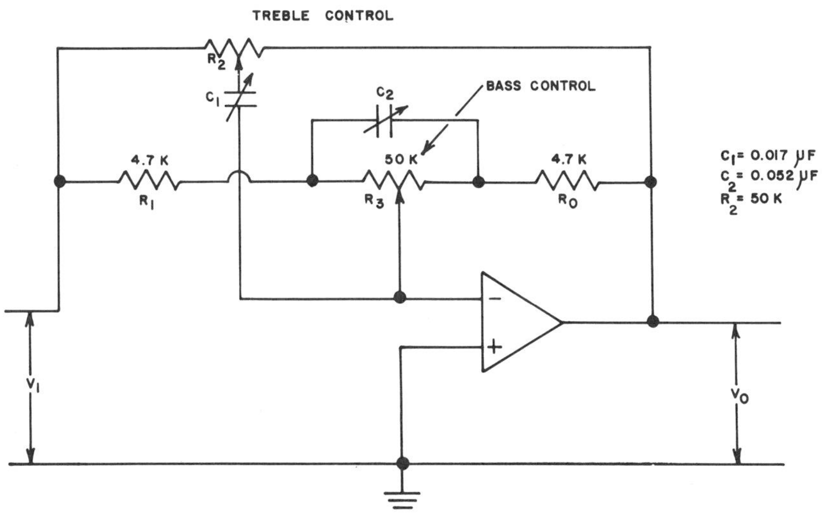

Figure 3-1. Amplifier with bass and treble control.

SINGLE OP-AMP AMPLIFIER WITH ADJUSTABLE BASS OR TREBLE CONTROL

This is a clever circuit that allows you to boost or cut either the low or the high frequency amplification. This amplifier, with controlled feedback, is shown in Figure 3-1.

In this circuit the treble control R_2 C_1 operates by passing a fraction of the input signal directly to the (-) input. The capacitor C_1 stops low frequency signals and passes high frequency signals. The bass control works on the principle that C_2 R_3 stops low frequencies and the variable resistor permits a controlled fraction of those frequencies to enter the amplifier. To try this out, you ought to use variable capacitors for C_1 and C_2, $R_1 = R_0 = 4.7\ \text{k}\Omega$, and resistor decade boxes (variable resistors) for the treble and bass controls. Hook up a signal generator to the input, an oscilloscope to the output, and have fun!*

OP-AMPS AS CURRENT SOURCES

Suppose we want to build a hot-wire anemometer resistance thermometer. The sensor is a 0.001 inch diameter platinum wire. We run a constant current through

*In your analysis of this circuit, it will help if you recall what the impedance of capacitors is like in terms of frequency and capacitance. If we make C_1 larger, the impedance will be smaller, and more of the treble signal will get to the op-amp input. The same thing happens if we move our point of contact to the left end of the resistor: we will send more treble signal to the op-amp input with the result that the treble input will be larger. The reader should try the same type of analysis with the bass control. How should the capacitor C_2 be set to put the maximum low frequency signal into the op-amp input?

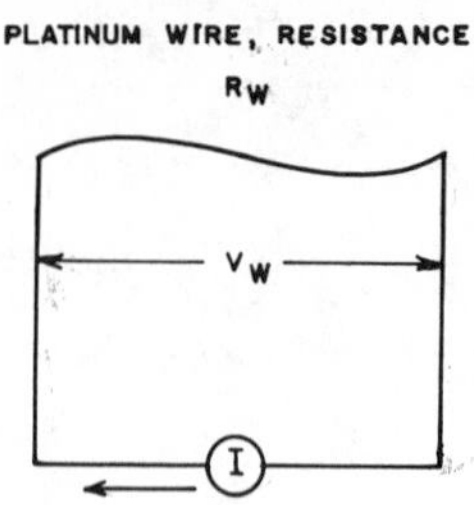

Figure 3-2. Hot-wire anemometer circuit.

the wire, and this heats the wire to some temperature, T_W. As the ambient temperature or the wind velocity changes, heat is gained or lost and the wire temperature changes. The relation between temperature and wire resistance is

$$R_w = R_0 (1 + \alpha [T_w - T_0])$$

Here $\alpha = 0.003$ (for platinum), and R_0 is the resistance at $T = T_0$ (usually 25°C).

The measurement circuit is shown in Figure 3-2. The voltage V_w is $I_w R_w$, or $IR_0 (1 + \alpha [T_w - T_0])$. Therefore, by measuring V_w, we can measure T_w, which is what we want to do. However, we must be sure that I is constant, so we need a *constant current source.*

The simplest constant current source is a battery and resistor (Figure 3-3). If R_w is about 1 ohm and R_1 is 1000 ohms, the current is controlled by R_1, and variations in R_w will have a very small effect on I. So we can write

$$V_w = \left(\frac{V}{R_1}\right) R_w = \left(\frac{V}{R_1}\right) (1 + \alpha [T_w - T_0]) R_0$$

This circuit works, but it is inefficient; most of the battery voltage is wasted in the power loss, $I^2 R$, due to the heating of resistor R_1.

So let's use an op-amp current supply. Our first thought is the current pump that was shown in Figure 2-13, which, in essence, is reproduced here as Figure 3-4. The current in R_0 is constant as long as V_1/R_1 is constant, so if we use our test wire for R_0, we are all set. The current through R_0 is fixed by the ratio of V_1/R_1, so V_0 (across R_0) must be $V_0 = (V_1/R_1) R_0$ or, in other terms, $V_w = (V_1/R_1) R_w$. You

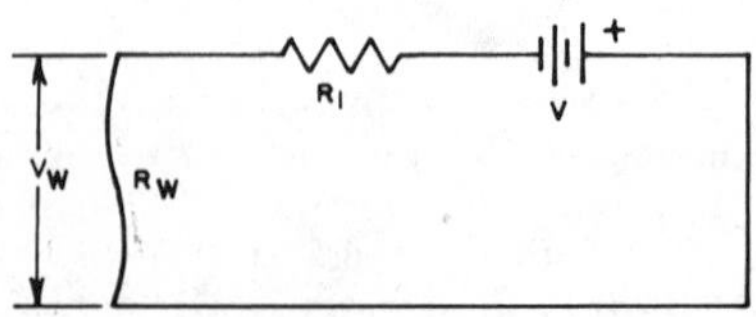

Figure 3-3. Simplest constant current circuit.

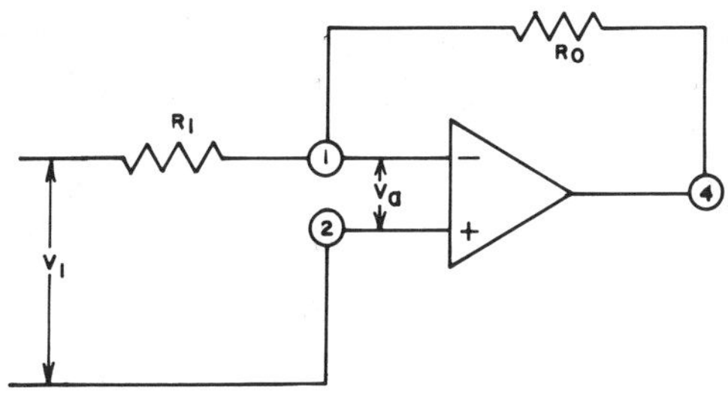

Figure 3-4. Op-amp constant current source.

should appreciate that we must measure V_0 (or V_w) between (4) and ground. The load R_0 (or R_w) is floating because (1) is only a "virtual" ground (recall that V_a = zero) due to feedback.

If the load R_w must be grounded at one end (real ground not virtual ground), we have to change the circuit, as is shown in Figure 3-5. Here R_0, R_1, R_2, and R_3 must satisfy the relation $R_1/R_2 = R_0/R_3$. The input voltage V_1 must be limited to a value that will not drive V_3 beyond the common mode limit specified by the manufacturer. This is a good hot-wire operational circuit. The hot wire is resistor R_L.

The analysis of this circuit is instructive. First, the op-amp terminals (1) and (2) are at voltage V_3 above ground. It follows that $I_1 = I_0$ and

$$I_1 = \frac{V_1 - V_3}{R_1} = \frac{V_3 - V_2}{R_0} = I_0 \tag{3-1}$$

Since no current flows in or out of the op-amp at (2), I_L must depend on the differ-

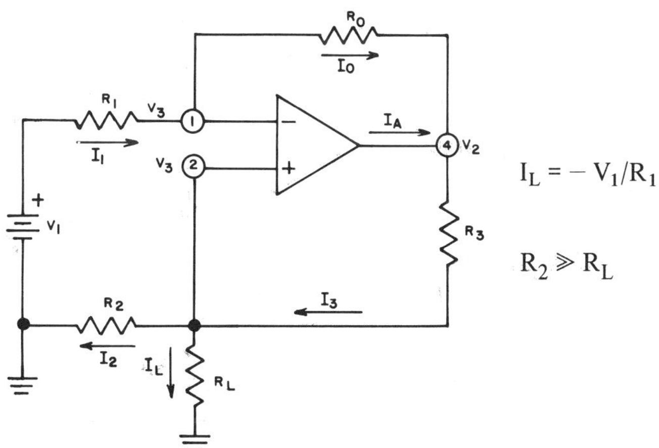

Figure 3-5. Current source with a grounded load.

ence between I_3 and I_2, $I_L = I_3 - I_2$. We note that $I_3 = (V_2 - V_3)/R_3$ and $I_2 = V_3/R_2$, so

$$I_L = I_3 - I_2 = \frac{V_2 - V_3}{R_3} - \frac{V_3}{R_2} \tag{3-2}$$

If we solve equation (3-1) for $V_2 - V_3$ and substitute the result in equation (3-2) we obtain

$$I_L = -\frac{(V_1 - V_3)R_0}{R_3 R_1} - \frac{V_3}{R_2} \tag{3-3}$$

This can be written as:

$$I_L = -V_1\left(\frac{R_0}{R_3 R_1}\right) - V_3\left(\frac{1}{R_2} - \frac{R_0}{R_3 R_1}\right) \tag{3-4}$$

Now if we choose our resistors wisely so that $R_3 R_1 - R_2 R_0 = 0$, then

$$I_L = -V_1\left(\frac{R_0}{R_3 R_1}\right) \tag{3-5}$$

and if $R_3 = R_0$,

$$I_L = -\frac{V_1}{R_1} \tag{3-6}$$

PLAYING GAMES IN THE FEEDBACK LOOP – OR HOW TO LOOK SMARTER THAN YOU REALLY ARE

Way back in Figure 2-13 we discussed the fact that the current through the feedback resistor R_0 is independent of the value of R_0. The op-amp builds up its output voltage V_0 until $I_0 = I_1$ and holds this condition to the limits of its ability.

There are all sorts of neat ways to make use of this capability. For example, given a 0–5 volt DC meter with a rating of 20 kΩ per volt, you should remember from Chapter 1 that the meter goes full scale if a 50 μA current passes through it (20 kΩ in series with 1 volt means a current of 1 volt/20 kΩ = 50 μA). So far all is well, but suppose you want to use this meter outside in upstate New York where it goes to (UGH!) -30°F in winter. You query the manufacturer and he says that the meter's internal resistance goes up and down with temperature. This will wreck your accuracy because the ohms-per-volt rating will go from 20 kΩ at 70°F to say 22 kΩ at -30°F (this is a bad case). What to do?

The solution is to put the meter in the feedback circuit of an op-amp, as shown

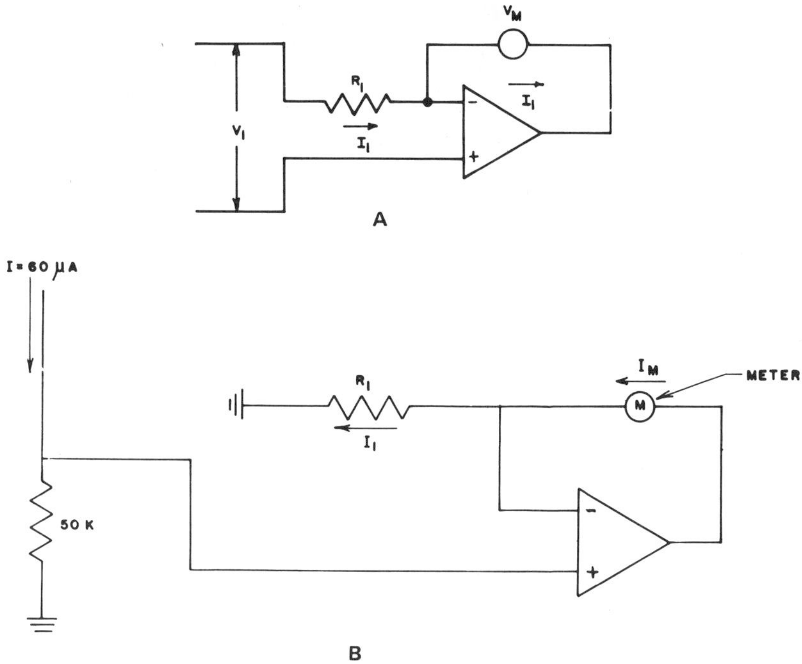

Figure 3-6. *A.* Making a poor meter look like an expensive one. *B.* Another meter improvement circuit utilizing an op-amp.

in Figure 3-6A. Now your unknown voltage generates a current, I_1. This current must flow through the voltmeter regardless of whether the temperature is 70°F or -30°F. The temperature-sensitive voltmeter is sensitive no more. The meter (V_M) goes full scale (5 volts) when I_1 is 50 μA. We can read V_M, calculate I_1, and obtain V_1 from $V_1 = I_1 R_1 = (V_M/5)(50) R_1$.

There are many variations of this game. For a choice selection, see J. I. Smith, *Modern Operational Circuit Design,* New York: Wiley, 1971.

Another way to play games with a voltmeter in the op-amp feedback circuit is shown in Figure 3-6B. This can make a rugged, low cost voltmeter work as well as an expensive laboratory device. As an example, consider the case in which we want to measure a 3 volt drop across a 50 kΩ resistor with a 0–5 volt voltmeter rated at 1000 ohms per volt. The 1000-ohm-per-volt rating tells us that this meter requires a 1 mA full-scale current, so if we hang it across the 50 kΩ resistor, we might upset the entire circuit that we are trying to measure. The problem is the low effective resistance of the voltmeter (i.e., 5000 ohms).

The solution involves taking advantage of the fact that a voltmeter is really an ammeter with a series resistance and that ammeters respond to current. We know that the voltmeter goes full scale at 5 volts and has a 5000 ohm resistance, so the maximum meter current is 1 mA. In Figure 3-6B we have an op-amp circuit that insures the current through the meter will be 1 mA when the input voltage is 5

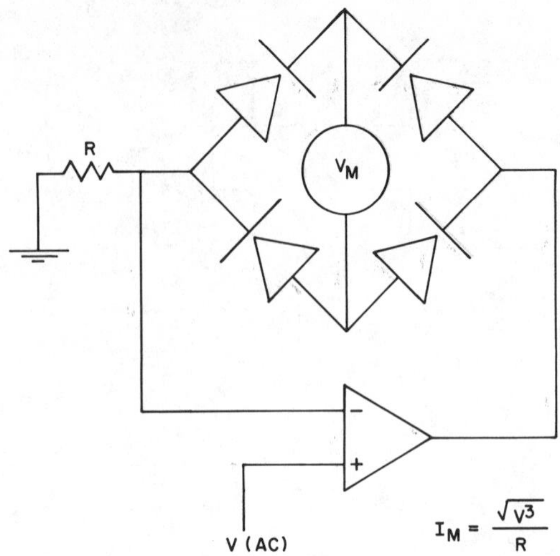

Figure 3-7. High-impedance AC voltmeter.

volts and that the meter will read 3 volts when the input is 3 volts. The trick is easy if R_1 = 5000 ohms (the meter resistance).

Suppose the input voltage is 3 volts, the op-amp has infinite resistance, so no current is drawn from the circuit under test. We have feedback, so the (−) input must also be at 3 volts, the current is $3/R_1$ and $R_1 = R_m$ so I_1 must be 3/5000 or 0.6 mA. For a meter that goes full scale (5 volts) with a current of 1 mA, a current of 0.6 mA will give a reading of (0.6) (5) = 3 volts. The trick is to make sure that R_1 equals the meter resistance.

Another application of this technique involves using an op-amp to build a high-impedance AC meter. AC meters actually read $\sqrt{V^2}$, and the fact that they have built-in diodes usually means that they are low resistance type devices. However by putting the AC meter in the feedback loop we can have it do its thing in a circuit with very high effective resistance (the input impedance of an ideal op-amp is infinite).

You will note in Figure 3-7 that we give the current through the meter as a function of the applied voltage V and the feedback resistance R. Remember that the meter itself is a *DC device;* the diodes are what let it read AC. If the voltage V goes from 0–6 volts, the current through the meter will be $6/R$; if it is a 1 mA meter movement (1000 ohms per volt), you will need an R of 6000 ohms, $6/R$ = 1 mA.

VOLTAGE OR CURRENT BOOSTER CIRCUITS

While we are on the subject of op-amps as current or voltage sources, we should mention the limitations of op-amps in this respect. (Have no fear; there are ways of getting around these limitations.) Typical op-amp output voltages are ±10 volts at ±10 mA. For more money, outputs of ±150 volts at ±5 mA or ±10 volts at ±100 mA

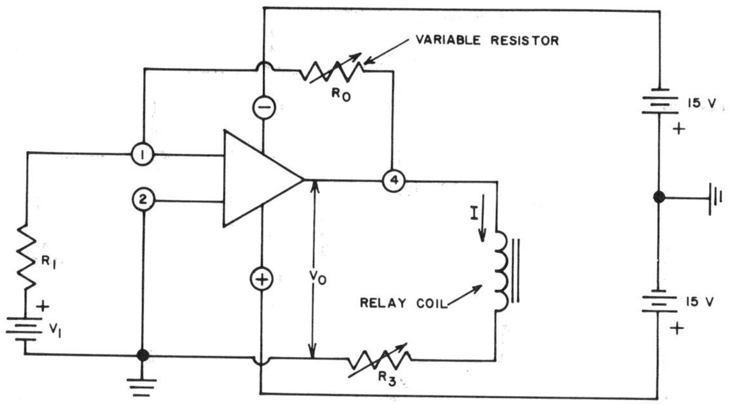

Figure 3-8. Relay control circuit.

may be obtained.* Usually it is better to use a conventional op-amp and then attach some sort of *current* or *voltage booster* circuit. The op-amp is a precision instrument that can be used to *control* large currents, but it should *not* be used to deliver maximum currents if accurate control is necessary.

Voltage Booster

A problem requiring a voltage booster is shown in Figure 3-8. In this circuit the objective was to open a relay when a variable resistor reached a certain value. Figure 3-8 shows our first design for this circuit. Things were great, except that the relay we found in the junk box took more voltage than the op-amp could deliver. To appreciate our predicament, let's do some figuring. Let's assume that the relay coil operates at a current I when a voltage V_0 is applied to the relay coil circuit. Note that the $V_0/V_1 = -(R_0/R_1)$, where V_1 is a fixed voltage generated by a mercury battery. We want the relay to operate when $R_0 = R_s$ (R_s is some pre-chosen value). *If* we could ignore the relay coil resistance, the relay current, I, would be $I = V_0/R_3$. Actually, we should also put in R_c, the coil resistance; then $I = V_0/(R_3 + R_c)$. Substituting for V_0, we have

$$I = -\frac{R_0}{R_1}\left(\frac{V_1}{R_3 + R_c}\right)$$

V_1 and R_1 are known, so we set $R_0 = R_s$ and adjust R_3 to provide the current needed for relay operation. We know when the proper value is reached because the relay *opens* at that value. Now we thought we were set: whenever $R_0 = R_s$, the relay would open. Our problem, however, was that the available op-amp had a maximum

*For those of you who are affluent, The KepCo people make op-amps with outputs up to 1000 amps, but these are expensive. KepCo's address is listed in the Appendix, so you can write to them for prices.

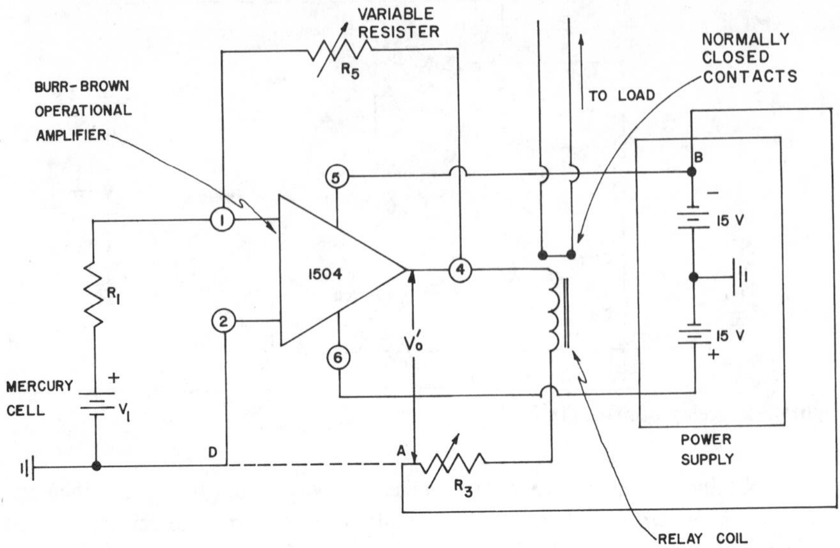

Figure 3-9. Revised relay control circuit with a voltage booster.

voltage output of 10 volts, and we had a relay that needed 12 volts to operate. Our op-amp could not supply sufficient voltage to drive it.

What could we do to change the circuit to supply more voltage? The revised circuit is shown in Figure 3-9. The voltage V_0' is $V_0 + 15$ volts, so if V_0 varies ±10 volts, V_0' ranges from +5 to +25 volts. This was quite adequate to drive the relay. Note that if we removed wire *A–B* and put in wire *A–D* (the dotted line), we would have our old circuit of Figure 3-8 again. This trick of using the power supply is a clever one: we can get voltages from +5 to +25 volts relative to the negative power supply voltage with a ±10 volt op-amp.

Current Booster

The next trick we would like to play with op-amps is to increase the output *current.* For example, we might want to use an op-amp with a ±10 volt output at 5 mA as a linear amplifier for use with a stereo system. Let us assume that we have a 7 ohm speaker that can be driven at about 1.2 amps maximum for a *peak* power output of 10 watts (we think, though some of our younger readers might disagree, that 10 watts peak is enough to deafen anyone). To produce 10 watts peak, we must increase our current by a factor of 240, and that will take two steps. We also want the system to stay linear so Bach doesn't sound like rock and roll.

What, you might ask, does *linear* mean? It means that the output is some *constant* times the input, i.e., $V_0 = -(R_0/R_1)\ V_1$. Op-amps are inherently linear, so we haven't discussed this concept very much. However, many transistor circuits are nonlinear, and we have to be aware of the term; for example, $V_0 = \log V_1$ is a *nonlinear* relationship.

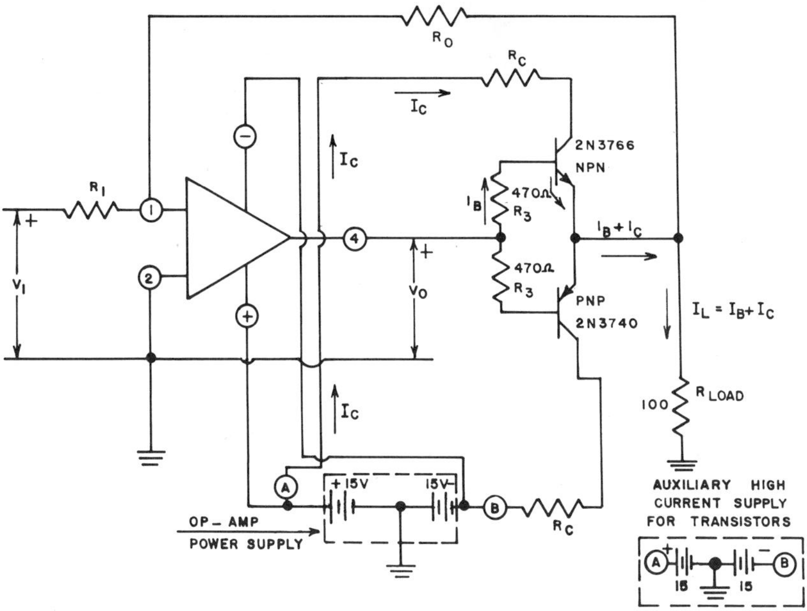

Figure 3-10. Current booster circuit.

In Figure 3-10 we show our first current booster circuit; it contains two transistors in the *common emitter* configuration (we will explain what "common emitter" means in Chapter 7). The op-amp itself is in the usual inverting amplifier hook-up. The output of the op-amp is used to drive *base current* into the two transistors. For the moment, you should think of a transistor as a black box having three leads: *collector, emitter,* and *base.* When current flows into the base, it turns the transistor "on" and allows current to flow from the collector to the emitter. If the collector current is greater than the base current, we define the *current gain* (A) as I_C/I_B, where I_C and I_B are the collector and base currents.

Before you go ape and start worrying about learning to work with transistors, you should appreciate that at THIS POINT in the book we are just showing you a useful transistor circuit in case you want to copy it and build an amplifier. For the moment, you should just think about a transistor as a circuit in which a variable resistor controls the current as shown in Figure 3-11A. The same situation, except with a transistor in the circuit, instead of a resistor, is shown in Figure 3-11B, in which the current I_c is controlled by the base current I_b. For the moment that is *all* you really need to know about transistors. (For those who just cannot wait, we have Chapter 7; go on and read it if you care to.)

To continue with our discussion of transistors, we note that there are two kinds of transistors: *PNP* and *NPN.* (Think of a PNP transistor as requiring a negative "turn on" signal and vice versa.) The PNP transistor requires a (–) base voltage relative to the emitter to "turn on"; the NPN requires a (+) base voltage signal to "turn on." We will discuss transistors in detail in Chapter 7. All you need to know at

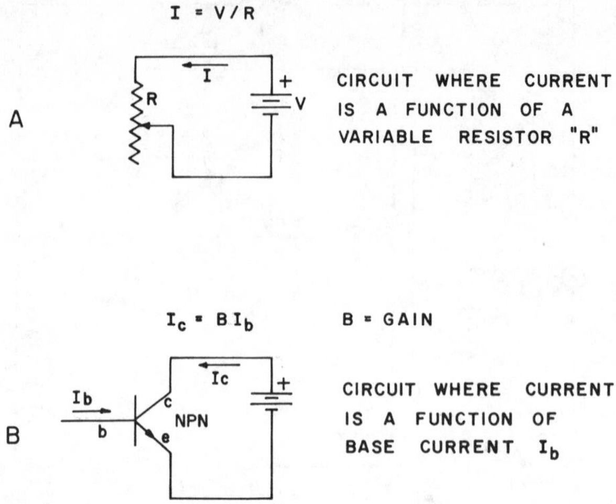

Figure 3-11. Comparison of resistor and transistor current circuits.

present is that in this case, the transistors are hooked up so that little or no collector-emitter current flows unless there is a signal of the proper polarity at the base.

In Figure 3-10 the transistors, an NPN and a PNP, are in what we call a *push-pull* circuit. In a sense, the common emitter is much like the voltage follower shown in Figure 2-15. The voltage follower had unity gain, but it served as an impedance matching device and, incidentally, provided considerable current gain. The transistor and resistor circuit of Figure 3-10 is designed to produce still more current gain. You will notice that the load resistance is 100 ohms rather than 7 ohms, which we had originally assumed for our speaker. This is because the op-amp has an output of +10 volts at 10 mA, and a minimum estimate of the current gain of the push-pull circuit is about 20. The available current is then about 20 × 10 mA, or 200 mA. If we are to use the 10 volt output of the op-amp and the 200 mA capability of the push-pull circuit, we need a load resistance of about 100 ohms. This means that at 10 volts, we will draw 100 mA (it is always best to be a little conservative and not drive components to their limits). If the maximum available current is 200 mA, the use of only 100 mA will give us a good safety factor.

Let's consider how the circuit of Figure 3-10 operates. Initially (when V_0 = zero), both transistors are *reverse biased,* i.e., no current flows through R_L. When V_0 goes (+), current from the op-amp flows into the base of the NPN transistor, removing the reverse bias and allowing a large current, I_L, to flow through the load. Typically, $I_B + I_C$ might be 100 mA, so if I_B = 5 mA, the current gain is 19. When V_0 goes (-), the NPN transistor remains off but the PNP transistor conducts; the current $I_B + I_C$ flows through the load as before, but in the opposite direction.

You should note that the power supply must provide the current for both the op-amp *and* the transistors. You might have to use an ausiliary supply for the transistors. An additional high-current transistor power supply is shown in Figure 3-10.

To use it, unhook the transistors at A and B on the op-amp supply and connect them to A and B on the auxiliary supply.

The resistor R_0 in Figure 3-10 should be adjusted in accordance with R_1 to find the proper value of V_0. The formula $V_0 = -(R_0/R_1)\ V_1$ still applies when the feedback is taken across the load, as is shown in Figure 3-10.

It is most important that the reader appreciate *why* the feedback is taken across the load instead of just being connected to the output of the op-amp. Putting the feedback connection across the load allows the op-amp to protect the circuit against thermal runaway. *Thermal runaway* means that if a transistor gets hot, its current I_c goes up; this generates still more heat (due to I^2R), and the transistor is quickly destroyed. With the op-amp on guard, any tendency to thermal runaway will be detected as an increase in load current that produces a larger negative feedback. This in turn will decrease the output of the op-amp and tend to turn off the transistor. This is a *most important* protection feature and should be understood before you go further.

Since you are still learning how to play this game, you should protect the transistors from an accidental zap if the output is short circuited. The resistors R_c are inserted for this purpose. They should have sufficient resistance to hold the collector current within the transistor limits if R_L goes to zero. You can leave them out if you are brave; if you are chicken (and welcome to the club) a good choice for R_c is 500 ohms. Of course, any power taken up in R_c is power that cannot be used in the load. While you are learning, a large resistor, such as 500 ohms, is good; it makes you feel confident. Later, you can use 10 ohm resistors for R_c without problems. As long as R_c is 500 ohms, you will be able to develop only about 3 volts across the load; when the current reaches 30 mA, the entire 15 volts is dropped across R_c.

You might wonder where the 470 ohm resistors in Figure 3-10 came from. This resistance was picked so that I_b will develop a small voltage to reverse bias the off transistor. If I_B is 5 mA, then 470 ohms times 5 mA equals 2.35 volts. This small reverse bias helps remove something called *crossover distortion* that would otherwise occur when V_0 swings from (+) to (-). (*Crossover distortion* refers to the troubles [e.g., bad sound] you get when the op-amp output goes from (+) to (-) or (-) to (+). For the moment, consider it a useful phrase to impress your friends with when discussing high-fidelity audio systems.) Of course, this 2.35 volts is lost as far as being available to put across the load. ("Every thorn has its rose.")

Darlington Amplifiers

This term is used for several different transistor circuits that really do the same thing; the current from the first transistor is used to turn on the second one. In the first circuit, Figure 3-12, we use two NPN transistors (two PNP types could be used instead). Looking at Q_1, a positive voltage turns on the transistor giving a larger current from the collector to the emitter. This emitter current becomes the base current for the next transistor. The only limit here is the power rating of the transistors (each one has to carry more current) and the power supply.

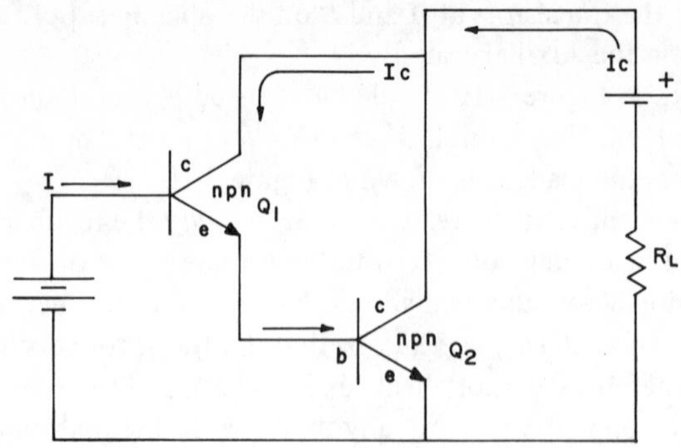

Figure 3-12. Darlington amplifier.

If we make use of a PNP and an NPN transistor (this is called a *complementary pair*), the circuit must look something like that shown in Figure 3-13. Notice that the first transistor has the emitter on top. When Q_1 turns on there is an increased current to the collector (as shown by the arrow I_{c_1}), and this current turns on the base of Q_2. The only thing to appreciate about this circuit is the way the reverse bias gets to the collector of Q_1. The emitter of Q_2 is forward biased with respect to the base, and current comes through this forward-biased junction to the collector of Q_1.

If you want to play this game with an NPN and PNP transistor pair you have to use the system shown in Figure 3-14 of the text. The emitter of Q_3 is forward biased with respect to the base, so current passes easily through this junction. However, the collector of Q_1 is reverse biased, and this makes Q_1 look like a high resistance. This in turn keeps a large (+) voltage on the base of Q_3, thereby keeping Q_3 off (not conducting). When Q_1 turns on, its resistance drops; this reduces the (+) voltage on the base of Q_3, thereby turning Q_3 on.

(One simple analogy to the Darlington system is a series of hydraulic tanks, each one 30 times larger than the one before. A man can open the first valve by hand but the second one is too much for him, so the flow from the first tank is used to "turn on" the second tank. The second turns on the third, and each time the rate of flow [the amplification] increases by a factor of 30. In a short time one person is controlling [via this multiplication system] a flow of millions of gallons via 50 inch valves, which even a gorilla couldn't turn by hand. If this is a hydraulic amplifier, the Darlington is a current amplifier.)

The thing to appreciate is that the Darlington system can lead to almost unlimited current gain by increasing the number of stages. The limit is set by the last transistor in the series because it must carry the maximum current. Typical current gain per stage is about 30; for two stages, the current gain would be 900. Present-day transistors have maximum current limits of about 350 amps.* Above 350 amps, nonlinear devices, such as silicon-controlled rectifiers (SCRs) and TRIACs can be

*Westinghouse markets 500 amp transistors as of 1980, and current levels are still going up.

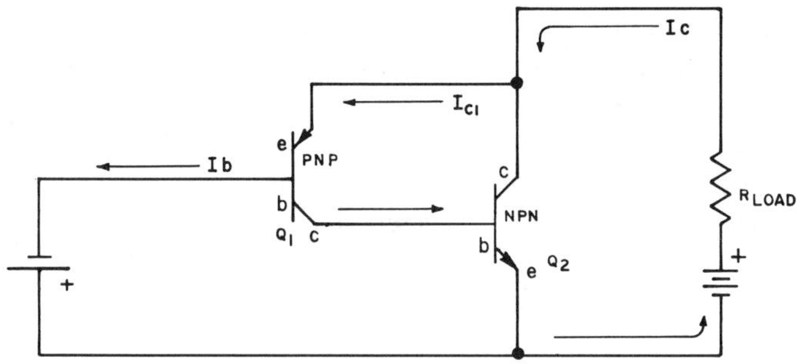

Figure 3-13. Darlington amplifier.

used to go up to hundreds of amps. SCRs and TRIACs will be covered in Chapter [illegible]. (Of course, with all these devices there are some limitations. Any leakage current through the first transistor will be amplified by the subsequent devices in the circuit. *Leakage current* is that current which flows between the emitter and collector when there is no input signal to the base. This leakage may limit the amount of gain any amplifying system can develop. For now you can ignore it. We designed this circuit so leakage would not bother you.)

To illustrate the use of our Darlington amplifier circuit of Figure 3-13, let's consider its application in the circuit of Figure 3-14. This is a practical audio amplifier that can drive a 6 to 16 ohm speaker.

We assume that the op-amp provides an output of up to 10 mA at ±10 volts.

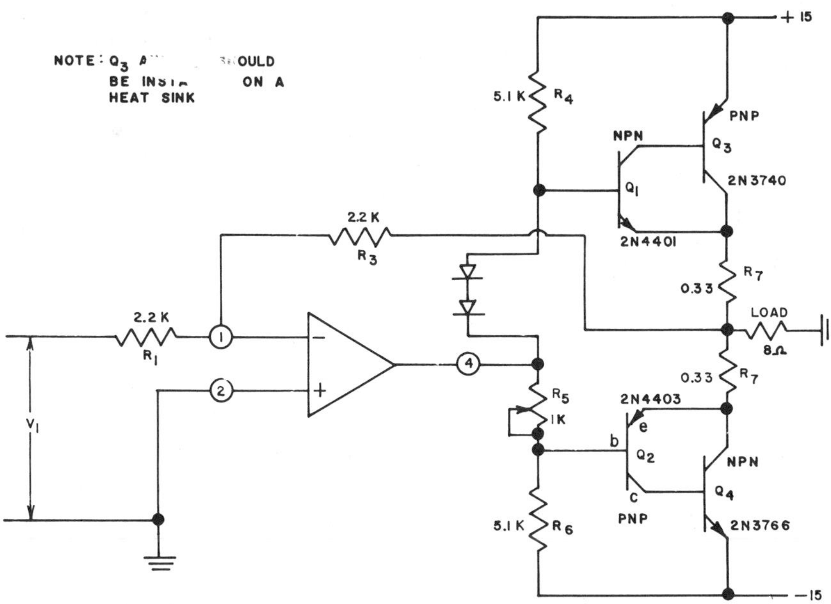

Figure 3-14. Darlington amplifier for an audio system.

Resistors R_4, R_5, and R_6 and the two diodes provide a small forward bias to the base of transistors Q_1 and Q_2. (Note that Q_2 and Q_4 correspond to Q_1 and Q_2 in Figure 3-12.) This prevents distortion when the output of the op-amp goes from (+) to (-). Without this arrangement, both sides of the push-pull circuit would be "off" for a portion of the voltage swing. The 0.33 ohm resistors (R_7) tend to stabilize this bias point. The value of 0.33 ohm is merely some convenient value much less than 8 ohms. (We will have a lot more to say about bias stabilization with emitter resistors in Chapter 7.)

The two diodes shown (recall Figure 1-2) should be made of silicon. They will each have a voltage drop of about 0.7 volt. The total voltage drop of about 1.4 volts will balance the two 0.7 volt emitter-bias voltages of Q_1 and Q_2. Any additional voltage dropped across R_5 will appear across the 0.33 ohm resistors (this voltage will be due to bias current flowing in Q_3 and Q_4). The 0.7 volt junction voltages are only approximate, so you will have to adjust R_5 to balance the system (if the music sounds good, it is balanced). Another way to balance the system would be to use a sine wave input and an oscilloscope on the output: a sine wave output would tell you that all is well.

Assuming that each pair of transistors has an average gain of 30 (a conservative value), the current gain will be 900. To evaluate the current that the op-amp must deliver, we divide the load into two parts. The 8 ohm load resistor is the first part of the op-amp load. The maximum voltage across the 8 ohm load resistor is the op-amp maximum (+10 volts) minus the base-to-emitter drop in Q_2 and the drop across R_7, which might be a total of 1 volt. The remaining 9 volts across the 8 ohm resistor generates a current of (9/8) = 1.13 amps. The output current required from the op-amp for this load current is 1.13 amps divided by the current gain of the circuit [(30) (30) = 900] : the net output current is then (1.13/900) = 1.26 mA. You should note that most of the current comes from the ±15 volt transistor supply. The op-amp provides the base current to make things go.

The second part of the current required from the op-amp is that which flows through R_5 when the op-amp output is +10 volts and the total voltage drop (i.e., 10 + 15 = 25 volts) is across R_5 and R_6. This additional current would be under 5 mA. Adding this to our previous result of 1.26 mA, we find the total required current to be 5 mA + 1.26 mA or about 6 mA. This is well within the 10 mA capability of the op-amp.

The output power in the 8 ohm speaker would be $I^2 R = (1.13)^2\, 8 = 10.2$ watts. This might well be the number that an unethical amplifier manufacturer might quote, but, to be honest, we should realize that the output is a series of sine waves, so the true output power is more like $(I_{rms})^2\, 8 = (0.707)^2\,(1.13)^2\, 8 = 5.1$ watts. This is a significant output if directed into a decent loudspeaker. If you want more power, you could use a 4 ohm speaker. The problem is the limited 10 volt op-amp output. If you put in a one-transistor voltage amplifier (Figure 7-11), you could easily get an output of 200 watts rms.

So this is how you would build an audio amplifier with an op-amp driver. You might wonder how we picked the values of R_4, R_5, and R_6. The resistances of R_4

and R_6 must be large enough so that excessive current is not drawn from the op-amp, yet they must pass enough current to drive the transistors, the combination of diodes, and R_5. We have already seen that the value of 5.1 kΩ for R_5 plus R_6 doesn't overload the op-amp. Choosing a 100 mA bias current requires a base current for Q_1 and Q_2 of 100 mA/900 = 0.111 mA. With no signal applied, the current in R_4 and R_5 is approximately 15 volts/5.1 kΩ = 2.94 mA. This could produce 2 to 3 volts across the 1 kΩ resistor, R_5. This is far more than needed, and we can be assured that some lesser setting of R_5 will provide the desired bias. The 2 to 3 volts is more than sufficient because Q_1 and Q_2 can be turned on by only 1.4 volts (2 × 0.7; recall that 0.7 volt is the emitter-base voltage of each transistor).

At this point you may complain that you really don't know how to pick out transistors and design an amplifier from scratch. We agree with your complaint, but part of our teaching technique is *not* to discuss a great mess of theory before telling you how to do something useful. *First* we show you how to build a working amplifier; *then* we discuss the operation in a qualitative way. In Chapter 7, we will discuss the theory of transistors and how to pick them. You don't have to like our technique, but we think it works. Engineers, when they're not being idealists, are pragmatists: if it works, it is good. If this reverse technique teaches you to like electronics and to want to learn more about it, we can't be too far wrong!

ISOLATING PARALLEL RESISTANCES WITH OP-AMPS

Perhaps once in a lifetime you will run into the problem of measuring one resistor in a parallel circuit. An example would be a problem in a geological experiment in which you have three electrodes in the ground and you want to measure the resistance R_0 between, say, points *A* and *B* without interference from path *A–C–B* (Figure 3-15). Please notice that here we are powering the op-amp with batteries or a floating source. Op-amp ground is on the batteries or the source, not Mother Earth.

In Figure 3-15, if V_1 is a fixed reference voltage, the output – when the switch is open – is

$$\frac{V_0}{V_1} = -\frac{R}{R_1}$$

where

$$R = \frac{R_0\,(R_2 + R_3)}{R_2 + R_3 + R_0}$$

When the switch is closed, R_2 and R_3 are *out* of the circuit. This is easy to see if we note that the op-amp input (1) is essentially at *ground potential,* so R_2 is grounded at both ends. Resistor R_3 is now a resistor between V_0 and ground. This resistor will

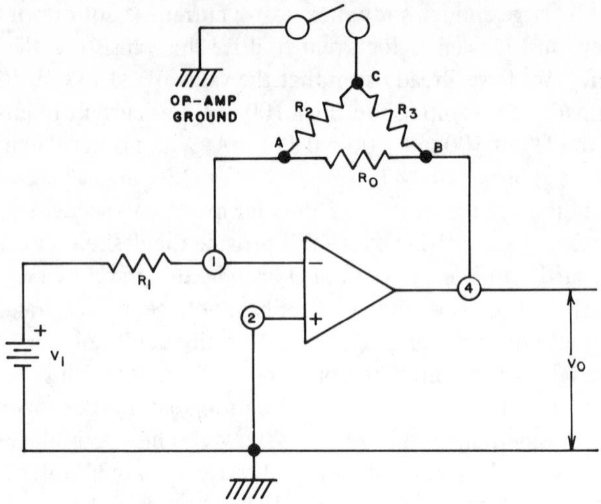

Figure 3-15. Resistance isolation circuit.

have no effect on V_0 unless I_0, which is equal to just $I_0 = V_0/R_3$, is greater than 10 mA. Assuming that I_0 is *less* than 10 mA, we have our old result

$$\frac{V_0}{V_1} = -\frac{R_0}{R_1}$$

which allows us to measure R_0 independently of R_2 and R_3.

We repeat that "ground" here is *on the op-amp, not* on Mother Earth. You can play this game with an isolated op-amp even if R_0, R_2, and R_3 are 1000 volts above earth ground! Of course, you do have to operate the op-amp on batteries or a power supply that is isolated from ground (batteries are safer). Remember to isolate yourself, too, or use a long stick to make adjustments. A system that is 1000 volts above ground can give you a "high" that you won't care to repeat. (Please note that this circuit is intended to measure the resistance of one leg of a three-leg delta* connection, when the circuit is *not* carrying current. You could even do this measurement with the whole system biased above ground, provided you adjust the op-amp system with a long stick.)

REFERENCE VOLTAGE SOURCES

In another section (p. 184), we will discuss the design of power supplies using op-amps. For the moment, our interest is in the design of *reference* voltage sources.

*A delta connected system is used on most three-phase motors where the three windings are connected in the form of a delta (or an equilateral triangle) with the power lines connected to the vertices (corners) of the triangle.

A voltage source is a low-current, regulated, adjustable source of voltage for reference and calibration. The source should remain constant over a reasonable period of time, vary only slightly with changes in ambient temperatures, and provide enough current to operate something like a VOM. A current of 1 mA at maximum output voltage is ideal.

The best DC voltage source is the well known, and seldom seen, *Weston cell.* Without discussing how a Weston cell works (see any textbook in physical chemistry for that), we can say that:

1. It has an output voltage of about 1 volt, which can be accurate to five decimal places if your budget will permit it.
2. Its variation with temperature is about 0.4×10^{-3} volts per degree centigrade near 25°C.
3. It is stable for long periods of time (years).
4. But you can't get current from it. A drain of 10 μA won't hurt, but you can't draw 100 mA and expect the voltage reading to be correct.
5. It costs about $150, so you can't use it carelessly.

As you might have guessed, an op-amp circuit provides the solution to problems 4 and 5. In Figure 3-16, we show the simplest circuit for this purpose. This is really our old friend, the *inverting op-amp circuit.* The capacitors C_1 and C_2 are there merely to knock out stray AC noise signals. If C_2 = 10 pF and $C_1 = 4\ C_2$, all will be well. Note that R_1 and R_2 are nothing but our old R_1 input resistance that has been divided into two parts to make room for C_1. The resistance of R_1 and R_2 should be about 100 kΩ to hold the current from the standard cell to 10 μA or less. The resistor R_0 should be a precision resistance box; then the output can be adjusted

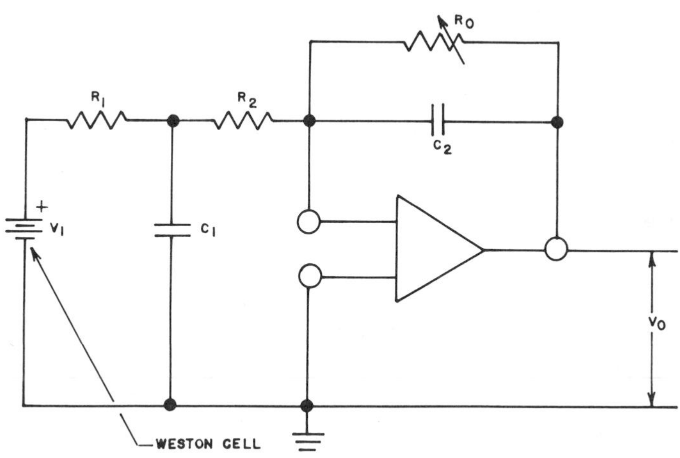

Figure 3-16. Op-amp voltage reference source using a Weston cell.

by varying R_0 all the way up to the maximum output of the op-amp. Of course,

$$\frac{V_0}{V_1} = -\frac{R_0}{R_1 + R_2}$$

You say that you can't afford a Weston cell. Peasant! To quote a well-known 19th century financier, "Anyone who has to ask about the cost of a yacht has no business owning one!" Having got that dig in, we must admit that in our lab there is a distinct shortage of Weston cells. We use mercury batteries for reference voltages. A mercury battery has:

1. A fairly stable voltage-time curve until it is depleted; then its voltage drops to almost nothing. (Note that zinc-carbon cells don't go that way, which is why mercury batteries are better.)
2. A small voltage variation with temperature; about 0.01 volt per degree centigrade near 25°C.
3. The advantage of being inexpensive; they cost about $1.50 each.

Now let us suppose that for every depth, there is one still lower, and all you have are zinc-carbon dry-cell batteries. What do you do next? Here we introduce a new gadget, the *zener diode.* You may recall that we discussed diodes in Chapter 1 and showed the diode *V-I* curve, which we now repeat in Figure 3-17.

In Figure 3-17, we have added a region beyond the breakdown voltage where most diodes die; this region is where the zener diodes live and flourish. To see why, notice what happens beyond V_Z: a slight increase in $-V$ means a huge increase in $-I$. To use this capability for regulation, we can set up a *regulator circuit* such as that shown in Figure 3-18.

The operation of a zener diode voltage regulator circuit is simple: if V goes up (due to a temperature change), the current through the zener diode goes up much

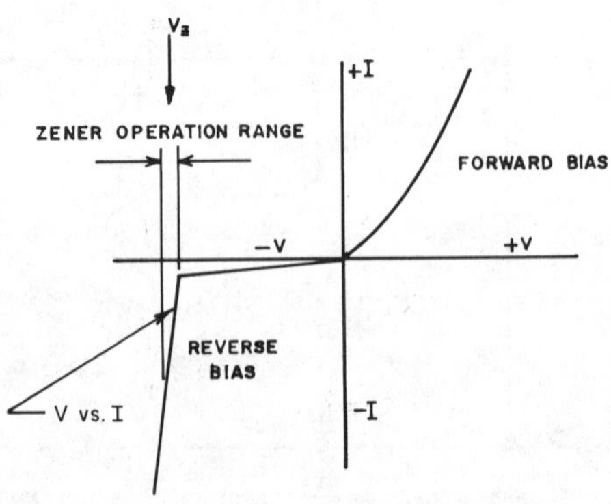

Figure 3-17. Zener diode *V-I* curve.

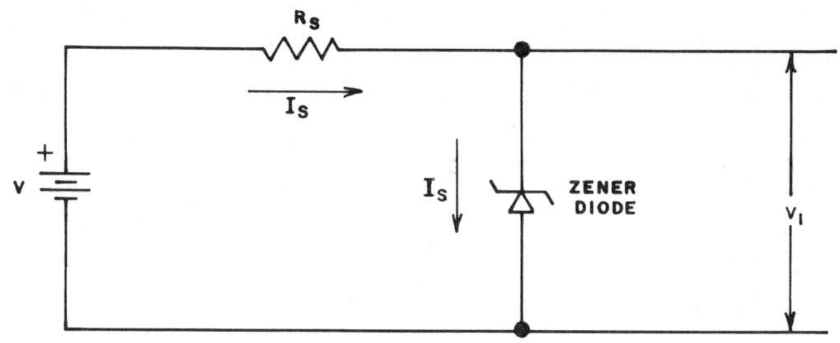

Figure 3-18. Zener-diode regulated voltage source.

faster ($I_s = [V - V_z]/R_s$). This increases the voltage drop across R_s, so V_1 is held close to the zener diode voltage V_z. The exact value of V_1 depends upon I_s, since $V - I_s R_s = V_1$. In practice, any change in V is greatly reduced when it appears at V_1. In fact, $V_1 = V_z$ as long as the battery has sufficient voltage to keep the zener diode voltage above the minimum required value.

To design a zener diode voltage regulator, we follow a simple sequence of steps:

1. What output voltage (V_1) do we want? Suppose we have an old 15 volt battery that is no longer useful for driving an op-amp but can still deliver 15 volts at about 3 mA for a reference source.
2. We decide what the maximum and minimum source voltages will be. In this case, V_{max} is 15 volts; V_{min} will be set by the zener diode we choose.
3. If the load impedance is about 70 kΩ (as it is in this case, since it will be an op-amp), we can ignore it because the current through the load is only a small fraction of the total current through R_s. Now we turn to the *Motorola Zener Diode Handbook* (if you don't have one, you can turn to the *Allied Electronics Catalog*). We decide that we want an output voltage of, say, 6.8 volts, so we pick a 1N957 zener diode rated at V_z = 6.8 volts, I_{zt} = 18.5 mA, and Z_{zt} = 4.5 ohms.* These are *average* conditions, not maximum or minimum values. The terms themselves mean that at a current of 18.5 mA, the zener diode voltage will be 6.8 volts, plus or minus some tolerance which may or may not be given. The zener diode resistance (4.5 ohms) is the slope of the *V-I* curve in the zener diode region.

Now to proceed to our design. As the battery voltage drops off, a point will be reached where the zener diode will no longer operate as a zener diode. Let's assume that this occurs at a battery voltage of 10 volts; when V = 10 volts, we throw away the battery. The zener diode voltage is 6.8 volts. The minimum zener diode current is 1 mA (this fact comes from the Motorola handbook or your knowledge that $I_{z(min)} = 5\% \, I_{zt}$). The value of R_s is given by the fact that the drop across R_s must be 10 volts - 6.8 volts, or 3.2 volts at a current of 1 mA. This means that R_s must

*R_{zt} would be consistent with our usage in this book, but Z_{zt} is used in the handbooks.

be 3.2 kΩ . . . . To be more accurate (nitpicker), you could add the 0.1 mA to the load and get R_s = 2.9 kΩ.

The next question is: What happens when V = 15 volts? Since the drop across R_s is 15 - 6.8 = 8.2 volts, the current I_s must be 2.6 mA. The power dissipation in the zener diode is the product of voltage times current, which, in this case, is 2.6 mA times 6.8 volts, or 17.7 milliwatts (mW). Our zener diode is a 400 mW type, so we are quite safe. (You should notice that we used $I \times V$ rather than $I^2 R$ for calculation of the power dissipation because with zener diodes, the value of R is subject to great variation as operational conditions change.)

We have succeeded in turning a discarded battery into a stable, ±5% voltage, source that can be used to drive an op-amp. This is an example of the economical use of material that is characteristic of successful laboratories.

At this point you might think that we have pretty well finished with the zener diode; however, there are a few more points yet to go. First, what about situations where you can't ignore the current to the load? A case of this type might arise when the load is a radio that requires, say, 50 mA at 6.8 volts rather than an op-amp that has an effectively infinite input impedance. To handle this situation we redesign the circuit of Figure 3-18 as follows:

1. The battery must be able to deliver considerably more current, as we shall see below.
2. The current through the load (the radio), call it I_L, must be significantly *less* than that through the zener resistor R_s. This will insure that the output voltage to the load remains constant even though the load resistance may vary somewhat. This will change our method of sizing the zener.

We begin by choosing the current I_s as some 10–15 times I_L; for this case 600 mA (0.6A) is the best choice. The lowest allowed battery voltage is again 10 volts so the voltage drop across R_s is 3.2 volts at a current of 0.6 amps. Ohm's law yields a resistance value of 5.333 ohms, which in turn indicates that at the maximum voltage (15 volts) the current will be 15 - 6.8 = 8.2 volts. Dividing this by 5.333 ohms yields 1.54 amps of current of which only 0.050 amps goes to the load. In this case we choose a 2 amp zener and the job is done. Again the worst case of power dissipation is 2 amps × 6.8 volts or 13.6 watts. This is a "big" zener but that is what we need for the job. The point to remember is that in this case we COULD NOT IGNORE THE CURRENT TO THE LOAD.

Another problem with simple zener diode circuits may arise as the ambient temperature changes, and we call it *temperature sensitivity*. Typically, V_z will change by 0.5 volt in the temperature range from 25°C–125°C. If this is a difficulty in a particular application, temperature-stabilized zener diodes are available (at higher cost), or you can use two zener diodes, as shown in Figure 3-19. In this circuit, both zener diodes change their V_z at the same rate (you hope), so the 2.1 volt reference value is probably good to 1% or better over a wide range of temperatures. The above circuit uses a 20 volt battery. However, the technique can be used with almost any voltage source.

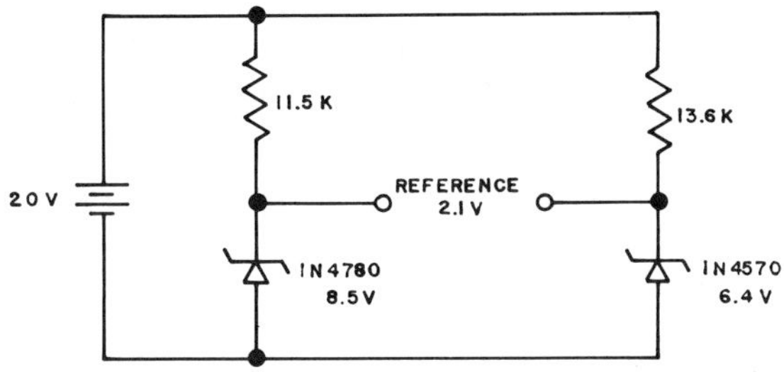

Figure 3-19. Temperature-stabilized zener diode reference circuit. (By permission of the Motorola Semiconductor Products, Inc., Phoenix, AZ 85008.)

Remember that with zener diodes you have available a stable source of reference voltage for most laboratory applications.

CURRENT REFERENCE SOURCES

Another useful gadget is the *current regulator diode.* It isn't really a "diode," but the expression has become established so we have to use it. A *V-I* curve for such a diode is shown in Figure 3-20. Within the operating range, the current remains constant while the voltage varies.

One application of this device is the stabilization of a zener diode voltage source; this is shown in Figure 3-21. If V varies with time or temperature, I_p stays constant so long as V stays in the plateau region of Figure 3-20. This means that I_z is constant, and, since $I_z = I_p$, V_z is also constant. This takes care of the tendency of the zener diode voltage to vary with temperature.

For use as a *current source,* the current regulator diode can be used either in series between the voltage source and the load, or as a current regulator to control the base current to a transistor (the transistor output would provide the current to the load).

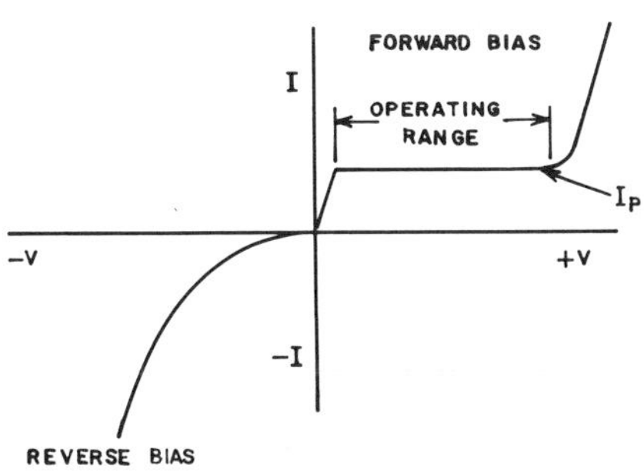

Figure 3-20. Current regulator diode *V-I* curve.

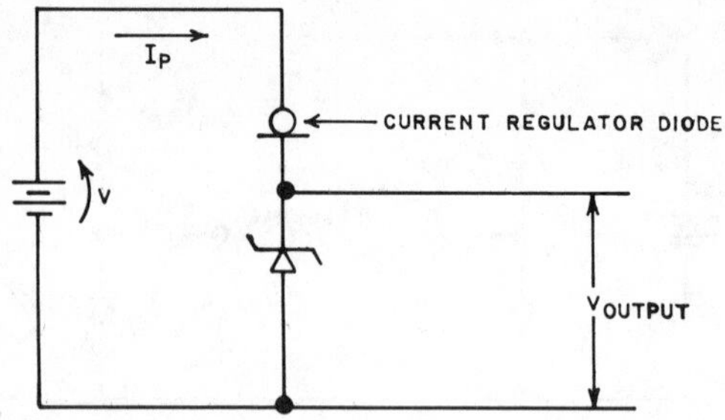

Figure 3-21. Temperature-stabilized zener diode voltage source.

VOLTAGE COMPARISON CIRCUITS

Occasionally we want to know when a voltage reaches some pre-chosen value. An elementary circuit for this purpose is shown in Figure 3-22. In this circuit, V_2 is our reference voltage. As long as V_1 is greater than V_2, V_0 is held at the maximum negative voltage of the op-amp. When V_1 is less than V_2 by as little as 0.2 mV, V_0 swings to the maximum positive output. V_2 can be set at almost any (±10 volt) arbitrary positive or negative voltage. This makes the circuit useful for things like signaling when a well is full or when an animal walks into a cage and closes a switch. This is a good "first circuit" for assisting you in thinking about *how* gate circuits *work.* However, it would not be a good one to build because the op-amp acts like a direct connection between the two voltage sources. You might end up exceeding the input voltage limit on the op-amp.

A much better circuit for this latter purpose is shown in Figure 3-23. The operation of this circuit involves setting the bridge so that R_1 is about 1 MΩ and $R_1 = R_2 = R_3 < R_x$.This unbalanced voltage puts a current through R_4 and applies a voltage to the op-amp input (1), which in turn produces an output voltage V_0. When an animal steps on two conducting strips, it produces a low resistance in parallel with R_x, which shifts the balance in the bridge and changes the polarity of the voltage applied to the op-amp.

What about R_5 and R_6? The resistor R_5 provides some positive feedback to snap V_0 from one polarity to the other as the input voltage changes. The silicon diode network and the resistor R_6 are not really needed; the circuit would work without them. You might be interested in looking at the discussion on page 132 to

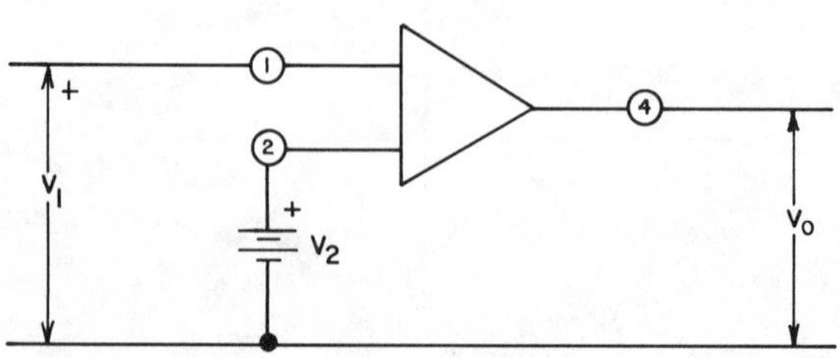

Figure 3-22. Voltage comparison circuit.

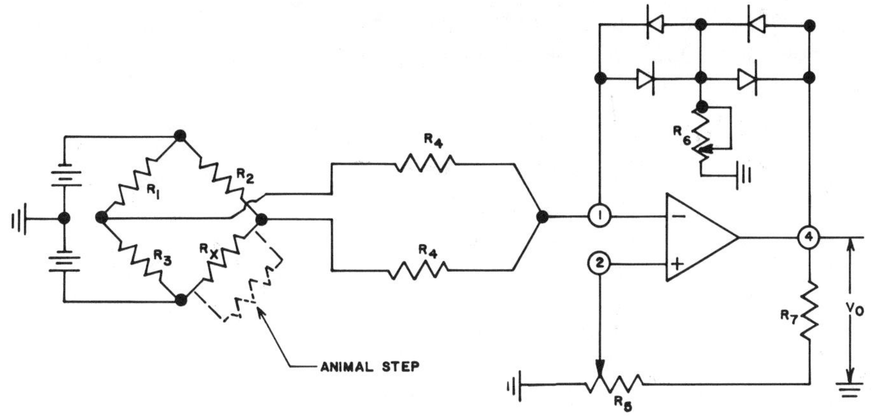

Figure 3-23. Voltage comparison detector.

appreciate that the diodes in Figure 3-23 act like ideal diodes. If you used the circuit without the diodes, the value of V_0 could swing from + 10 volts to −10 volts. This might induce op-amp saturation, which is bad (it makes for slow response). The diode network and R_6 provide some feedback that keeps V_0 from swinging the full ±10 volts. This allows quick response to changes in the input voltage. Among the professionals, this technique is called "bounding the circuit"; it is well worth knowing. (More on this point later.)

CURRENT-TO-VOLTAGE CIRCUIT

Many types of instruments generate a current signal, but most meters and recorders measure voltage. A single op-amp circuit can be used to give a voltage signal that is proportional to a current. Such a circuit is shown in Figure 3-24. Again, it is our old friend, the inverting op-amp circuit.

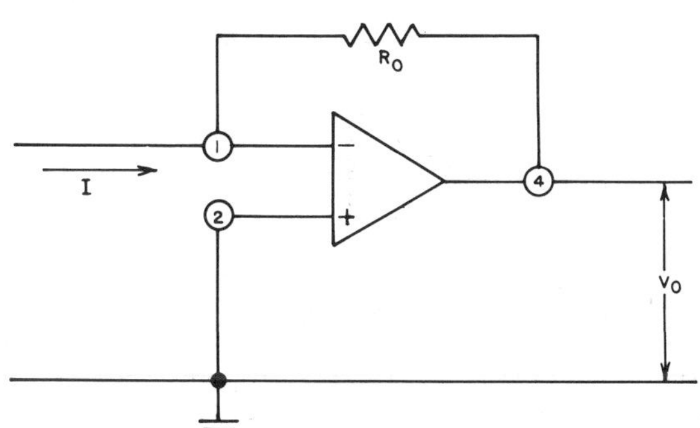

Figure 3-24. Current-to-voltage circuit.

The voltage drop in the current loop is essentially *zero* in this circuit. The relation between the output V_0, the current I, and the feedback resistance R_0 is

$$V_0 = -IR_0$$

For example, if $I = 0.01\ \mu A$ and $R_0 = 10\ M\Omega$, the output, V_0, is 0.1 volt with at least several milliamperes of current capability.

The analysis of this circuit is quite easy. If no current flows into the op-amp at (1), then any current that flows into (1) must flow out through R_0. The voltage drop across R_0 must be IR_0, and this is the value of V_0. The op-amp can supply the necessary current to keep $V_0 = -IR_0$, even when the current to the load is as high as 5 mA.

There are a number of applications for a circuit of this type in Figure 3-25. In Figure 3-25A we have put an ammeter in the feedback loop of an op-amp, and as a result we can measure the current without introducing any voltage drop due to the ammeter resistance (in this case the input impedance of the op-amp circuit is zero.) Another trick of this type is shown in Figure 3-25B, where we have put in an ammeter and two resistors; the current through the meter is larger than the current from the source, which might make it easier to measure, but the input impedance of the op-amp circuit is still zero.

If you want to derive the formula in Figure 3-25B, just ignore the meter for a moment and realize that the current in R_1 must be just I_1 but that the meter has to deliver the current for R_2 as well. If the op-amp output voltage is $V_0 = -I_1R_1$ as usual, then the current I_M follows from:

$$I_M = I_1 + I_2 = \frac{V_0}{R_1} + \frac{V_0}{R_2}$$

$$I_M = -I_1R_1\left(\frac{1}{R_1} + \frac{1}{R_2}\right)$$

$$I_M = -I_1\left(1 + \frac{R_1}{R_2}\right)$$

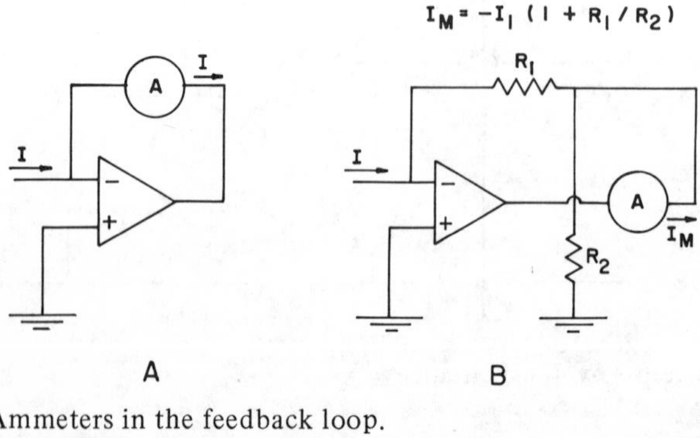

Figure 3-25. Ammeters in the feedback loop.

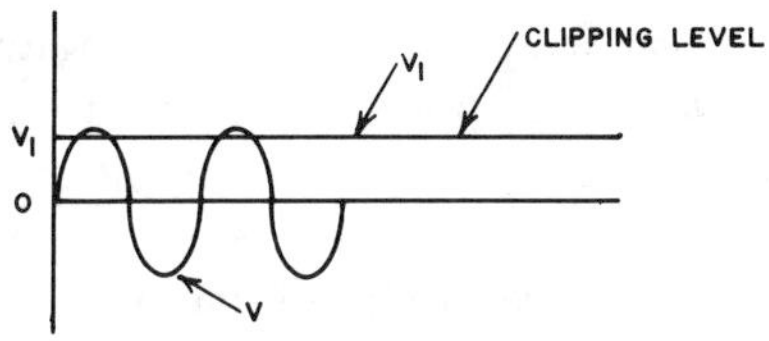

Figure 3-26. Clipping circuit input voltages.

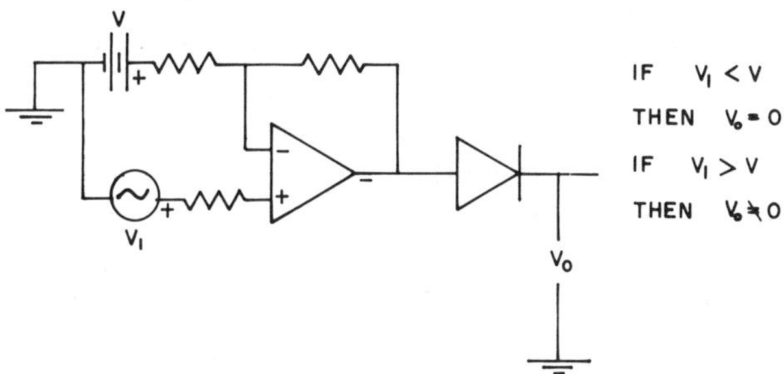

Figure 3-27. Op-amp clipper circuit.

OP-AMP CLIPPER CIRCUITS

Suppose we want an op-amp circuit that cuts off all voltages below a certain level, V_1. For example, if we want to amplify only the voltages above level V_1 (Figure 3-26), we can use a circuit like that shown in Figure 3-27. As long as V_1 is less than V, the voltage V_0 is zero. When V_1 is greater than V (and has the polarity shown in Figure 3-27), the output V_0 is not zero.

GATING CIRCUITS

A *gate circuit* is a useful device that acts as a switch which provides a specific output upon receiving the proper signal. A simple gate circuit is shown in Figure 3-28. If V_1

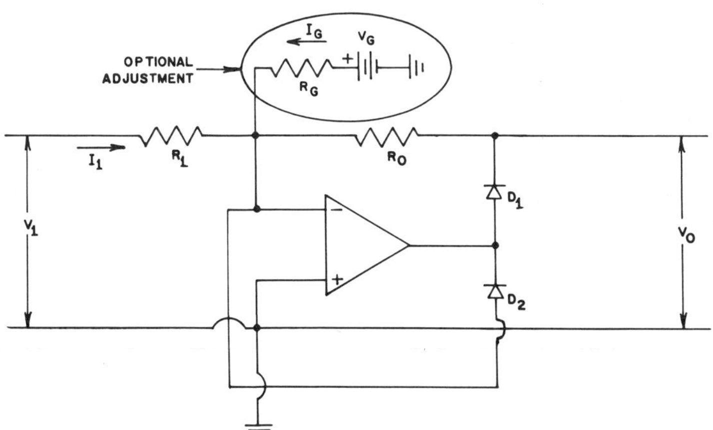

Figure 3-28. Negative gate circuit.

goes (+), V_0 goes to zero because D_1 will not pass the negative voltage coming out of the op-amp. This negative voltage, however, passes through diode D_2 to provide the usual negative feedback signal to the (-) input of the op-amp. If V_1 goes (-), the system is just our inverting op-amp circuit. The point to appreciate is that this is a gate circuit that passes only (-) signals to the output, where, of course, they appear as (+) signals. No (+) input signals ever get to the output, because they are blocked by the diode D_1. (Here again D_1 and D_2 act like ideal diodes.)

We can change this gate circuit from one that switches its output as V_1 changes from (+) to (-) to a gate that switches at some arbitrary voltage V_G. This is shown in Figure 3-28, where we have put in a battery and a resistor indicated by V_G and R_G. The response of the circuit depends on the sum of two currents: $I_1 = V_1/R_1$ and $I_G = V_G/R_G$. If V_1 goes (-) with respect to ground, the circuit switches when I_1 is numerically greater than I_G, that is when $|V_1| > (R_1/R_G)\,|V_G|$. This is, in effect, an *addition circuit;* the input signal goes (-) when $I_1 + I_G < 0$ (we will tell you in more detail what an addition circuit is in a little while – see Figure 3-30 if you're impatient).

We have discussed this device as though it were only a current-crossing detector, but you should realize that it can also be a voltage-crossing detector if that's what you want it to be. Look over Figure 3-28 once more and convince yourself of this; we will be using this idea again.

You might be interested in noting that in the circuit of Figure 3-28, the real diodes, D_1 and D_2, act like ideal diodes, in contrast to those in the circuit of Figure 3-27. (If you have forgotten the meaning of "real" and "ideal" as applied to diodes, go back to Chapter 1 and look it up.) In the circuit of Figure 3-28, when D_1 is in forward bias and D_2 is in reverse bias, the forward-bias resistance of D_1 is much less than R_0, so the system acts as though the forward resistance were "ideal" or zero. The reverse-bias leakage current that flows through D_2 is so much less than the forward-bias current through D_1 that it can be considered zero by comparison. Thus, D_2 acts like an ideal or zero-leakage diode; when the output polarity reverses, D_1 and D_2 change from forward to reverse bias and vice versa, but they are still effectively ideal.

A modified version of the circuit shown in Figure 3-28 is shown as Figure 3-29. Here, instead of two diodes, we use four with a grounded center section. (If you plan to use circuits of this type at frequencies above 50 Hz, you should use what are called "instrument" diodes or "switching" diodes. They work better in the feedback loop because of their low capacitance.) This improves the operation, especially for high speed switching. Figure 3-29 also shows a voltage or current-crossing detector circuit, but the use of the diode feedback network requires some explanation. Suppose that initially V_G and V_1 are (+), a current $(I_G + I_1)$ is generated. This puts a (+) voltage on the inverting op-amp input and the output goes (-). Now diodes D_3 and D_4 are reverse biased, while diode D_1 goes into forward bias while carrying current $(I_G + I_1)$. Some of this current goes to ground via the 1 kΩ resistor R_2, and the rest goes into the op-amp at the negatively biased output (remember current flows from (+) to (-)). Since the output voltage is negative, the feedback current goes through diodes D_2 and D_1. Some of the feedback current is lost through R_2, but the remain-

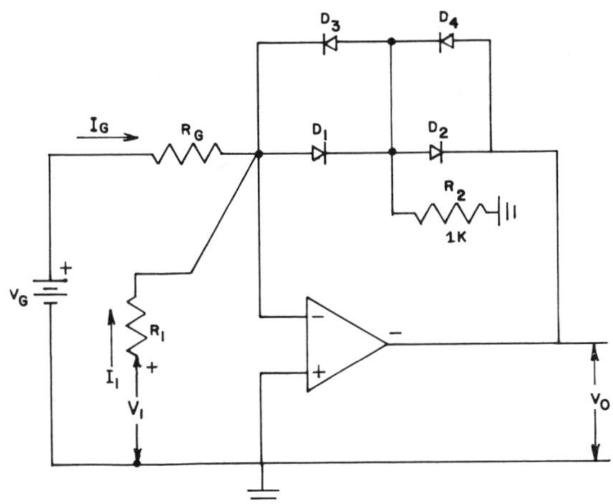

Figure 3-29. Improved zero-crossing detector (modified negative gate circuit).

der provides the necessary feedback to the (-) input of the op-amp. What keeps V_0 from going too high? The answer is easy: if V_0 gets above 1 volt, the feedback through D_1 and D_2 reduces the input signal to zero. The result is that V_0 is held at about -1 volt. (If you need a larger V_0, reduce the 1 kΩ resistance to 0.5 kΩ.) The circuit will now switch sharply when V_1 goes negative and becomes numerically large enough to satisfy the relation $|V_1| > (R_1/R_G)\,|V_G|$. Putting it in terms of current, we can say that switching will occur when $I_G - I_1 < 0$. This is a "fun" circuit – build it and see it work.

If you want to play games with this circuit, you can put zener diodes in place of diodes D_2 and D_4. Consider the result of placing 3 volt zener diodes at these positions. Diodes D_1 and D_3 will act exactly the same way as they did before. However, the zener diodes won't go into their zener mode and conduct current until V_0 reaches about 3.4 volts. This circuit "snaps" even faster than the simple diode circuit of Figure 3-29, and it gives you a larger V_0. (We will leave it as an exercise for you to figure out which way the zener diodes should face.)

OP-AMPS AS COMPUTING ELEMENTS

The op-amp has a number of applications in analog computation; in fact, op-amps were first developed by the analog computer people. The mathematical operations that can be performed by an analog computer include addition, subtraction, multiplication, and division, as well as the operations of integration and differentiation. With this set of operations, many tedious and difficult problems may be solved quite easily using a computer.

In this chapter we shall show a few simple examples using op-amps as a tool for solving equations. Op-amps can be used to perform these mathematical operations if we represent *number values* by *voltages* and *currents* (this is where the term "analog" comes in).

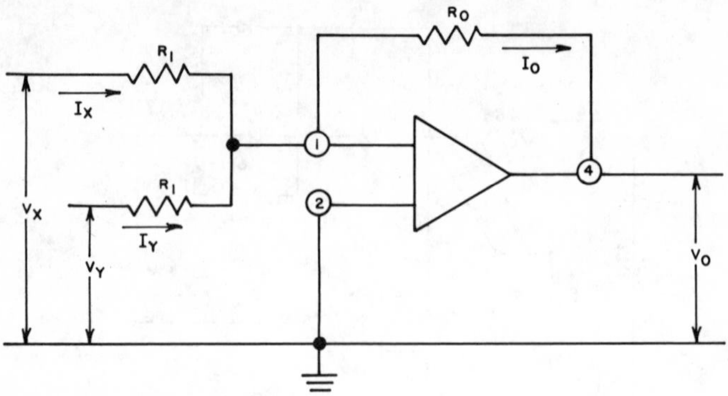

Figure 3-30. Op-amp addition circuit.

Addition Operation

The operation of addition is simple if we regard two numbers, say 10 and 12, as voltages, say $V_x = 5$ volts and $V_y = 6$ volts. To add them, we use the simple circuit shown in Figure 3-30.

The gain law can be obtained as follows:

$$I_x + I_y = I_0$$

since

$$I_x = \frac{V_x}{R_1}$$

$$I_y = \frac{V_y}{R_1}$$

$$I_0 = -\frac{V_0}{R_0}$$

then, by substitution,

$$\frac{V_x}{R_1} + \frac{V_y}{R_1} = -\frac{V_0}{R_0}$$

or

$$V_0 = -\frac{R_0}{R_1}(V_x + V_y) = -\frac{R_0}{R_1}(11)$$

It should be clear that since V_x and V_y represent numbers, we have *performed the operation of addition.* Our addition system can also handle negative numbers quite easily. If our first number were -10 instead of +10, V_x would be -5 volts and V_0 would be $V_0 = -(R_0/R_1)(V_y - V_x) = -(R_0/R_1)(1)$.

The application of op-amps to mathematical operations is a testimony to their linearity and stability. You should be aware that the accuracy of an op-amp analog computer is limited. Under normal conditions, one should not expect better than ±1% accuracy. This limitation is due to a combination of op-amp drift and changes in external parameters that occur because of aging and deterioration ("Moth and rust doth corrupt," as the Bible puts it). Within this limitation, op-amps can be used for all sorts of interesting and useful studies. A few of them are listed here; many more are given in the Teledyne Philbrick book listed in the Appendix.

Our op-amp addition circuit has other useful properties. We can change the resistor R_1 to weight the value of V_x and V_y. For example, if V_x and V_y represent quantities of water used by a farmer and V_x costs twice as much as V_y, the total cost for this quantity of water is

$$V_0 = -R_0 \left(\frac{V_y}{R_y} + \frac{V_x}{R_x} \right)$$

The cost per gallon is represented by the *inverse* of the resistance involved. For the case in which $V_x = 5$ volts and $V_y = 6$ volts, we might choose $R_x = 1\ \text{k}\Omega$ and $R_y = 2\ \text{k}\Omega$ (remember that cost is inverse to resistance), thus obtaining

$$V_0 = -R_0 \left(\frac{5}{1 \times 10^3} + \frac{6}{2 \times 10^3} \right)$$

or

$$V_0 = -R_0\,(8 \times 10^{-3})$$

Since V_0 is a measure of the total pumping cost, by adjusting R_0 we can bring V_0 to any value up to 10 volts. Once R_0 is fixed, this system can be used to investigate all sorts of interesting problems.

We will give an example below that involves what is called *sequential addition.* The objective here is to show what op-amps *can* do; the nuts and bolts of *how to do it* were discussed in the first edition of this book. Try your local college or university library or the book by J. Finkel, *Computer-Aided Experimentation,* N.Y.: Wiley, 1975.

Suppose that a farmer has two wells, *X* and *Y*, and that pumping *X* costs 6 cents per 1000 gallons for the first 1500 gallons, 3 cents per 1000 gallons for the next 1500 gallons, and 2 cents per 1000 gallons after that. Pumping well *Y* costs 10 cents per 1000 gallons for the first 2000 gallons; then the cost drops to 1 cent per 1000 gallons. Given that 10,000 gallons are needed, what is the optimum *lowest*

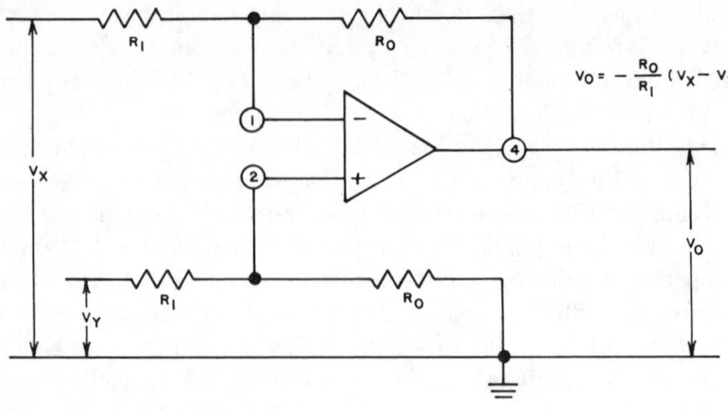

Figure 3-31. Op-amp subtraction circuit.

cost pumping schedule? To compute this by hand would be tedious, but our simple analog computer can do it by simply setting water quantities equal to voltages and costs inversely proportional to resistances. Then it can simply add up every possible combination of the use of wells X and Y to yield the proper total amount of water. The combination that yields the lowest value of V_0 is the optimum scheme.

Subtraction Operations

Subtraction with op-amps is easy. We could always invert a voltage and then run it into terminal (1) of the op-amp for subtraction, but it is easier to show how to subtract two positive voltages, say $V_x = 5$ volts and $V_y = 6$ volts. A typical circuit for subtraction is shown in Figure 3-31. You should note that this circuit is merely the old common mode circuit of Figure 2-16 in another application. The analysis is just the same.

Integration and Differentiation Operations

Earlier we mentioned the word *integration.* This word raises fears all over the place: for certain citizens, integration means racial problems; to many Life Science people, integration means calculus and they're terrified. Take heart, integration in the mathematical sense, however, is nothing to fear, no matter what your views on the former issues may be.*

Suppose you have a plant growing in artificial light and you are using varying light intensities and times. You want the *total* intensity-time sum from the intensity-time plot that is shown in Figure 3-32. What you are looking for is the total area

*If you are totally blanked out by calculus and think that you ought to know something more about it, we recommend that you purchase a copy of W. A. Granville's *Elements of the Differential and Integral Calculus,* New York: Ginn, 1934. It's an old-fashioned book, but it, or something similar from that era, will present the material you need in a simple and straightforward manner. It keeps what a friend once called the "delta and epsilon nonsense" down to an absolute minimum.

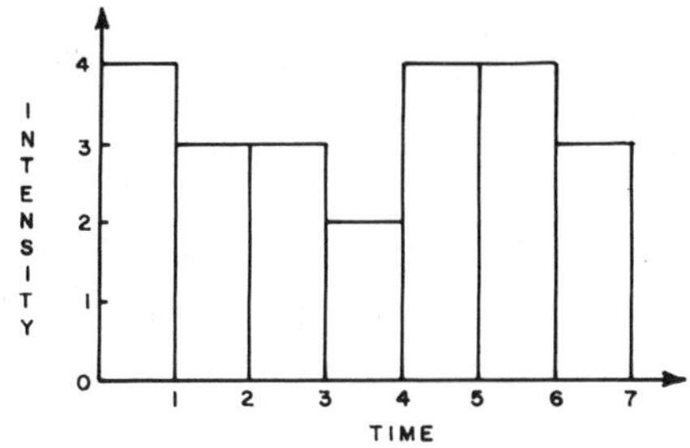

Figure 3-32. Intensity-time plot.

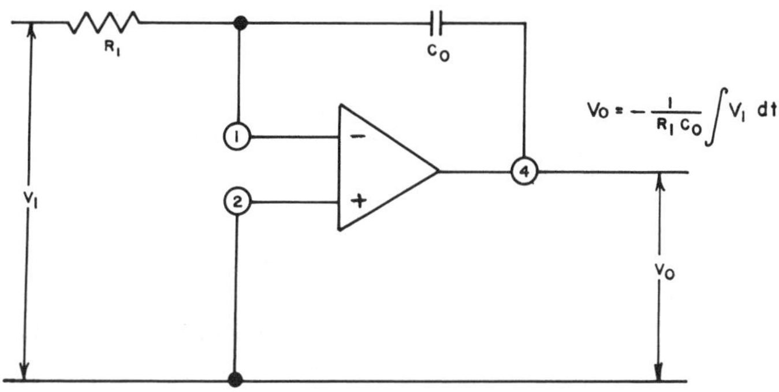

Figure 3-33. Op-amp integration circuit.

under the curve, or 4(1) + 3(1) + 3(1) + 2(1), and so on. Mathematicians write this as $\sum_{i=1}^{T} (It)_i$.

The point is that if the intensity, I, can be represented by a voltage, this summation can be done with the op-amp circuit shown in Figure 3-33.* (If you have drift problems with this circuit, a 10 MΩ resistor across C_0 will help a lot.) In this circuit, the intensity is modeled as V_1, the input voltage. The symbol $\int$ simply means that we do the sum in a series of infinitesimally small intervals, but to all intents and purposes the result is the same as if we did the addition in a series of finite steps: 4(1) + 3(1) + 3(1) + . . . , and so on. (Any mathematician who reads this book is welcome to froth at the mouth as he reads this section, but here we are concerned with *practical applications,* not theoretical subtleties.)

Differentiation using op-amps is more difficult than integration, but it is still possible. The meaning of the term *differentiation* can be demonstrated by considering the growth of a population in the presence of unlimited food. Assuming the usual gloomy Malthusian statistics, we show in Figure 3-34 the population increase

*In Figure 3-33 we are introducing another convention often used in circuit diagrams – no ground symbol is shown, the reader is *supposed* to KNOW that V_1 and V_0 are measured with respect to ground. We will use this convention now and then so that you will get used to it. Remember all voltages are with respect to ground unless otherwise indicated.

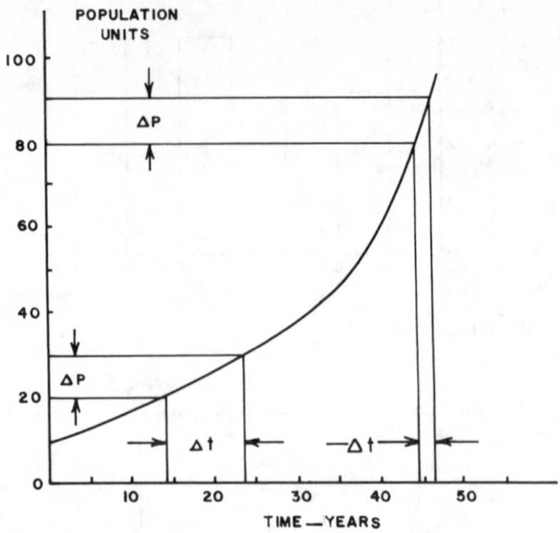

Figure 3-34. Population growth versus time curve.

versus time of a typical *small* planet. Note that in Figure 3-34, equal population jumps occur in shorter times as the overall time increases. We say that $\Delta P/\Delta t$ is the rate of growth or the *slope* of the curve. If we measure $\Delta P/\Delta t$ over smaller and smaller intervals, the ratio $\Delta P/\Delta t$ approaches dP/dt. We call dP/dt the *derivative* of P with respect to t; it is simply the rate at which P changes with time.

When we discuss analog computers in more detail, we will return to this topic. At the moment, we only want to point out that if P can be represented as a voltage, the derivative dP/dt can be obtained continuously by the circuit shown in Figure 3-35. The operation of this circuit is straightforward. AC signals pass through the capacitor C_1 and appear at the op-amp input (1). The op-amp then amplifies these signals via the usual feedback resistor R_0. The important point is that rapidly changing signals get through the capacitor C_1 easily, slowly changing signals less easily, and DC signals not at all. The system responds only to the *rate of change* of V_1, not V_1 itself. Thus, it functions as a *differentiator.*

A practical circuit for differentiation involves at least one more component to ensure stability. The modified circuit is also shown in Figure 3-35, where a small capacitor C_R has been added across R_0. This makes $R_0\ C_R$ into a filter that takes out some of the high frequency noise that gets into the circuit through C_1. We still have a differentiator, but it rolls off (recall Chapter 1) at some frequency $\omega = 2\pi f = (1/R_0\ C_R)$.

Building good differentiator circuits takes a little practice. If you have access to a ramp generator, you should hook it to the circuit of Figure 3-35. (A *ramp generator* is just a voltage source whose value changes with time in some pre-chosen way. Sometimes it is called a *staircase generator.*) Investigate the effect of varying C_R and R_0. If the input looks like the curve in Figure 3-36A, the output should look like that shown in Figure 3-36B. If your wave shapes look bad, put a 100 μF capacitor across R_0. That might help, but good differentiating circuits *are* somewhat difficult

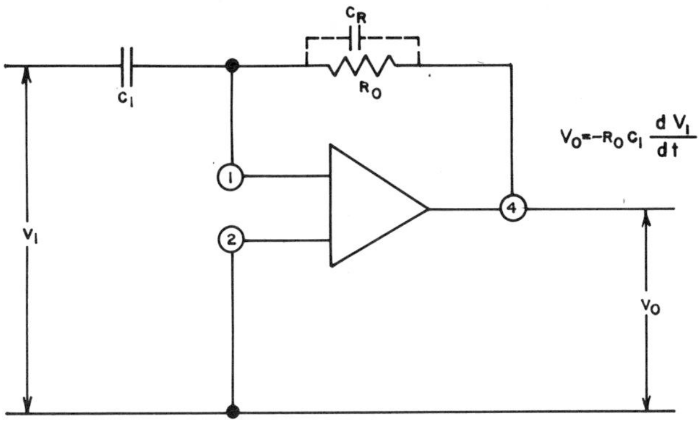

Figure 3-35. Op-amp differentiation circuit.

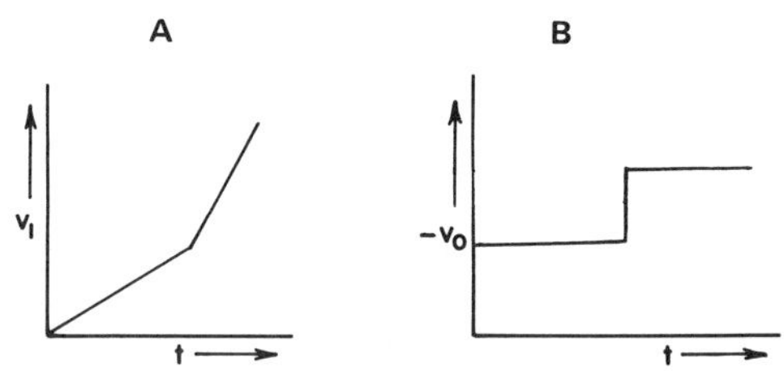

Figure 3-36. Differentiation circuit voltages. *A*. Input. *B*. Output.

to build. If you have instability or noise problems, put a variable resistor in series with C_1 and a variable capacitor C_R across R_0. Then you can start trying to find some combination that works; this is called "breadboarding the circuit." If you can't make the damned thing work for your application, get out your Burr-Brown or Teledyne Philbrick catalog (see Appendix) and buy one.

Another variation on the circuit of Figure 3-35 involves putting in a diode network in place of R_0, as shown in Figure 3-37. Note the similarity between this diode network and that of our zero-crossing detector (Figure 3-29). In the circuit of Figure 3-37, R andC act as a capacitor-resistor differentiation circuit, or a high-pass filter. When V_1 is put into the switching circuit through this differentiator, the circuit will switch when dV_1/dt changes sign. This is why it is called a *change-of-slope detector.* When the input signal V_1 is increasing, then $dV_1/dt > 0$, the input signal is effectively (+), and V_0 is (-). When V_1 begins to decrease, then $dV_1/dt < 0$, and V_0 immediately goes (+). You can use regular diodes here and get an output swing of ±1 volt; with zener diodes, you can swing as much as ±10 volts—don't forget V_1, V_0 are measured with respect to ground.

Another application of this circuit is as a "peak detector," since it switches whenever the input signal stops rising and begins to decrease. One student used this to provide a signal when the flow of water in a desert wash stopped rising and began

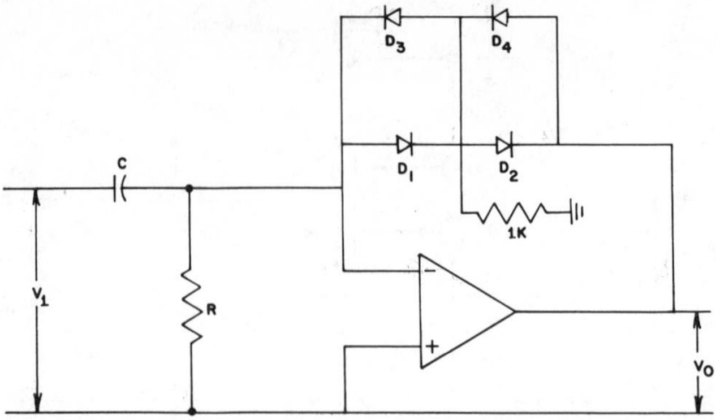

Figure 3-37. Change-of-slope detector.

to go down. (His detection system was a series of resistors that the rising water shorted out. The output current from the circuit was a function of water level.)

Multiplication, Division, and Square Root Operations

Operations that involve taking products have been difficult to accomplish with op-amps. Until very recently, multiplier circuits have been complex, and their accuracy has been limited. Better op-amp multipliers are now available, and the use of a so-called *quarter-square multiplier* will be discussed in detail. Mechanical or Hall effect multipliers are also available but will not be discussed here.

Given two voltages V_1 and V_2, we can write the product $V_1 V_2 = ¼ [(V_1 - V_2)^2 - (V_1 + V_2)^2]$. The term *quarter-square* comes from the factor 1/4 ahead of the bracket.

An op-amp multiplier is a special type of device. They cost about $7.50 (as of January 1980). They have *no* gain; in fact, the input signals are attenuated so that the gain is less than one. The multiplier output must go to an op-amp amplifier to bring the gain back up. The complete circuit is shown in Figure 3-38. (The devices involved happen to be Motorola products because this company developed the circuit.) The output of a multiplier circuit with an op-amp to provide gain can be the direct product of the input voltages ($V_0 = V_1 V_2$).

There are other applications of multipliers. If $V_1 = V_2$ and both are sinusoidal, the multiplier functions as a *frequency doubler,* i.e.,

$$(\cos 2\pi f t)^2 = \frac{1}{2}(1 + \cos 4\pi f t)$$

If the multiplier is in the op-amp feedback loop as shown in Figure 3-39, the result is a *division circuit.* It is important to note that the circuit of Figure 3-39 is a functional diagram rather than a complete circuit that shows every pin and solder con-

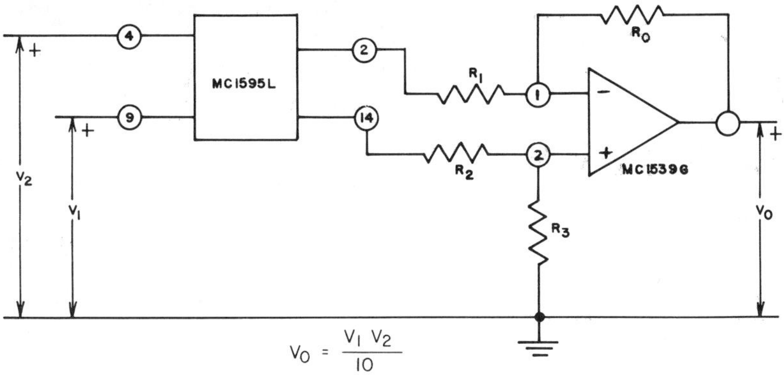

Figure 3-38. Op-amp multiplier circuit.

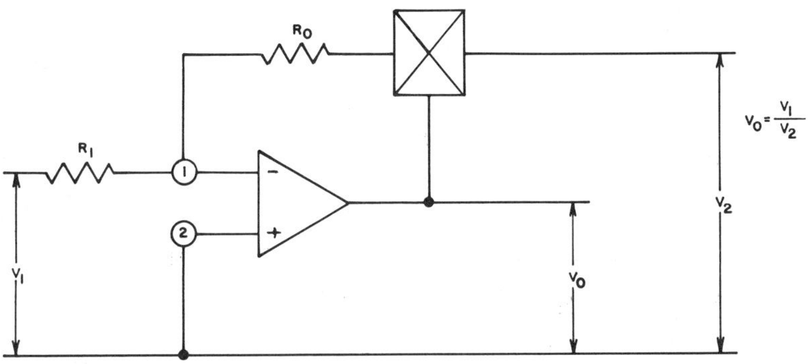

Figure 3-39. Op-amp division circuit.

nection. In most cases, people buy – rather than build – circuits of this type. If you want to build such a circuit, it is best to contact Motorola for the detailed application notes.*

Similar comments apply to the application of multipliers as *square root circuits* (Figure 3-40A) and as *mean-square function generators* (Figure 3-40B); gadgets like these are too difficult for most people to build. You should plan to buy them if you need them.

SIGNAL AVERAGING

The word *averaging* is a tricky one, and one must be aware of the kind of signal to be averaged and what sort of average is wanted. For example, if the signal is *pure AC,*

*For application notes on multipliers, write to Motorola Semiconductor Products, Inc., P.O. Box 20912, Phoenix, AZ 85036. Ask for *AN 489, AN 490,* and *AN 261.*

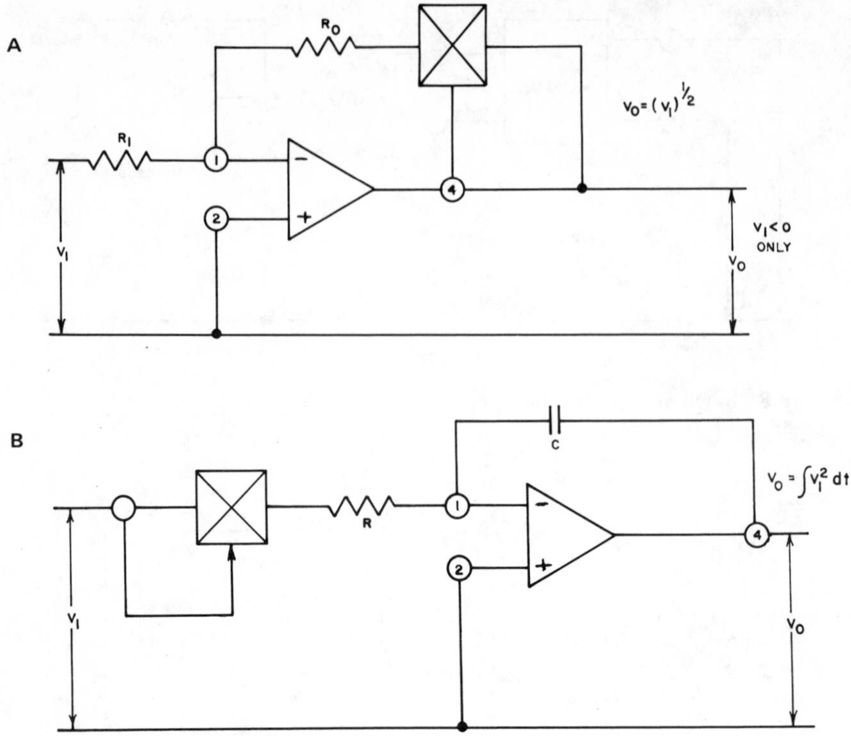

Figure 3-40. *A.* Square root circuit. *B.* Mean square circuit.

the average is *zero.* If the signal is *AC and DC,* the average is the *DC signal only.* If the signal is really a noisy one but not a sine or cosine wave, the average will be a true measure of the net signal. You can think of this as *short-term averaging.* These three types of signals are shown in Figure 3-41.

In the case of Figure 3-41A (a sine or cosine signal), the average is

$$\frac{1}{T}\int V dt = 0$$

so we work with the *rms average,* which is defined as

$$V_{rms} = \left(\frac{1}{T}\int V^2\ dt\right)^{1/2}$$

The V_{rms} is *not* zero for sine or cosine waves, which is why we use it; another reason is that Ohm's law ($V = IR$) yields the same result with *AC* or *DC,* if we use rms.

Our first averaging circuit will not be an rms type, but rather the type of circuit

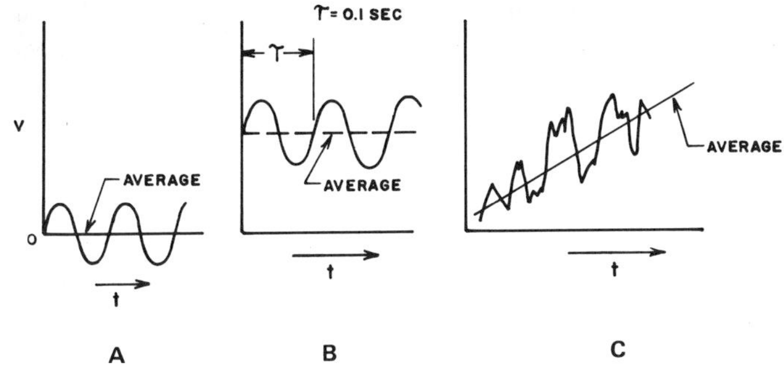

Figure 3-41. Signal averages. *A*. AC only. *B*. AC and DC. *C*. Noisy signal.

you would use for signals such as those shown in Figure 3-41B or 3-41C, where a *true average,* $V_{avg} = 1/\tau \int V dt$, does *not* yield a zero result.

The first question is: What period do we average over? In Figure 3-41B it is clearly 1 cycle or 0.1 second. (You might ask, why not average over 10 cycles? The answer is that it won't make any difference in the final result, and a longer averaging time makes the circuit sluggish in reaching the proper value.)

In Figure 3-41C the period is not defined; in fact, there isn't a true noise period. All we can do is average over a time that is relatively *long* compared with the average noise period. You might have to do some trial-and-error work and vary circuit parameters to get the right averaging time.

In a typical case, the signal shown in Figure 3-41C might be the output of a thermocouple immersed in a turbulent airstream. The average airstream temperature is rising steadily, but this may be hard to see because of the turbulent fluctuations. Let's assume that we read the output every 10 seconds and that the noise period is 0.1 second. The averaging is done with a modified integrating op-amp circuit, as shown in Figure 3-42.

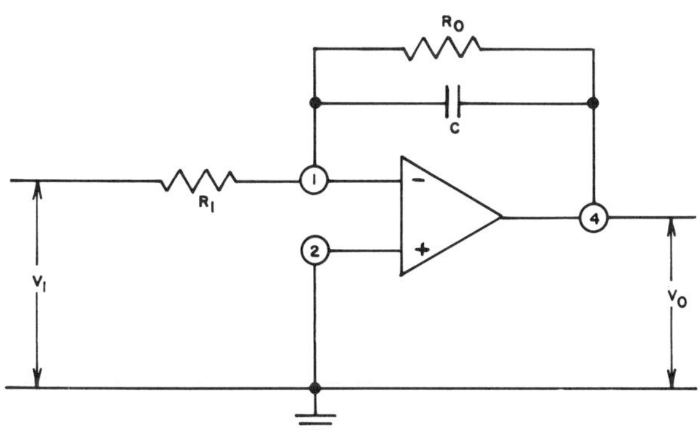

Figure 3-42. Op-amp averaging circuit.

Our original integrating op-amp (see Figure 3-33) had either no R_0 resistor or a very large one across the capacitor. The output was simply the input integrated over time:

$$V_0 = -\frac{1}{R_1 C}\int V_1 dt$$

The effect of R_0 is twofold: (1) it helps filter out short-period noise signals (recall our *RC* filter in Chapter 1) and (2) it gives some DC gain. For example, if we know that the level of the input signal V_1 will vary from 1 mV to 10 mV due to the increase in airstream temperature, we choose a gain of 300. This means that when $V_1 = 10$ mV, the output V_0 will equal 3 volts. The noise has an average frequency of, say, 10 Hz (which is typical for turbulent eddy flow), so we want our $R_0 C$ circuit to cut off everything above, say, 5 Hz. We choose $f = 1/2\pi R_0 C$, that is, if $R_0 = 10\ \text{k}\Omega$ and $C = 3.2\ \mu\text{F}$, then the relation $2\pi f = 1/R_0 C$ is satisfied. If a signal of the type shown in Figure 3-41B or C is the input signal, the output V_0 would look like that shown in Figure 3-43, in which the straight lines represent the *average* of each type of signal.

For those of you who may have trouble understanding the integration circuit, we suggest that you think of a water jet filling a big tank. If the water flow rate varies, this variation will change the rate at which the tank fills; but, in a real sense, the tank integrates (adds up) the flow over a period of time. If we do not empty the tank it will eventually be filled. This is an example of "saturation."

If we put in a valve so that the water can run out of the tank at a rate which prevents overflow, the level of water in the tank will reach some point determined by the average rate of flow in and the rate at which the water runs out through the valve. If the tank is large enough, the level of water will remain almost constant even though the input flow rate may vary wildly over short periods of time. We could call this system a "leaky integrator."

Now let us return to the problem of finding the rms average of pure AC signal. If our input signal is V_1, we define

$$|V_1| = (A^2 \sin^2 \omega t)^{1/2}$$

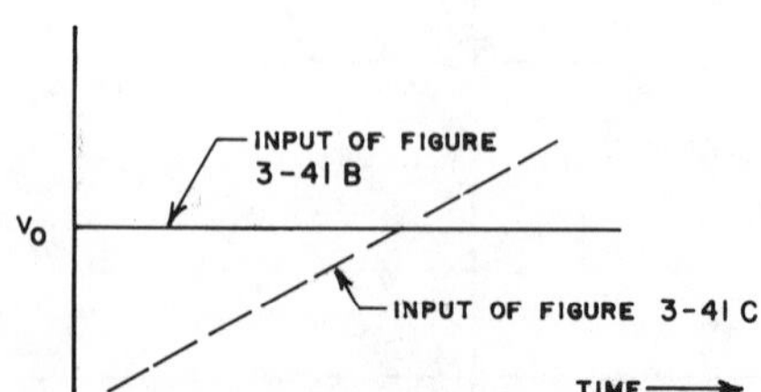

Figure 3-43. Outputs of an averaging circuit.

The vertical bars indicate that the absolute value of V_1 is to be used (whether positive or negative). Note that V_1^2 is *always positive,* but it does vary with time as $\sin^2 \omega t$. (Recall that ω is just a shorthand way of writing $2\pi f$, where f is the frequency in Hz or cycles per second.) To get V_{rms}, we must average this varying value of V_1 over some interval to obtain

$$V_{rms} = \left(\frac{1}{T}\int_0^T V_1^2\, dt\right)^{1/2}$$

So our root-mean-square circuit must have three parts: (1) a precision rectifier to turn AC into pulsating DC, (2) a square root circuit, and (3) an averaging circuit to obtain the final V_{rms} value. The circuit for this three-step process is shown in Figure 3-44.

You might wonder whether all AC meters have this much circuitry in them. The answer is "not quite," but if we want to build a *true* rms circuit, we must process the signal through squaring, averaging, and square-rooting circuits. Common AC voltmeters don't really read rms, even though their scales are clearly labeled "rms." VOMs usually have simple diode bridge rectifiers that produce a voltage *related* to the *average* of the rectified AC input. Their scales are calibrated to read in rms for a *sine wave only;* any other waveform would not be read as a rms voltage. A VTVM has an AC amplifier that raises low voltages to levels suitable for rectification by diodes (you need several volts or more to get a useful output from a diode bridge).

Figure 3-44 shows that a precision AC averaging circuit can be built. With the

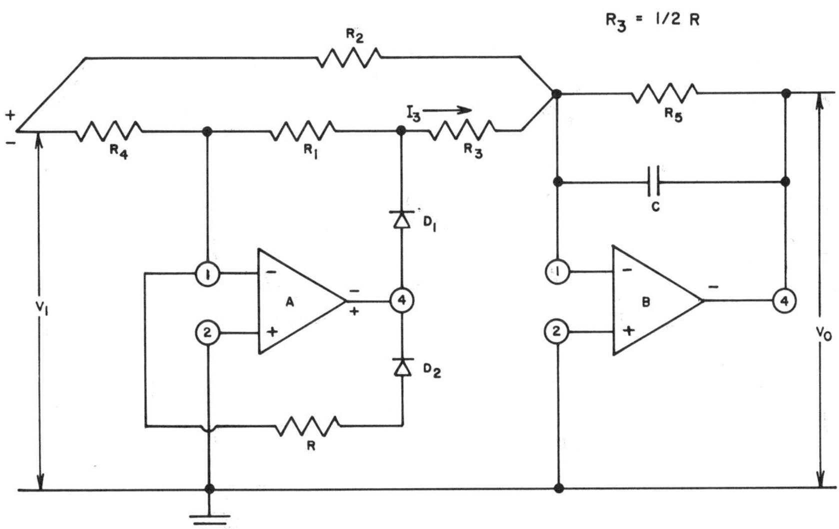

Figure 3-44. Precision AC averaging circuit. (If you really need something of this sort, it might be better to buy than build. Check the Burr-Brown catalog for prices.)

capacitor C across the output, we obtain a DC voltage equal to the rectified average of the input waveform

$$\left(\frac{1}{T}\int_0^T V_1^2\,dt\right)^{1/2}$$

To understand this circuit, we have to take it in steps. First, the resistances are all equal in magnitude except that $R_3 = \frac{1}{2}R$, and they are numbered *1, 2, 3, 4,* and *5* for purposes of explanation only. Second, D_1 and D_2 are diodes. (Do you remember about diodes? If not, *stop* right here, go back to Chapter 1, and review the section on diodes. Otherwise you are lost.)

Assuming that you know about diodes, let's return to Figure 3-44. When V_1 is (+), the output from (4) on op-amp A is (-), so diode D_1 is cut off (reverse biased). Diode D_2, however, is conducting, so the voltage at (1) on the op-amp A is brought to zero. In contrast, the output from op-amp B is $-V_0$. This is clear if we note that $I_3 = 0$; therefore R_2 and R_5 act like an inverting op-amp circuit with op-amp B. There is unity gain, so the output of op-amp B is $-V_1 = V_0$.

When V_1 is (-), the output of op-amp A is (+), which passes current through diode D_1 (diode D_2 is now cut off). This (+) signal is presented at the input of op-amp B via R_3 ($R_3 = \frac{1}{2}R$), giving a gain of 2 and an inversion to generate an output of $V_0 = -2\ V_1$. The voltage V_1 also appears at op-amp B by another route, namely via R_2, and, since $R_2 = R_5$, there is a unity-gain signal at V_0 of $+V_1$. The net output of op-amp B is $V_0 = -V_1$.

Thus we have a full-wave rectifying circuit, and we can say that $V_0 = -|V_1|$ or $-\sqrt{V_1^2}$. The next step is to average V_0, which is done by the capacitor C on op-amp B. The net result is an *ideal* averaging circuit. (The term *ideal* is used because the diodes in the output of op-amp A act like ideal diodes in spite of their real diode characteristics. However, this is not always the case. You should recall that a real diode has both reverse current leakage and forward-bias voltage drop.)

OP-AMPS AS FILTERS

In Chapter 1 we discussed the use of a filter as a device for separating voltages of different frequencies. (There are amplitude filters, too; they were called "gate circuits" earlier.) We defined *high-pass filters* as those that stop frequencies below some value and pass frequencies above that value. *Low-pass filters* do just the opposite, and *slot filters* pass a band of frequencies. There are also *band-attenuation filters* that pass all but a band of frequencies.

Band-Pass Filters

The characteristics of typical band-pass filters can be defined in terms of the curve of Figure 3-45. To discuss Figure 3-45, we need a way to characterize filters and

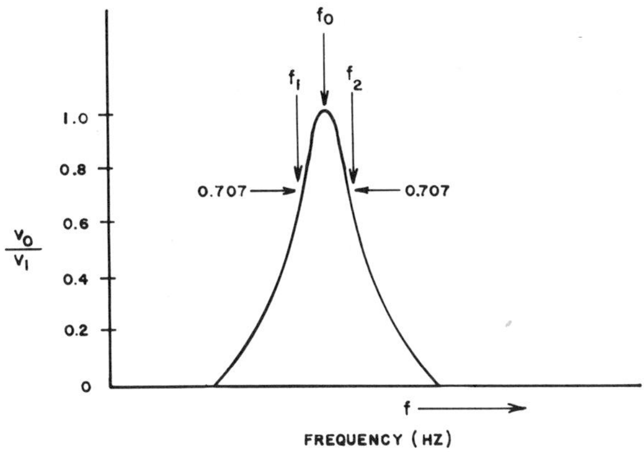

Figure 3-45. Slot filter Bode plot.

compare them. For this type of analysis, it is common to define the *one-half power points* where

$$\frac{V_0}{V_1} = \frac{1}{\sqrt{2}} = 0.707$$

You might wonder why $V_0/V_1 = 0.707$ is chosen. It comes from the fact that the power into a resistor is $I^2R = V^2/R$. If the power into R is V^2/R at the frequency $f = f_0$, then at $f = f_1$ and f_2, it will be

$$\frac{1}{R}\left(\frac{V}{\sqrt{2}}\right)^2 = \frac{V^2}{2R} = \frac{1}{2}\left(\frac{V^2}{R}\right)$$

This is why the term *one-half power* is used.

If we take the frequencies f_1 and f_2 where $V_0/V_1 = 0.707$, we can calculate a *quality factor*

$$Q = \frac{f_0}{f_2 - f_1}$$

This gives us a *measure* of the filter. In general, the *larger* Q is, the *better* the filter. Of course, there are times when a low Q circuit is necessary. When we talk of a Q factor, we are usually referring to a circuit tuned to one center frequency f_0. All frequencies other than f_0 are attenuated to some degree. If the signal source does not have a highly stable frequency, a large Q filter would result in an unstable amplitude in the filter output. (The precise definition of Q [at resonance] is

$$Q = \frac{2\pi \text{ (energy stored in circuit)}}{\text{(power dissipated in one cycle)}}$$

This can be applied to any circuit; however, the formula given in the text is a useful approximation that can be applied to band-pass filters.)

Other types of filters exist that do not follow this simple definition of Q. For example, you might want to pass or stop all frequencies in a specific band, e.g., 100 Hz to 800 Hz. Filters of this type are called Chebyshev, Butterworth, or slot filters. These will be discussed later (p. 152).

RC Filters

The notation above is not easily applied to high-pass, low-pass, or broad band-pass filters. For devices of this type, we use the roll-off in decibels (dB) per octave or decibels per decade. (If you are vague about this, admit it and go back to Figure 2-7 of Chapter 2 for an explanation of roll-off.)

Typically, a simple *RC* filter will roll off 6 dB per octave, two *RC* filters in series roll off 12 dB per octave, and on. It is not practical to put many simple *RC* filters in series, since the signal loss would be too great. But, as you might expect, op-amps come to the rescue. Op-amp active filters can filter as sharply as six or seven *RC* filters in series without significant loss of signal strength.

To appreciate these considerations, let's look at an old friend, the *RC* filter shown in Figure 3-46. You should recall from Chapter 1 that the reactance of a capacitor is

$$X_C = \frac{1}{2\pi fC}$$

As the frequency *increases,* the value of X_C *decreases,* indicating that more of the input signal will be bypassed to ground through the capacitor. Conversely, as the frequency *decreases,* X_C *increases,* and more signal is delivered to the output.

The simple *RC* filter is very sensitive to the load that you place on its output. To measure V_0, you must use a voltmeter with an input resistance that is large compared to *R*. Another problem with simple *RC* filters is the loss of signal strength

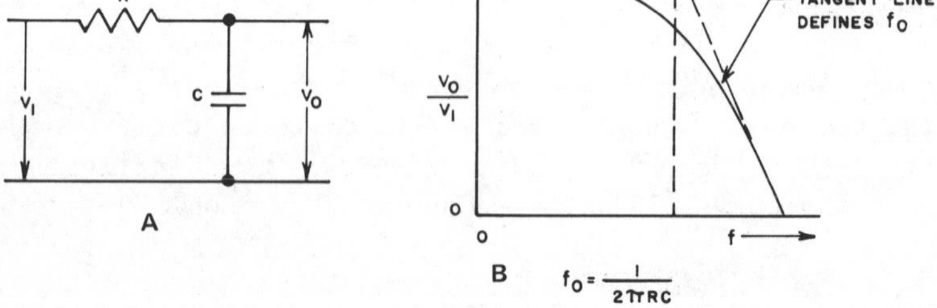

Figure 3-46. Low-pass *RC* filter. *A. RC* filter circuit. *B.* Bode plot.

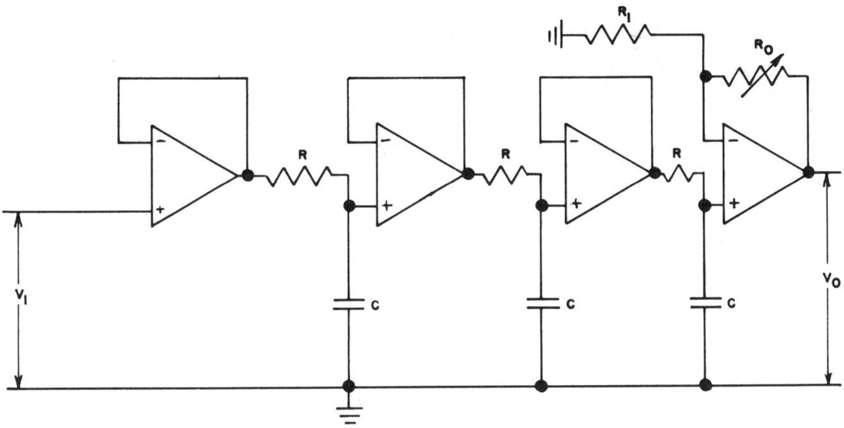

Figure 3-47. Third-order *RC* filter.

because of the resistor R. If we tried to get a sharper cut-off by using two of these *RC* filters in series, the output signal might be too small to use.

The solution is (as always) one or more op-amps. In Figure 3-47 we show three *RC* filters in series with noninverting op-amp buffers. The buffers isolate each filter from its load, and the last op-amp provides any gain that may be needed. This is a simple *third-order low-pass filter.* This filter is expensive in terms of op-amps, but it gives very sharp cut-off: as much as 18 dB per octave. A good experiment for learning about active filters would be to build this circuit and see it work. Try measuring the roll-off as you add each section to the filter string.

Op-Amp Active Filters

LOW-PASS FILTER

A *single op-amp, second-order, low-pass filter* is shown in Figure 3-48. The R_1C_1 section acts as a filter as before – remember that point (1) is at ground potential – but there is more to the story: low frequency signals are stopped by C_1 and are passed through the bypass directly to the output. Some high frequency signals pass through C_1 and appear at (1), where they become an input signal for the op-amp. The op-amp inverts this signal and delivers it to (4). This inverted signal is combined with the original high frequency signal that went to (4) via the bypass. If the signals are of *opposite polarity* and *equal amplitude,* the net result is *zero* for all high frequencies. (Clever, ain't it!) [This circuit is instructive, but it is not simple. To appreciate it assume the V_1 is (+) with respect to ground, and this makes the signal that goes through the bypass lead positive; but the signal that goes to the inverting input of the op-amp is (-) because of the capacitor C_1. The output of the op-amp is (+) but the capacitor C_2 changes the signal to a (-) signal. If R_0 is adjusted so that the gain is about 1, the high frequency signals that get through the bypass will be just the opposite of the signals that come out of the op-amp. This is a *low-pass*

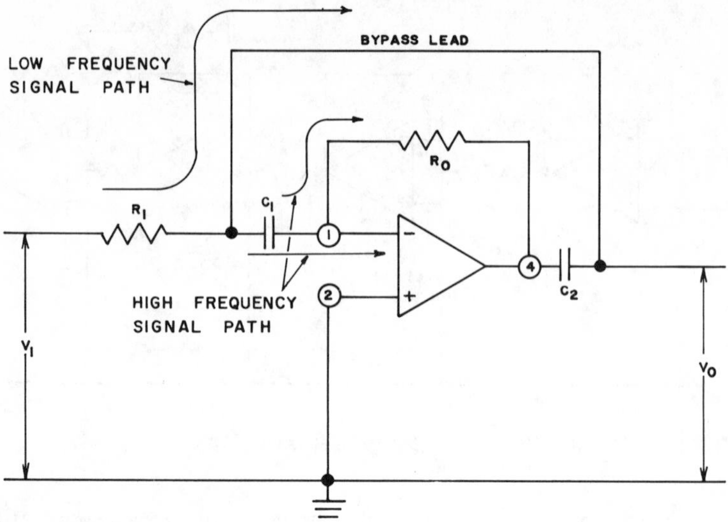

Figure 3-48. Low-pass op-amp active filter.

filter. The trick here is to remember that if you have a (+) voltage on one side of a capacitor, the other side must be (-).]

You should build this circuit (for best results make $R_0 \approx R_1$) and try to convince yourself that it works. The cut-off frequency is controlled by $f_c = 1/(2\pi \sqrt{R_1 R_0 C_1 C_2})$. Note that the roll-off of this filter is about twice as sharp as the *RC* filter alone. This filter has one problem: its DC output impedance is not decreased by the op-amp; it is equal to R_1. This occurs because there is a capacitor in the op-amp output circuit and a direct connection exists between R_1 and the output. If this causes problems, a buffer op-amp can be used to follow the filter circuit.

HIGH-PASS FILTER

High-pass *RC* filters were discussed in Chapter 1. An *active high-pass filter* is shown in Figure 3-49B. This active filter has the same Bode plot curve as the passive filter. The op-amp is introduced to make the filtering action *independent of the load on the output.* In contrast, the passive *RC* filter is *very* load-dependent.

The high-pass filter shown in Figure 3-49 can provide gain, but its major function is not gain but buffering, for which $R_0 = 0$ and $R_1 = \infty$ should be used.

SLOT FILTER

If high-pass and low-pass filters are available, *slot filters* to pass or stop certain frequency bands are also possible. A typical slot filter to pass and provide gain at 1000 Hz is shown in Figure 3-50. Although detailed discussion of this circuit is out of place here, we can indicate that C_1 provides low frequency cut-off, while the C_2R_2 network acts like a high resistance at a frequency f_0. This produces the gain at $f = f_0$

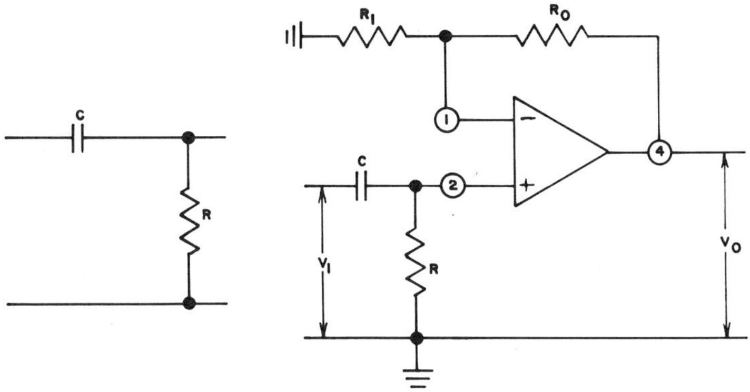

Figure 3-49. High-pass filters. *A. RC* passive. *B. RC* active.

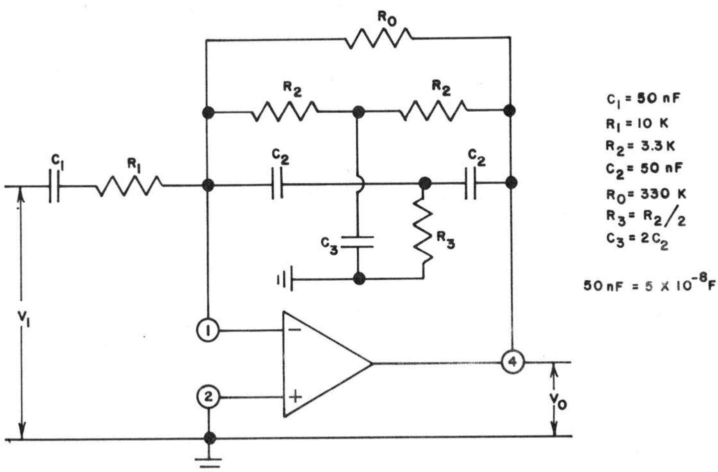

Figure 3-50. Slot filter.

(Figure 3-51). The circuit of Figure 3-50 has a frequency peak at f_0 at 1 kHz. To adjust the frequency peak, we use the relation

$$f_0 = \frac{1}{2\pi R_2 C_2}$$

and adjust C_1 so that $C_1R_1 > 2C_2R_2$. The gain can be adjusted by changing R_0. The resistance R_1 should be kept under 100 kΩ.

BUTTERWORTH, CHEBYSHEV, and BESSEL FILTERS

After having discussed filters at such great length, we must admit that in many cases it is more convenient to *buy* active filters than to try to build them. This is especially

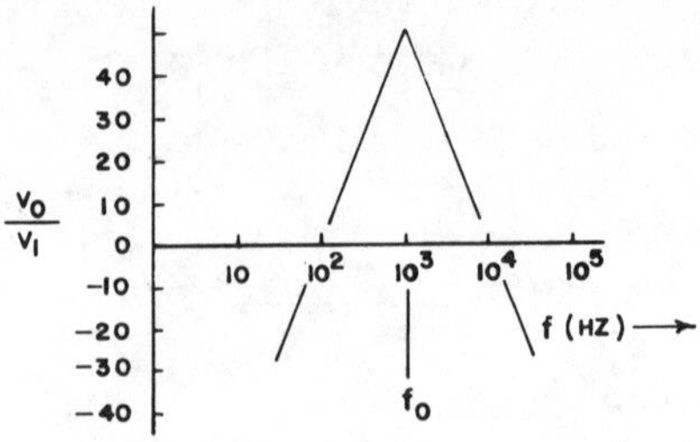

Figure 3-51. Frequency response of a slot filter with gain.

the case when high values of Q (recall Figure 3-45) or sharp cut-offs are needed. In order to purchase active filters, you must first know the nomenclature of these devices. There have been many Ph.D. man-years of effort expended in this area, and present-day active filters are not easily explained in terms of simple *RC* concepts.*

The two most common types of active filters are called *Butterworth* and *Chebyshev filters* after the men who contributed to the development of the art. The Butterworth filter has a very flat pass band, as shown in Figure 3-52. The Chebyshev filter has more ripple in its pass band, but it is sharper in cut-off than an equivalent Butterworth filter (see Figure 3-52).

Another type of filter is the *Bessel filter.* This filter has a linear roll-off as the frequency increases beyond the cut-off frequency. The Chebyshev filter has the sharpest roll-off, but, besides having ripple in its pass band, it also suffers from poor transient response. This means that if the signal is applied suddenly, the output of the filter will exhibit a certain rise time. (*Rise time* is simply the time for the output to reach its final value after the input signal has been applied.) When the signal is removed, continued *ringing* (see Figure 3-63) may occur. For this reason, Butter-

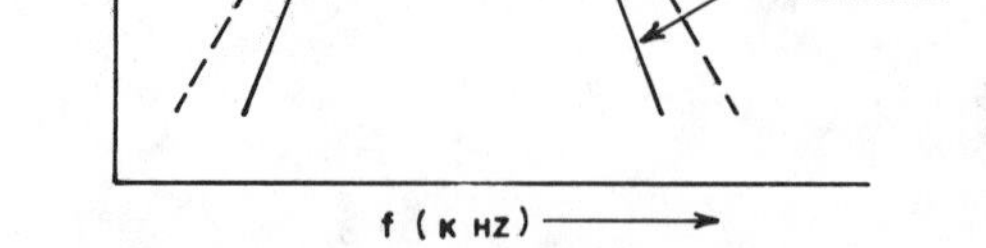

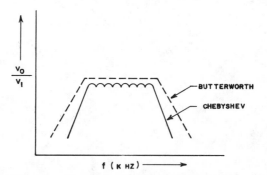

Figure 3-52. Active filter band-pass characteristics; Butterworth versus Chebyshev filters.

*Some good references on active filter design are D. E. Lancaster, *Active Filter Cookbook,* Indianapolis, Howard W. Sams, 1975; J. Hilbrun and D. Johnson, *Manual of Active Filter Design,* New York, McGraw-Hill, 1974; P. R. Geffe, *Simplified Modern Filter Design,* New York, Ryder – a division of Hayden Publications, 1963.

worth and Bessel filters are used when transient response to rapidly changing signals is important.

Commercial active filters are rated in terms of their order: second, third, fourth, and so on. In general, the higher the order, the better (but, of course, the cost is higher, too!). The term *order* comes from certain functions that define the filtering capability of the devices (we refuse to open this can of worms and try to explain in detail what is meant by "orders"). A fourth-order filter has a faster roll-off than a second-order one, and so on. If you want to learn more about active filters, you can start taking EE courses. For now, "Let the Poles Fall Where They May" (that's an EE joke, not an ethnic one).

HIGH QUALITY FILTERS

For those of you who want to play games with *high quality filters,* we will show you a few circuits (see Figures 3-53 and 3-54) that you might want to build and try out. We will indicate a few component values for you to try, but you should expect to do some experimentation yourself. Use variable resistors and capacitors at one or two spots, hook up a frequency generator to the input, and have fun. This process is called "breadboarding" and electrical engineers do it all the time. In recent years, computers have been used to test out hypothetical circuits, but the old-fashioned oscilloscope and signal generator have *not* been forgotten.

The two filters shown in Figures 3-53 and 3-54 are worth discussing because they illustrate what might be called a *complementary pair.* Note that in both cases, the feedback network is the *same circuit* but the input filter *changes.* If you put a high-pass filter and a low-pass filter together, you get – if you pick your f_0 values right – *a notch filter.* Its characteristics are shown in Figure 3-55.

Other handy filter circuits for fun and profit are shown in Figures 3-56 and 3-57.

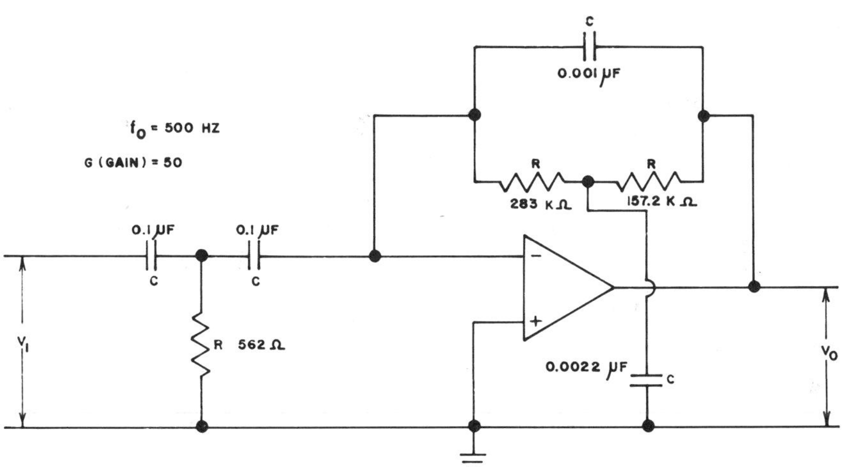

Figure 3-53. Second-order high-pass filter.

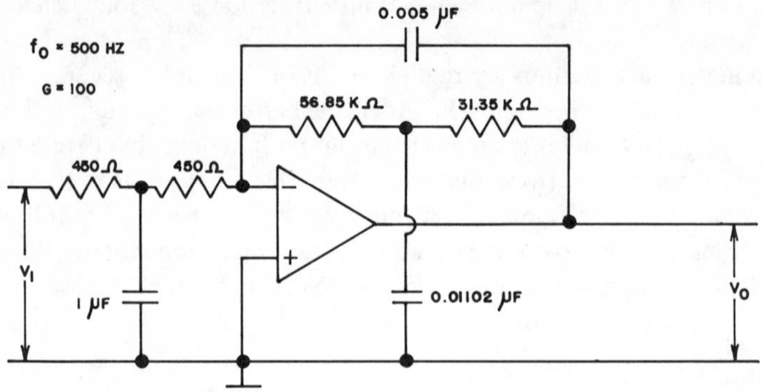

Figure 3-54. Second-order low-pass filter.

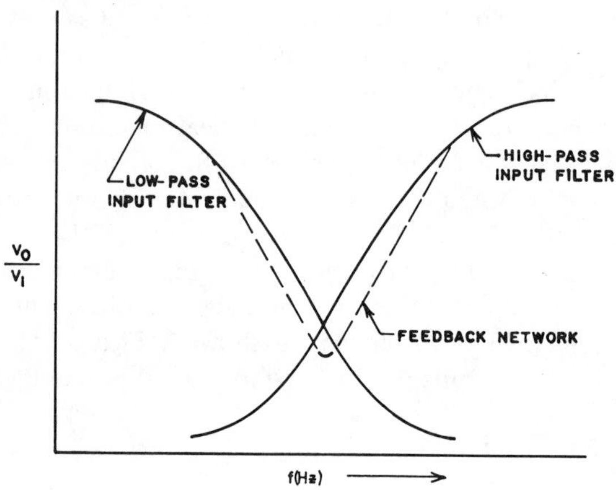

Figure 3-55. Notch filter characteristics.

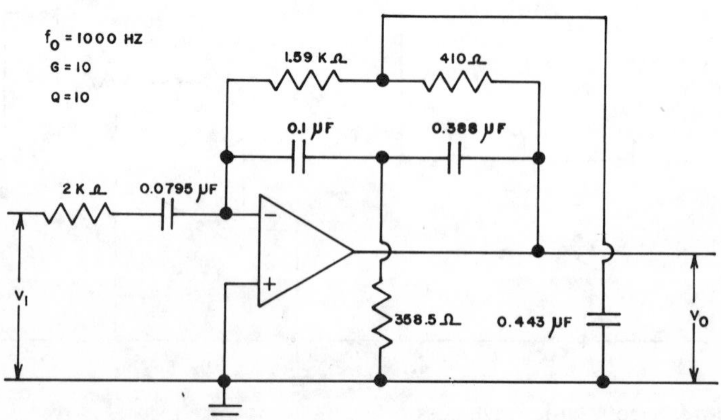

Figure 3-56. Second-order band-pass filter.

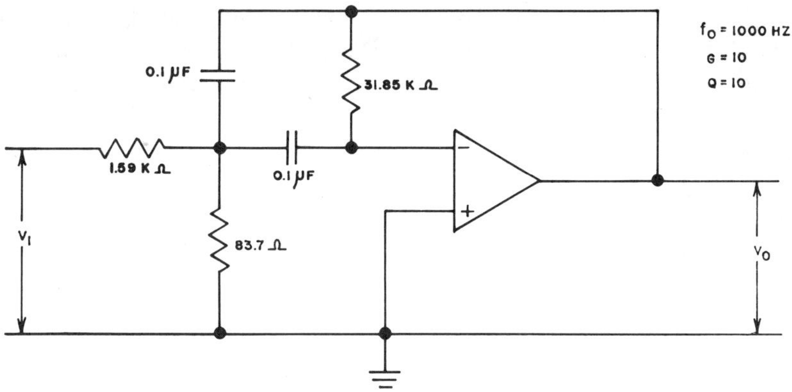

Figure 3-57. Simplified second-order band-pass filter.

Twin-Tee Filter

No discussion of filters would be complete without a discussion of the *twin-tee filter,* which is designed to remove 60 Hz noise picked up from power lines or through cheap power supplies. Of course, the first thing you should do to reduce noise is to use shielded wire, but it this doesn't cure the problem, a filter will help.

The basic twin-tee circuit configuration* is shown in Figure 3-58. This circuit is characterized by almost infinite rejection at the null frequency f_0. A twin-tee filter of this type is manufactured by the Heath Company for its recorders. Heath suggests $C_1 = 0.5\ \mu F$, $C_2 = 1\ \mu F$, $R_1 = 5.6\ k\Omega$, and $R_1 = 2.7\ k\Omega$. These values can be obtained from the equations given in Figure 3-58. You should be aware that there is a strong phase shift through the network as the null frequency is approached; it changes from plus 90° to minus 90° as the null point is passed. A rejection of not more than 45 dB should be expected because of the component value variation that makes it difficult to obtain perfect balance in the circuit.

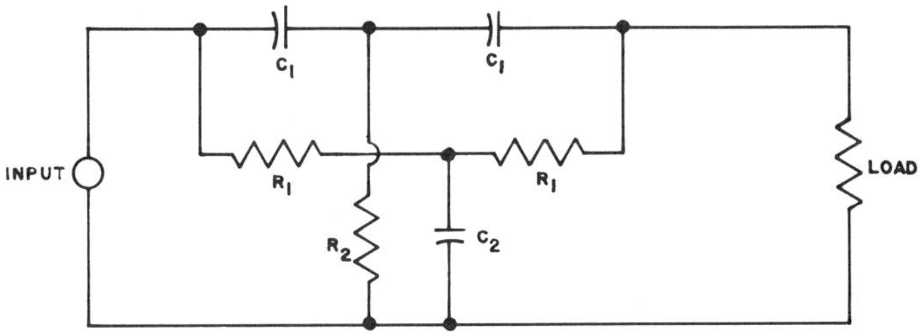

$$f_0 = \frac{\sqrt{H}}{2\pi R_1 C_1} \qquad H = \frac{2C_1}{C_2} = \frac{R_1}{2R_2}$$

Figure 3-58. Twin-tee filter.

*Contributed by Mr. Michael Pomeroy when he was in the University of Arizona Electrical Engineering Department.

Component values for a 60 Hz rejection filter are easily calculated. First, we make $C_1 = C_2$ for simplicity, and pick a value of 0.1 μF for C_1. We define a term H for this circuit as:

$$H = \frac{2\,C_1}{C_2} = \frac{R_1}{2\,R_2}$$

and

$$f_0 = \frac{\sqrt{H}}{2\pi\,R_1\,C_1}$$

Since we chose $C_1 = C_2$, then $H = 2$. The frequency we want to reject, f_0, is 60 Hz. So, by substituting values, we obtain

$$60 = \frac{\sqrt{2}}{2\pi\,R_1\,(0.1 \times 10^{-6})}$$

Solving for R_1 yields $R_1 = 37.6$ kΩ. From the definition of H, $R_2 = R_1/2\,H$. Therefore, $R_2 = 37.6/4 = 9.4$ kΩ. Using these values, we draw up the 60 Hz twin-tee filter shown in Figure 3-59.

If maximum attenuation is not needed, you can substitute a 3.6 kΩ, 5% resistor for R_1 and a 0.910 kΩ, 5% resistor for R_2, and you can then forget about the potentiometers shown in Figure 3-59. (Remember that all your output power goes through resistors R_1. If you have a 1000 watt system and an 8 ohm speaker, the rms current is 7.9 amps, which could be considered a source of noise pollution!) For best results, all capacitors should be matched. An impedance bridge, if you have access to one, will do just fine (an *impedance* or *RLC bridge* is a device for measuring the value of resistors, inductors, or capacitors). To "tune" the twin-tee filter, you can use an AC signal generator set at 60 Hz as the input signal, and observe the output on an oscilloscope while adjusting the potentiometers for minimum signal.

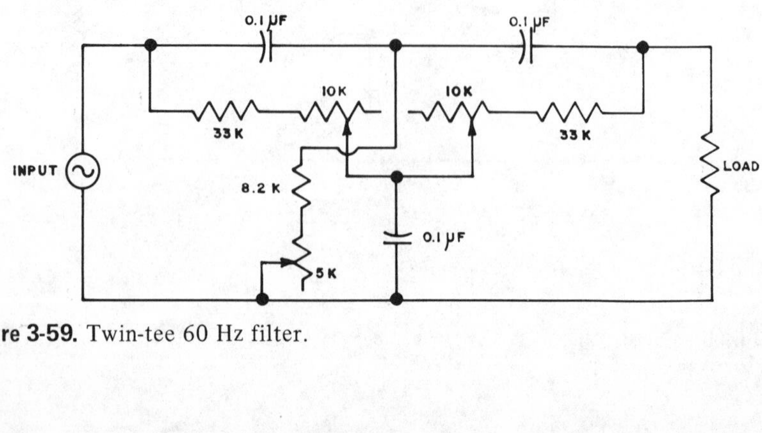

Figure 3-59. Twin-tee 60 Hz filter.

You can put several filters in series – i.e., hook the input of a second filter to the output of the first filter – for even better filtering. Of course, there is signal loss in each stage because of the resistors, so too much filtering can be as bad as too little.

Twin-tee filters can be designed for almost any frequency by using our handy formulas. If you build one, you should try the effect of putting it in the feedback loop of an op-amp. In this case it becomes a band-pass filter with a central pass frequency of f_0.

In Figure 3-60 we show a variation of this filter circuit with a Motorola MC1533 op-amp for passing $f_0 = 400$ Hz. To pick the component values for this circuit, we use the following rules:

1. The parallel resistance of $2R_1$ and R_L should not draw enough current to overload the amplifier.
2. R_2 should be a value close to R_1 (at least within a factor of 5).
3. We obtain H as before, and then calculate the ratio C_1/C_2.

This circuit can give you a 40 dB rejection without much effort or special tuning.

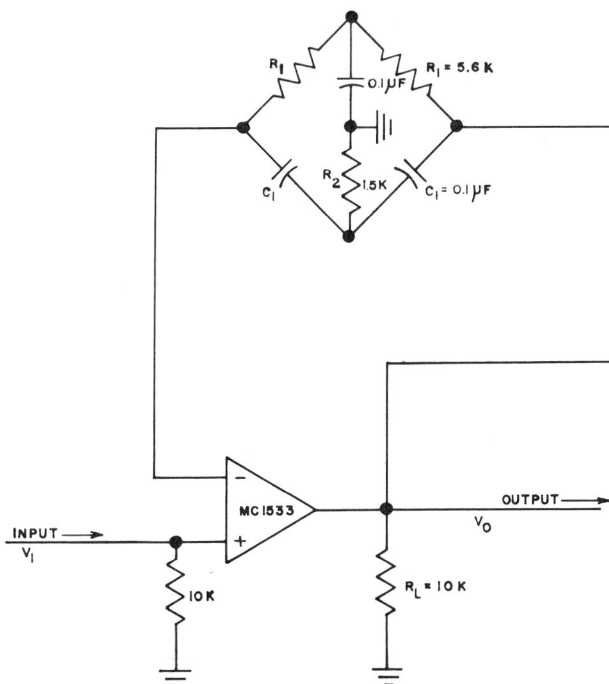

Figure 3-60. Twin-tee op-amp band-pass filter. (From *Motorola Application Note 248;* courtesy of Motorola Semiconductor Products, Inc.)

OP-AMP OSCILLATOR CIRCUITS

Earlier we mentioned that negative feedback stabilizes op-amp circuits; conversely, *positive feedback destabilizes* the op-amp circuit and induces oscillation. We should study this process because uncontrolled oscillations can be a problem, but controlled oscillations can be quite useful.

Negative feedback implies that if the inverting op-amp input goes (+), the output goes (-); this minus voltage is fed back to the input, thereby reducing the original (+) voltage. With positive feedback, a (+) input again yields a (-) output, which is inverted and fed back to the input to *increase* the input voltage. This in turn increases the output voltage and the process continues until the op-amp is driven to its maximum output voltage. This is called *saturation,* and it is not quite what we want. We can improve the situation by providing positive feedback but arranging things so that *before* the op-amp saturates, the input signal reverses in sign. This forces the op-amp to change its output voltage from positive to negative in a cyclic fashion: such a device is called an *oscillator.*

Phase-Shift Oscillator

Our first oscillator, shown in Figure 3-61, makes use of a phase-shift network to ensure that positive feedback will occur. In a qualitative sense, we can assume that if a (+) input is present at (1), the (-) output from (4) is phase-shifted by 180° in the feedback network. Thus, the signal fed back to (1) is (+) when (1) is (+), and vice versa. The system continues to oscillate at a frequency *f:*

$$f = \frac{1}{2\pi RC}$$

where

$$10 \text{ Hz} \leqslant f \leqslant 10 \text{ kHz}$$

This circuit is tuned into oscillation by the variable resistance $R/2$. The frequency of oscillation can be controlled by changing the two 40 kΩ resistors (R) or the two 1000 pF capacitors (C). (To appreciate how this circuit operates, you might want to look back at page 27 where phase shift was discussed. A single capacitor shifts the phase by 90°, while two capacitors in series generate 180 degrees of phase shift. That is just enough to provide the "positive feedback" we need to induce oscillation.)

The phase-shift oscillator is a good circuit to start with. It will oscillate at a particular frequency and at a constant amplitude for long periods of time. There are a few things to note about this circuit. First, the output of this simple oscillator will be a mixture of several frequencies rather than a pure sine wave. To get a fairly good sine wave, you have to use a more complex circuit, which is shown in Figure 3-62.

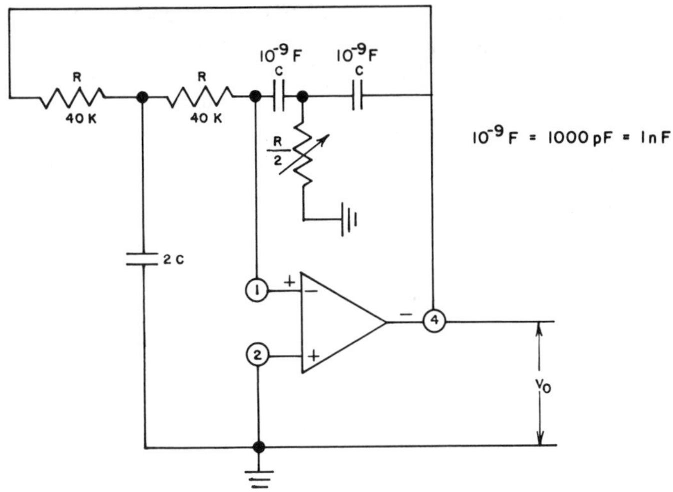

Figure 3-61. Phase-shift oscillator.

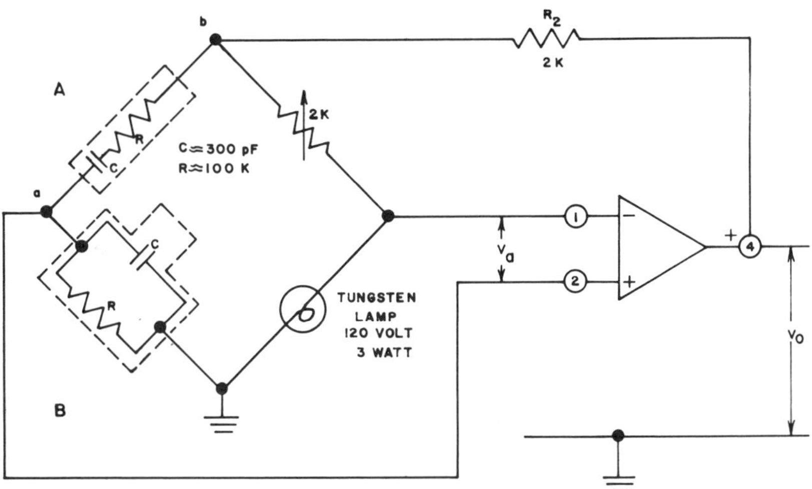

Figure 3-62. Wien bridge oscillator.

Second, in the circuit of Figure 3-61, the coarse frequency adjustment is set by *R* and *C*. Fine adjustments can be made by playing with the *R*/2 resistance. Third, the output impedance of this oscillator is relatively high, and you might want to follow it with another op-amp that uses V_0 as its input if you need current and high stability. Of course, it is important to keep in mind that the operating frequency of this circuit is limited by the bandwidth of your op-amp.

The phase-shift oscillator is an interesting circuit, and it can teach us some important things about how circuits work. Suppose input (1) is momentarily (+). This makes the output (4) negative, and this (-) signal comes back to (1) by *two* paths. One path is through the two capacitors (C), which introduces a 180° phase shift and a (+) feedback. This positive feedback drives the op-amp toward saturation, but at the same time, the (-) signal goes to (1) via the two resistors (R). This is negative feedback because no phase shift is induced by the resistors. Initially, positive feedback exceeds negative feedback because most of the negative feedback is shunted to ground by the capacitor 2C. However, as capacitor 2C is charged to some voltage near saturation, it blocks the flow of negative current to ground. All the negative feedback signal must then go to (1). This overwhelms the positive feedback from C + C, and the op-amp output begins to decrease. Eventually, (1) becomes slightly (-), the op-amp starts going toward the (+) saturation point, and the capacitor 2C begins to charge up with the opposite polarity.

This process continues indefinitely. As stated above, the frequency of oscillation is controlled by $f = 1/2\pi RC$, and the system is tuned into oscillation by $R/2$.

Wien Bridge Oscillator

To obtain a very pure sine wave, we must use the Wien bridge oscillator that is shown in Figure 3-62. Here again we have a phase-shift network, but it is somewhat more complex than that used in the circuit of Figure 3-61.

To understand the Wien bridge oscillator, we first must recall that we require positive feedback for oscillation. *Positive feedback* means that when the output of an amplifier increases, a portion of this output is tapped off and returned to the amplifier input in such a way as to drive the output more positive. This process cannot continue until the amplifier saturates, because if it did, the result would be a messy square wave or no signal at all!

To appreciate how the Wien bridge prevents saturation, we must go through the circuit in some detail. Let's assume that the op-amp input (1) is (-). The output, then, is (+) and is returned as negative feedback via the 2 kΩ resistor and the 2 kΩ potentiometer. These two resistances act like the resistor R_0 in our basic circuit, and they give the circuit some gain: without gain, there would be no oscillation. The (+) signal from the op-amp output (4) passes through the R-C series combination and then through the R-C parallel circuit.

This series-parallel network forms a phase-shift circuit. At some frequency f_x, where $f_x = 1/2\pi RC$, the voltage at point *a* is in phase with the voltage at point *b*, and the network behaves like two resistors in series. At the frequency $f_1 < f_x$, the voltage at *a* leads the voltage at *b*, whereas at the frequency $f_2 > f_x$, the voltage at *a* lags behind the voltage at *b*. At the frequency f_x, the voltage that is sent to (2) is in phase with the voltage at (4) and about one-third as large (remember that resistors don't induce any phase shift). The (+) signal that goes to (2) drives the op-amp toward saturation.

The next question is, what stops it from going all the way to saturation? Well, notice the 120 volt, 3 watt tungsten lamp in the circuit. When V_0 is small, the negative feedback is small, and the lamp has a low resistance because the filament temperature is low. As V_0 goes up, so does the feedback current, which heats up the filament by an amount equal to I^2R. The resistance then increases with temperature: $R = R_0(1 + \alpha[T - T_0])$. This in turn means a larger negative feedback voltage at (1) that stops V_0 from leading to saturation, and it starts the system going toward its (-) output phase. Once again the excursion will be limited, and we have a system that oscillates at a *single* frequency without saturation.

SPECIAL FUNCTION GENERATORS

We have discussed two sine wave generators – the phase-shift and the Wien bridge oscillators – but there are other types of signals that you may find useful. Among these are *square waves, saw-tooth waves,* and *programmable signals.*

Square Wave Generators

Square waves are used for testing and calibrating amplifiers. (Have patience, we will show you how to build square wave generators in the section entitled Multivibrators.) Figure 3-63 shows a typical square wave and the appearance of the square wave after passage through a defective amplifier. The defect in the amplifier can be determined from the change in the wave shape. Note that the peak-to-peak value of the square wave can be used to calibrate the amplifier. The actual wave shape shown on the oscilloscope screen is often used as a measure of the response of the amplifiers in the oscilloscope system. Figure 3-63B shows two possible types of poor high frequency response. Figure 3-63C shows the result of poor low frequency response.

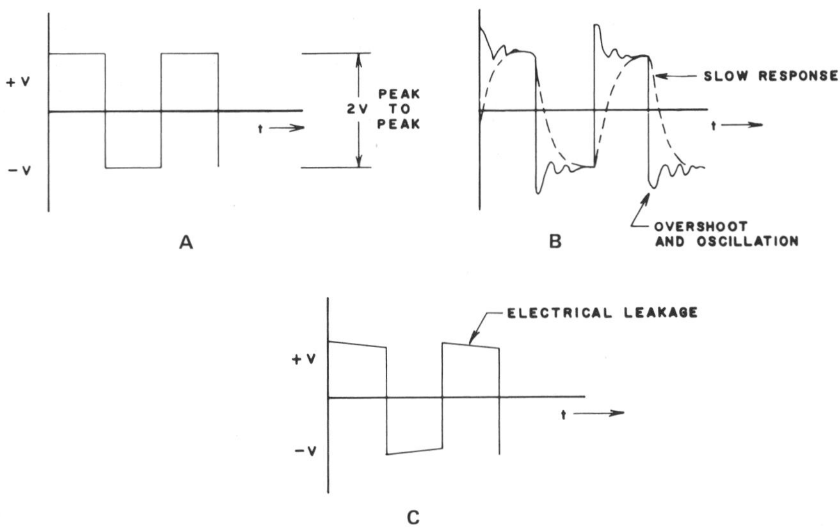

Figure 3-63. Square wave amplifier testing. *A*. Perfect square wave. *B*. Defective square wave (poor high frequency response). *C*. Defective square wave (poor low frequency response).

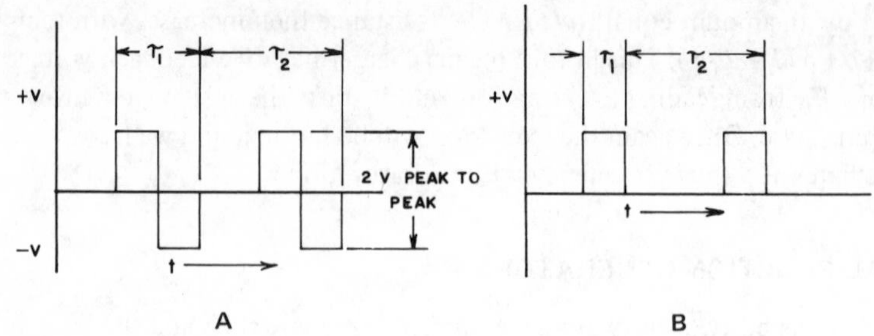

Figure 3-64. Square wave pulse generator outputs.

Square wave or pulse generators are versatile gadgets, and the waveforms shown in Figure 3-64 are often useful. The point to note in Figure 3-64A and B is that τ_1 is the *pulse width,* whereas τ_2 is the *pulse period,* and they are *not* the same. The reciprocal of τ_2 is the *pulse frequency* in cycles per second, or hertz:

$$f = \frac{1}{\tau_2}$$

The output shown in Figure 3-64B is called a *DC pulse signal.* The term *DC* means that the voltage doesn't go *negative;* it does *not* mean that the voltage is constant. It *does* mean that the voltage *never* changes sign.

Saw-Tooth Wave Generators

Saw-tooth wave generators are used for the horizontal sweep drive on oscilloscopes. For the moment, we should note that the output of a saw-tooth wave generator may be symmetrical or asymmetrical, as shown in Figure 3-65A and B, respectively. Both the amplitude and duration of such waveform's are usually adjustable.

Building a simple saw-tooth generator is easy. The procedure involves providing a constant voltage (V_1) to an integrator circuit to get $V_0 = V_1\ t/k$, where k is an arbitrary constant. If a symmetrical wave is desired (Figure 3-65A), V_1 is switched from (+) to (-) at the proper time, and the integration proceeds in the other direction.

If you want a saw-tooth waveform like that shown in Figure 3-65B, you have to

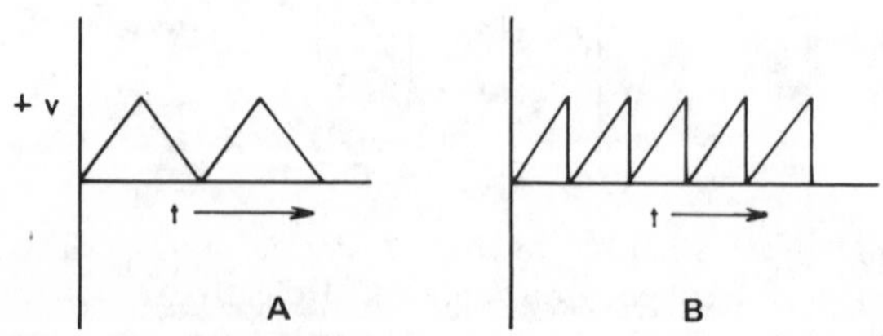

Figure 3-65. Saw-tooth signals.

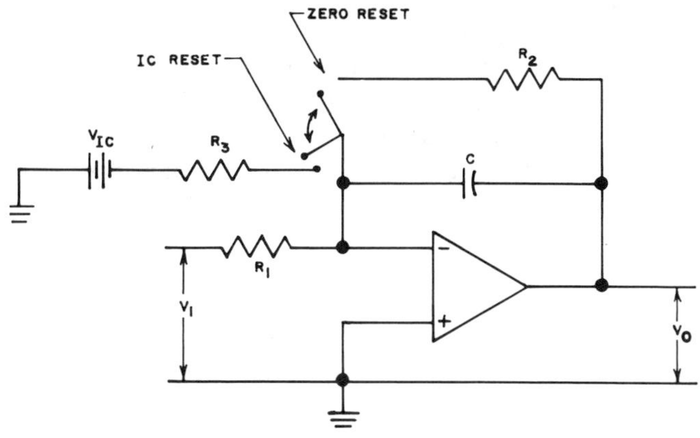

Figure 3-66. Op-amp zero reset or initial condition circuit.

short circuit the output of the integrator periodically to bring V_0 to zero. This is called *resetting* the integrator, and it is worth saying a few words about. The best thing for this purpose is a *dry-reed relay*. The term *dry reed* indicates that it is a fast-acting device that requires only a small coil voltage for actuation and will handle only small (~200 mA) currents at the contacts. The coil voltage and current present a problem when you try to drive the reed relay with an op-amp. Some reed relays have 6 volt coils, but it hard to find one that operates on 10 mA. If worse comes to worst (as it usually does), you can build a current booster using a one-transistor amplifier, as we showed previously (see Figure 3-10). If you feel rich, there are electronic switches for sale (Teledyne Philbrick and Burr-Brown) that will do the job quite easily.

A typical *reset circuit* is shown in Figure 3-66. When the zero-reset switch closes on the upper contact, the voltage across the capacitor discharges through R_2, so the decay time τ is approximately $R_2 C$. The resistance R_2 should be as small as possible for fast zeroing, but the current must not exceed the limits of the relay contacts. If an initial condition voltage is needed, the initial condition (IC) switch should be used after the zero reset. The time for the initial condition voltage to be set on the capacitor is given by $\tau \approx R_3 C$. Be sure to disconnect V_1 while you use the *IC* reset circuit.

You may wonder how time can be equated with the product of resistance times capacitance. It so happens that in terms of *fundamental dimensions*, resistance is defined as $R = \mu l/t$ and capacitance as $C = t^2/\mu l$, where μ is magnetic permeability, t is time, and l is length. (What is "magnetic permeability"? Look it up in a physics textbook if you're that curious.) The point is $(\mu l/t)(t^2/\mu l) = t$.

Ramp Generators

Last, but by no means least, are the *ramp* or *function generators* that produce a particular nonlinear signal. Now and then a circuit of this type is needed to provide the input to an analog computer, which is why the output of such a circuit is sometimes called *programmable signal.* The ramp generator will be discussed in some detail since it allows us to introduce some new concepts.

Many electronic devices are nonlinear. Diodes are a typical example; we mentioned this in Chapter 1. In forward bias, the diode current-voltage relation is

$$I = A(e^{BV} - 1)$$

where A and B are constants at a given temperature and e is the base of natural logarithms, a constant equal to 2.718. If the first term is much greater than 1 (as it usually is) we can write

$$I = Ae^{BV}$$

This means that the V-I curve of a diode will be an exponential one. If we can control the voltage at which a series of diodes goes into forward bias (and if we choose diodes wisely), we can generate a whole class of special voltage-versus-time curves. These curves would be the sum of the several diode exponential curves. As an example of this process, let's look at Figure 3-67, which shows the prototype curve generator.

To understand this circuit, let's suppose that initially V_1 is short circuited (i.e., $V_1 = 0$), and V_2 keeps diodes D_1 and D_2 in reverse bias. The voltage V_a may be obtained by noting that the current (when D_1 is off) is

$$I = \frac{V_2}{R_1 + R_2}$$

so

$$V_a = IR_1 = \frac{V_2 R_1}{R_1 + R_2}$$

Initially, $V_1 = 0$ and V_a is negative, so both diodes are reverse biased. As V_1 increases, there will be a point at which diode D_1 will begin to conduct. From that point on, the input to D_1 (i.e., V_a) will be at ground potential (ignoring the diode drop itself) because of feedback in the op-amp.

We want to determine the value of V_1 that will start D_1 conducting, but this is difficult. Therefore, we work a simpler problem: we find out what value of V_1 is required to bring V_a to zero. Obviously, any voltage greater than that value for V_1 (ignoring the 0.7 volt forward-bias voltage that is required to turn the diode "on") will start D_1 conducting. To find this critical value of V_1, we note that when $V_a = 0$ the voltage drop across R_2 must be exactly V_2. In this case, the current through R_2

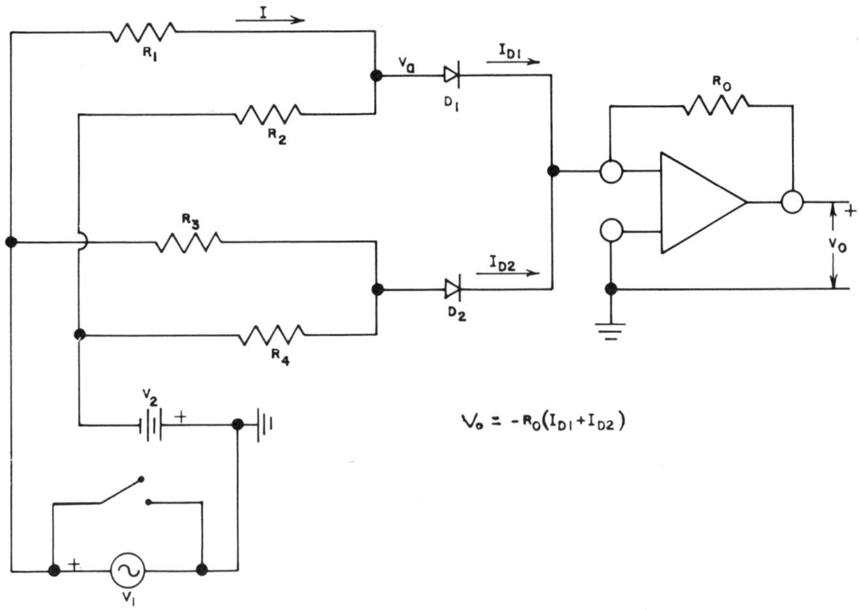

Figure 3-67. Voltage-time generator.

is $(V_1 + V_2)/(R_1 + R_2)$. Hence, the voltage drop across R_2 is $(V_1 + V_2)R_2/(R_1 + R_2)$. This voltage is equal to V_2, so we have

$$V_2 = \frac{(V_1 + V_2)R_2}{R_1 + R_2}$$

Solving this for V_1, we obtain

$$V_1 = V_2 \frac{R_1}{R_2}$$

If V_1 is greater than this value, D_1 is turned on and conducting.

Returning to the important point – the output of curve generators – we see that the current to the op-amp is zero until the condition $V_1 > V_2(R_1/R_2)$ is satisfied. After that, the voltage increases with current according to the usual $I = Ae^{BV}$ diode curve. At some chosen value of V_1 that is greater than $V_2(R_3/R_4)$, diode D_2 starts to conduct. The resultant current is the sum of the diode currents and is transformed by the op-amp into a voltage.

Figure 3-68 shows a typical case in which the input signal is linear ($V_1 = At$), and we want an output signal that looks like the curved line in the illustration. We

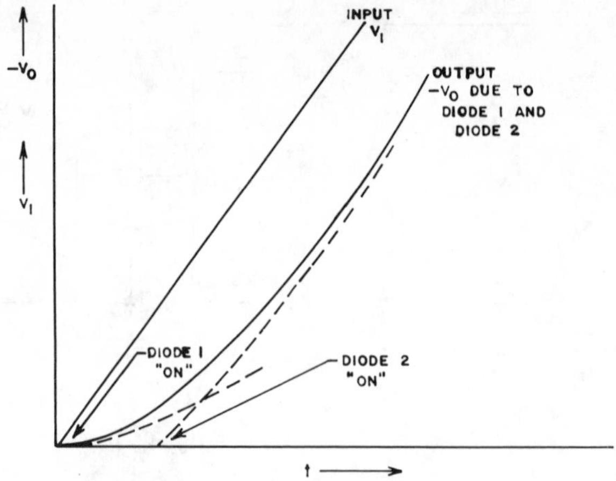

Figure 3-68. Input-output signals from curve generators (programmable signals).

set up the circuit as described for Figure 3-67. In this case, until the proper value of V_1 is reached to turn on diode D_1, there is no output, V_0. When D_1 turns on, the output begins to rise with a slope given by R_0/R_1. When D_2 turns on, the slope changes to R_0 divided by the parallel combination of R_1 and R_3, or

$$\frac{R_0(R_1 + R_3)}{R_1 R_3}$$

You should notice that in Figure 3-68, V_1 and $-V_0$ are plotted in the same direction, but V_0 increases negatively when V_1 is a positively increasing voltage.

The variety of diode, resistor, and op-amp combinations in such a circuit is almost unlimited. Given enough time and money, you could produce almost any voltage-time curve you want. This type of circuit, however, critically depends on the diodes you use and is limited to low-frequency operation by diode capacitance. The break points will not be as sharp as we have drawn them, and the break voltages will depend upon the properties of real diodes, which are temperature-sensitive. Nevertheless, there are times when a circuit like this can get you out of a real bind.

MULTIVIBRATORS

The term *multivibrator* covers a variety of circuits that can be used for all sorts of useful projects. In principle a multivibrator circuit has two possible outputs, say, +5 volts and -5 volts. If it jumps back and forth from +5 volts fo -5 volts continuously at some frequency *f* without any input signal, it is termed a *free-running multivibrator* or *astable vibrator.*

If the circuit has one stable state (say +5 volts output) and, upon receiving a signal, it changes to -5 volts output and stays that way for a predetermined time, before automatically going back to +5 volts, it is called a *monostable multivibrator* circuit.

A *bistable multivibrator* circuit is one that has two stable states (e.g., ±5 volts output) and switches back and forth when a signal of proper polarity comes into the input. This is often called a *flip-flop circuit.*

Free-Running Multivibrator

A free-running multivibrator is often used as a square wave generator. We will see this type of circuit again when we discuss telemetry (p. 208). The circuit for a free-running multivibrator is shown in Figure 3-69. Let's assume that the circuit is in the (+) output mode. This implies that the voltage at input (1) of the op-amp is (-). The capacitor C charges through the 250 kΩ resistor until the (+) voltage across C is larger than the (-) voltage originally present at (1). When this occurs, the output goes (-) and the capacitor begins charging up the other way. The process of switching is continuous, and the output oscillates between the plus-to-minus saturation voltage of the op-amp at a frequency $f = 1/[2\,RC \ln(1 + 2\,R_1/R_2)]$. It should be noted that in this formula, "ln" is in the *natural logarithm,* i.e., the logarithm to the base e, not to the base 10 ($\log_{10}$).

This circuit can also be driven from one state to another by an input pulse V_1. For example, if V_0 is positive and V_1 is a positive signal, the output will immediately switch to the negative value and stay there until V_1 goes to zero. At this time the system begins to oscillate freely again.

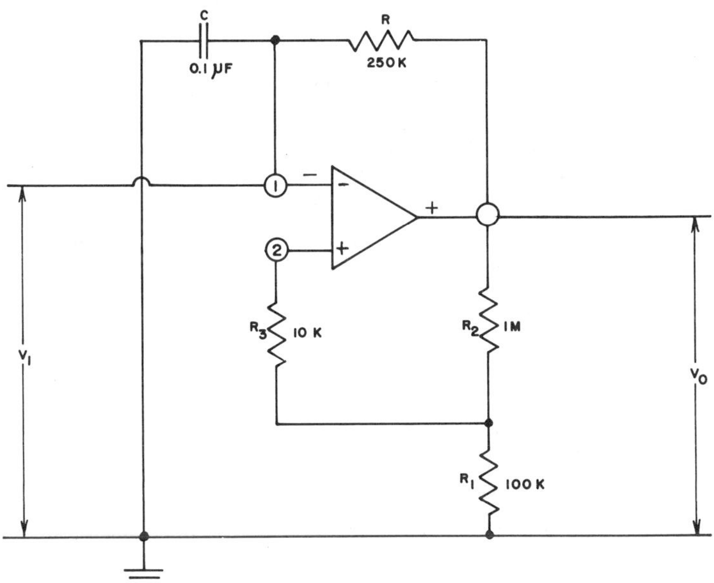

Figure 3-69. Op-amp free-running multivibrator.

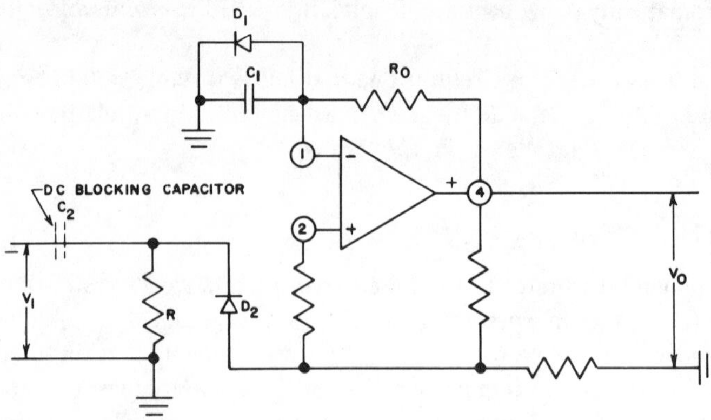

Figure 3-70. Monostable multivibrator.

Monostable Multivibrator

A circuit of this type is shown in Figure 3-70. The stable state is a positive output of, say, 10 volts (i.e., the saturation voltage of the op-amp). Initially, the op-amp is in its +10 volt output state. If V_1 – the trigger signal – goes (-) with respect to ground, this puts a (-) signal at input (2) on the op-amp. The op-amp output (4) goes (-) and C_1 begins to charge up through R_0. Eventually, the (-) signal at input (1) is greater than that at (2), so the op-amp switches back to +10 volts again. (Remember that the op-amp has gain; so long as V_1 is less than V_0, switching will occur.) Diode D_1 keeps the circuit from oscillating.

Of course it is easier to switch this circuit if the voltage across C_1 is not the full +10 volts of the op-amp. This is achieved by the diode D_1, which limits the (+) voltage across C_1 to, say, 0.5 volt. This means that any signal V_1 that is greater than 0.5 volt will induce switching. If V_1 is less than V_0, the circuit switches back to its stable state (as described above) in a time determined by R_0 and C_1.

You should note that the switching cycle occurs *only* if V_1 is a pulse. If V_1 is a step function, the circuit will be locked in the -10 volt output mode until V_1 goes to zero. This, however, can be useful in digital circuits. It is also important to note that a (+) V_1 voltage doesn't switch the circuit of Figure 3-70; such a signal would be blocked by D_2.

Bistable Multivibrator

This is a device that has *two* stable states. It flips from one to the other when a signal is received, which is why it is called a *flip-flop* circuit.

A typical circuit of this kind is shown in Figure 3-71. The op-amp is held at its saturated output (+ or -10 volts) because of the positive feedback via the resistor R_2. To see how switching occurs, assume that the output V_0 is (+). This maintains a positive input at (2), so the op-amp remains at saturation. Note that the voltage at (2) is not the full 10 volt output of the op-amp; diode D_1 is conducting and the

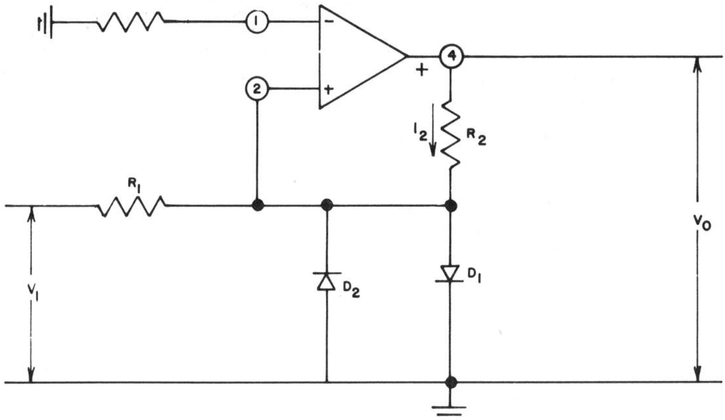

Figure 3-71. Bistable multivibrator.

voltage at (2) is $V_0 - I_2 R_2$. The current I_2 is set by $V_0 - V_d$, where V_d is the voltage across the forward-biased diode.

Switching occurs if V_1 is negative with respect to ground because it makes (2) negative, which in turn makes V_0 negative, and feedback via R_2 holds the circuit in the negative mode. Switching will occur again only if V_1 becomes positive with respect to ground.

The voltage necessary for switching will depend upon the resistance R_2. The larger R_2, the smaller the voltage fed back to (2), and the more easily switching can occur.

You may be interested to learn that a variant of this circuit is used in digital systems for a divide-by-two circuit. If the input is a square wave signal at, say, 200 Hz, the circuit will change its output state only when the input signal swings positive, thereby producing a 100 Hz square wave. By cascading these circuits, you can divide by two as often as you like. A good circuit for this purpose is shown in the Teledyne Philbrick manual (see the Appendix).

SCHMIDT TRIGGER CIRCUIT

This is a useful gadget that has the kind of output V_0 shown in Figure 3-72. Note that the voltage V_0 decreases along one path and increases along another. The values of V_0 trace out a closed loop that is sometimes called a *hysteresis loop* in analogy to a phenomenon observed in the study of magnetic materials.

An op-amp circuit to produce this effect is shown in Figure 3-73. One application of a Schmidt trigger is the generation of clean square waves from a messy sine wave signal. Other applications exist in digital circuits and in circuit test procedures.

Let's assume that initially, $V_2 = +15$ volts. This is above the lower trigger level (-3.8 volts), and the output V_0 must be -10 volts. Thus, the voltage drop across the 75 kΩ and 27 kΩ resistors must be set by the -10 volts at the top of the string and

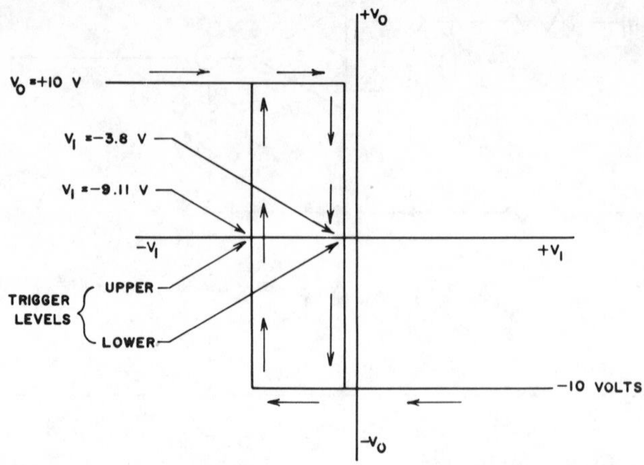

Figure 3-72. Schmidt trigger circuit input-output voltages.

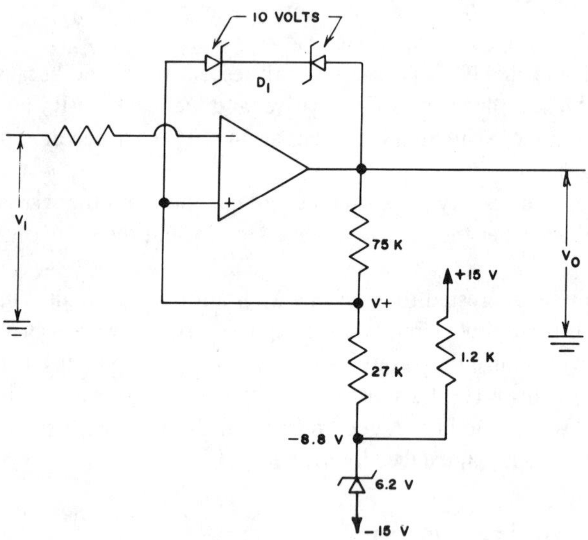

Figure 3-73. Schmidt trigger circuit.

the –8.8 volts at the bottom. The difference (10 – 8.8 = 1.2) fixes the current at I = 1.2 volts/102 kΩ = 11.8 μA, and the voltage drop across the 75 kΩ resistor is 11.8 X 75 X 10^{-3} = 0.885 volt. This means that V_+ must be -9.11 volts.

When V_1 drops, nothing happens (i.e., V_0 = -10 volts) until V_1 = -9.11 volts. At that point the input voltages are equal, the op-amp output is trying to go in both directions at once, and, as soon as V_1 becomes slightly less than -9.11 volts, the output switches to V_0 = +10 volts. Now there is 18.8 volts across the two resistors, and the drop across the 75 kΩ resistor is (18.8 volts/102 kΩ) X 75 kΩ = 13.8 volts.

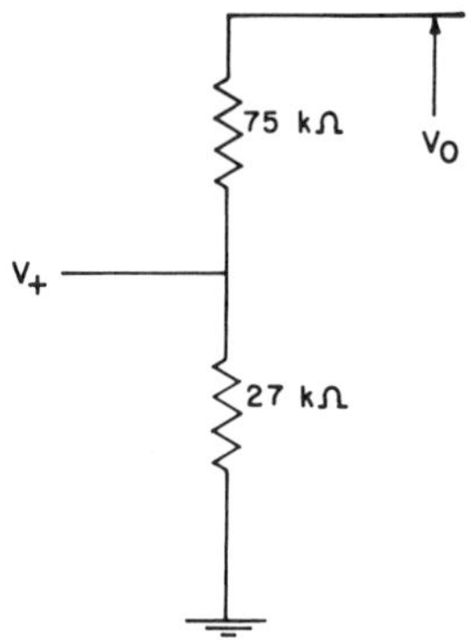

Figure 3-74. Modified Schmidt trigger circuit.

V_+ must be -3.8 volts, and the output remains at the V_0 = +10 volt condition until V_1 increases again and passes through the -3.8 level. At this point the system switches back to the V_0 = -10 volt condition. The 10 volt zener diodes shown in Figure 3-73 are there to keep the op-amp from going into saturation.
You can shift the hysteresis loop about by changing the voltage on the resistor string and removing the zener diode. Suppose you wanted to make the loop symmetrical about the vertical axis, you would change the system as shown in Figure 3-74.

To see how this works, assume that V_1 (the signal at the inverting input) is +15 volts. Now V_0 = -10 volts and the current through the resistor string is

$$\frac{10\,V}{27\text{ k} + 75\text{ k}} = 9.8 \cdot 10^{-5}\text{ A*}$$

The voltage V_+ is $(9.8 \cdot 10^{-5}) \times (27\text{ k}\Omega) = -2.65$ volts. Now V_1 begins to change and eventually goes negative to V_1 = -2.65. The moment V_1 becomes slightly more negative (to, say, -2.6501 volts), the op-amp output switches to V_0 = +10 volts and V_+ becomes +2.65 volts. This condition will persist until V_1 starts to go positive again and reaches, say, +2,6501 volts. At this point the output switches to V_0 = -10 volts. The loop is symmetrical about the vertical axis with switch points at ±2.65 volts.

APPLICATIONS OF OP-AMPS TO POWER SUPPLIES

The laboratory power supply is probably the most useful black box in the house. If you haven't already, you will soon be purchasing, specifying, or using these gadgets.

*You will notice that in many equations we have begun to leave out the Greek omega (Ω) to indicate ohms. This is not a mistake on the part of the author or the printer. In most EE circuits the omega is left out; the reader is expected to "know" that a resistor value is involved. Part of our job in this book is getting you ready for the "real world".

It is, therefore, a good time to discuss the jargon used in describing power supplies and how to use op-amps to make cheap power supplies work better. You can even build high current or voltage supplies with op-amps, but we assume that you aren't rolling in money and might want to do things inexpensively.

Evaluating Power Supplies

In evaluating DC power supplies the critical terms are:

1. Voltage range
2. Current output
3. Ripple
4. Line and load regulation
5. Short-circuit protection
6. Current limiting
7. Output (+) or (-), or (+) and (-) at the same time, with respect to ground.
8. Voltage and/or current meters

The terms *voltage range* and *current output* are fairly obvious. A constant *voltage supply* applies a given voltage to a load: if the load resistance changes, the current also changes, but the voltage does not. A constant *current supply* sends a given current through a load: if the load changes, the voltage changes in such a direction as to keep the current constant. (Constant voltage supplies are generally less expensive than constant current supplies.)

The word *ripple* refers to the 60 Hz or 120 Hz AC signal that appears at the output. If the DC output is 0–100 volts and the ripple is 1%, that's 1 volt rms. If you set the output for 1 volt DC, you still might have 1 volt rms of AC ripple on top of it. This *might* not hurt, but you had better know it is there because your DC output will really look something like that shown in Figure 3-75, and this may come as an unfortunate surprise! Actually, no reputable manufacturer would make such a supply (we hope). Remember 1 volt rms is 1.41 volts peak.

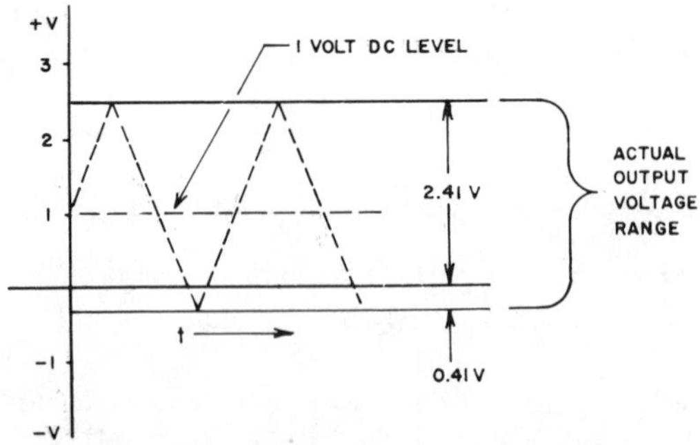

Figure 3-75. Power supply output with ripple.

All electronic gadgets are affected by *line-load changes*. The question is, how much? As an example, suppose you have a 10 volt, 10 amp voltage supply adjusted to deliver 5 volts at 4 amps to a load. Suddenly the load requires twice the current: how long does it take the supply to settle down to delivering 8 amps, and, when it does, what is the output voltage? Will it be 5 volts, 4.9 volts, or even less?

Line voltage changes occur whenever a large load (e.g., machinery) on the same line is switched on or off. This problem is usually at its worst in those universities that build instant buildings in Contemporary Cracker-Box. Someone turns on a compressor downstairs, and momentarily your voltage drops to 90 volts (rms). The compressor then shuts off, and your voltage jumps to 150 volts (rms) for a few seconds. How does the power supply respond to this? A *good* power supply soaks up these line transients and doesn't deliver them to your load. Daily line voltage variations are also common at universities because of hour-to-hour load variations. At 3:00 P.M. on a dark winter day, you may get 106 volts (rms); at 3:00 A.M., it may be 125 volts (rms).

If you have an experiment that will be disturbed by these line voltage changes, you must build or buy a power supply that will not transmit them to your load. A simple system for this purpose will be shown later in Figure 3-82. If your requirements are more complex, look at the Lambda, KepCo, or Sorenson catalog (their addresses are given in the Appendix).

Short-circuit protection may be necessary for both the power supply and the load. If a problem occurs, you may not want to zap the power supply, and if the load is, say, a $150 op-amp, you don't want to zap it, either! The best way to avoid the latter is to use a *current-limiting circuit*, which limits the current to some maximum value. We will show you some simple current-limiting circuits in a subsequent section on protection circuits.

Outputs on a power supply can be either (+) or (-), or (+) and (-) at the same time. Op-amps, for example, need *both* (+) and (-) voltages at the same time.

Meters for voltage or current are nice, but there are always questions about how accurate they are and whether they show any AC ripple. Most meters have poor response to transient line or load changes. If you want to check for sudden output variations, you should use an oscilloscope.

Buying Versus Building Power Supplies

Our attitude toward buying versus building power supplies is that it is always cheaper to buy milk than to keep a cow. Of course, if you are lacking in funds, a poor cow and free grass is better than no milk at all. Hell, if you were rich, you would have hired an electrical engineer and not have bothered reading this book at all. Since you are probably poor (having read this far), we will provide you with some suggestions. First, check with your nearest military base regarding the possibility of obtaining surplus or obsolete equipment. Get catalogs from various surplus sales outfits. Also, don't be afraid to write letters to large companies and beg for outmoded or secondhand equipment (see Chapter 8 and the Appendix).

In any case, arm yourself with a good supply of manufacturers' catalogs. (You can learn a lot for cheap just by looking through company catalogs; the company *has an interest* in your getting the message from their catalog.) If you haven't already done so, you should write for catalogs to:

Burr-Brown Research
The Heath Co.
Hewlett-Packard Co.
John Fluke Co.
KepCo
Lambda Co.
Teledyne Philbrick
Sorenson

The addresses of these companies are given in the Appendix. These catalogs, especially KepCo's, will explain power supplies in more detail than almost anyone needs. You should realize that *many other* manufacturers of power supplies exist and have excellent products; those mentioned here merely represent *our* experience.

Our experience, which has been mostly in poverty-stricken universities, is that the Heath Company's kits are easy to put together, low in cost, and very reliable. Of higher quality is the Hewlett-Packard line; they provide instruments and power supplies that are almost indestructible. KepCo gear is comparable to that of Hewlett-Packard. The highest quality – but high cost – power supplies come from Lambda or John Fluke.

The first type of power supply that you might consider is one that delivers ±12 or ±15 volts to drive an op-amp. Batteries are fine for demonstrations, but you might want to invest in a Burr-Brown or Teledyne Philbrick power supply to start with; that way you will have something that *works* while you are just starting out. (Burr-Brown, Teledyne Philbrick, and other manufacturers make power supplies that are especially designed for use with op-amps and that are so constructed as to provide some protection against what we call the "student demolition derby.") Later on, we will discuss home-made op-amp power supplies.

Hooking up the Power Supply

Most op-amp power supplies will drive two or more op-amps. How, then, do you hook them up to the power supply? Figure 3-76 shows two techniques for hooking up the *ground connection.* In one case, a *bus-bar* ground is used. (A *bus bar* is just a heavy wire that serves as current carrier for several op-amps.); in the other, *direct* ground wires are run. What is the difference? Well, if the distance between points *A* and *B* in Figure 3-76 is 20 inches and the wire is No. 22 copper, the resistance is about 0.005 ohm. If the ground current is 10 mA, the reference ground potential for op-amps A and B differs by $(10^{-2})(5 \times 10^{-3})$ or 50 μV. This *might* be enough voltage to give you fits if you don't know it is there.

Protection Circuits

Now, how about an op-amp controlled power supply circuit? Let's suppose you have a typical laboratory setup with an inexpensive DC power source like the Heath

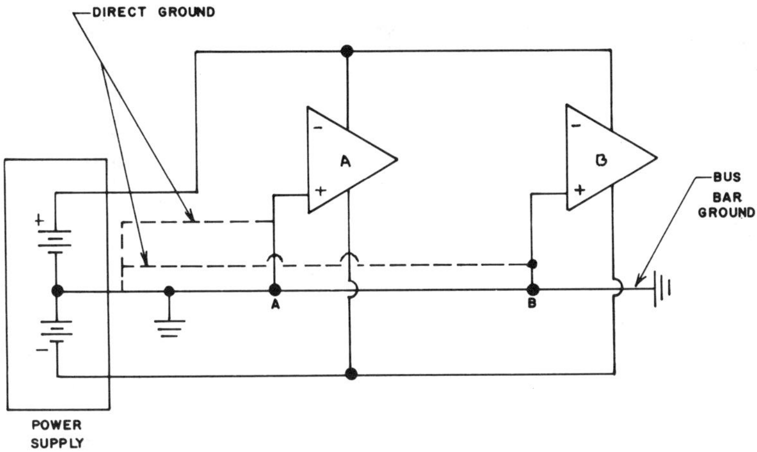

Figure 3-76. Op-amp power supply with two op-amps: bus-bar ground and direct ground connections.

IP-12 battery eliminator. The supply puts out 0–6 volts at 10 amps or 0–12 volts at 7.5 amps. The output has 0.3% ripple; this is fine, but you need *protection for the load,* which could be damaged by an internal short-circuit. Suppose the normal operating condition for this load is 12 volts at 2 amps; then if the current goes above 3 amps, you're in trouble!

The solution is (surprise) an op-amp and a device called a *silicon controlled rectifier* (SCR).

CROWBAR CIRCUIT

This combination of an op-amp and an SCR is called a *crowbar protection circuit.* It *can* short circuit the power supply, but Heathkit supplies are very rugged and easy to fix. Usually, only the power supply fuse blows, and that is a lot cheaper than a $50 integrated circuit. The operation of the system depends upon the SCR, which is a diode that passes only leakage current in reverse bias. In forward bias, the current is very low until the gate receives the proper signal. When the gate signal is received, the device "turns on" and passes current like a diode in forward bias. The *V-I* curve for an SCR is shown in Figure 3-78.

The protection circuit is "set" by adjusting the voltage to op-amp input (1) to some value V_1. This holds the output negative, but the (–) signal does not get through the diode to the SCR gate. Under normal operation, the voltage drop through the sensing resistor R_s is less than V_1 and the op-amp output is unchanged. Now if the load current goes up, the drop across R_s goes up to a value above V_1. This switches the op-amp to a (+) output, which passes through the diode and provides a (+) gate voltage for the SCR. The SCR turns on and short circuits the power supply to protect the load.

To pick the component values for this circuit, let's assume that the normal output is 12 volts at 2 amps with a shut-off at 2.5 amps by the crowbar circuit. We

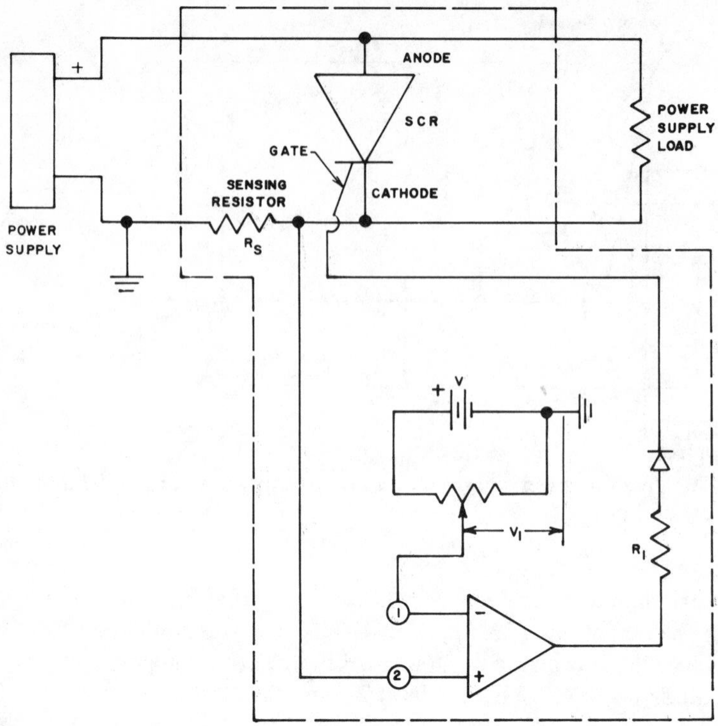

Figure 3-77. SCR crowbar protection circuit.

choose a 0.1 ohm, 1 watt sensing resistor: at 2 amps, the drop across it is 0.2 volt. We set V_1 at -0.24 volt to allow the op-amp to switch when the current rises to 2.4 amps. The diode must withstand a peak inverse voltave (PIV) of 50 volts and pass a current of 12 mA; a Motorola MR 1120 will do just fine and costs only about $1.00 as of 1980.

The SCR must hold off a 12 volt potential with the gate off, and it must carry a current of, say, 10 amps (or enough to blow the fuse in the power supply) when it gets the gate-on signal. A Motorola 2N6160 (costing about $5.00) will be more than adequate here: it holds off about 200 volts and has a 30 amp surge-current rating. It requires about 2 volts and 200 μA of gate signal to turn on, and our op-amp can provide this signal without trouble. The resistor R_1 in Figure 3-77 is used to limit the turn-on current if necessary; try 10 kΩ for a start.

CASCADED SCR CIRCUIT

Now the problem is, what if you wanted to use a bigger SCR that required, say, 200 mA of gate current? The op-amp won't supply 200 mA, but fear not, we still have some tricks up our sleeve. You can do one of two things. You can be crude and use two SCRs in series, as shown in Figure 3-79. In this circuit, SCR-1 is fired by the op-amp just as before, and it in turn fires SCR-2, which could be a 70 amp,

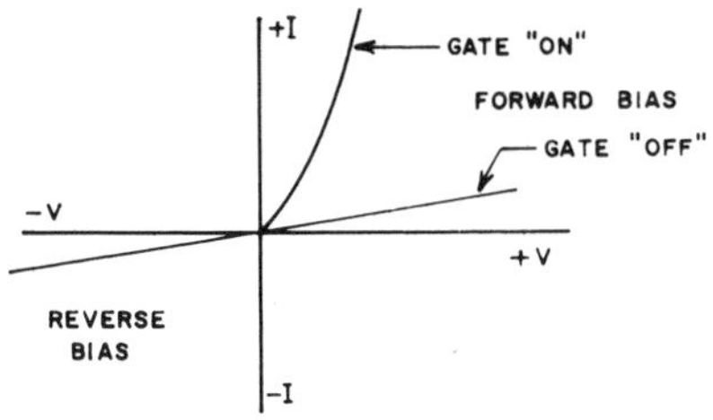

Figure 3-78. SCR *V-I* curve.

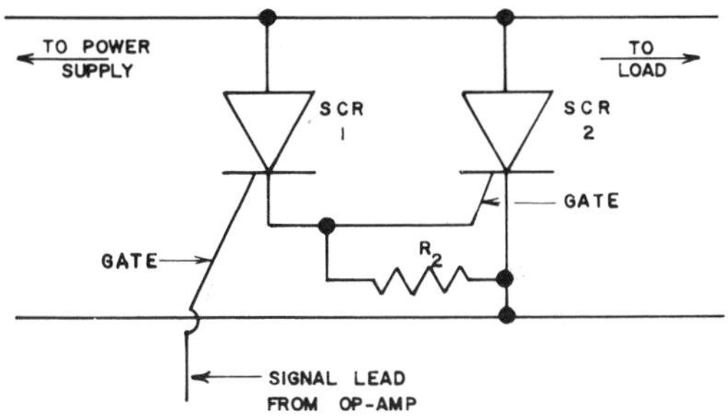

Figure 3-79. Cascaded SCR circuit.

or even a 150 amp, device if you needed it. The resistor R_2 is just there – *if you need it* – to keep leakage current through SCR-1 from firing SCR-2 accidentally. If this does happen, you will have to find the right value of R2 by trial and error: start with R_2 = 10 kΩ and see what happens. You may want to put a gate-current limiting resistor in the lead that goes to the gate of SCR-2. If so, start with 10 kΩ.

DIAC CIRCUIT

We admit that this use of cascaded SCRs as a crowbar circuit is an inelegant solution to our problem (most electrical engineers would cut their throats first). Another, and a more up-to-date, solution involves introducing a new type of device called a *DIAC.* This little jewel has a *V-I* curve as shown in Figure 3-80.

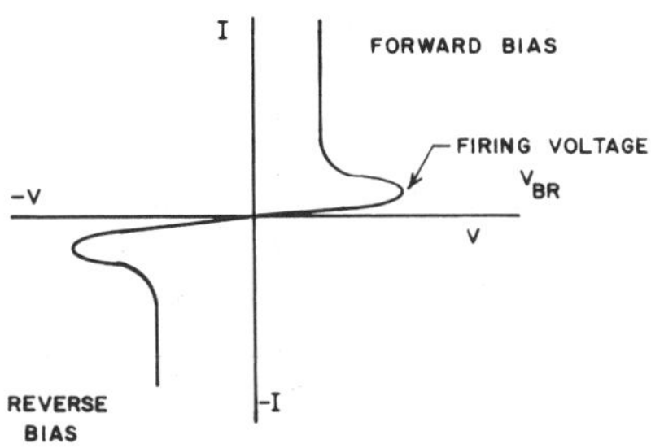

Figure 3-80. DIAC *V-I* curve.

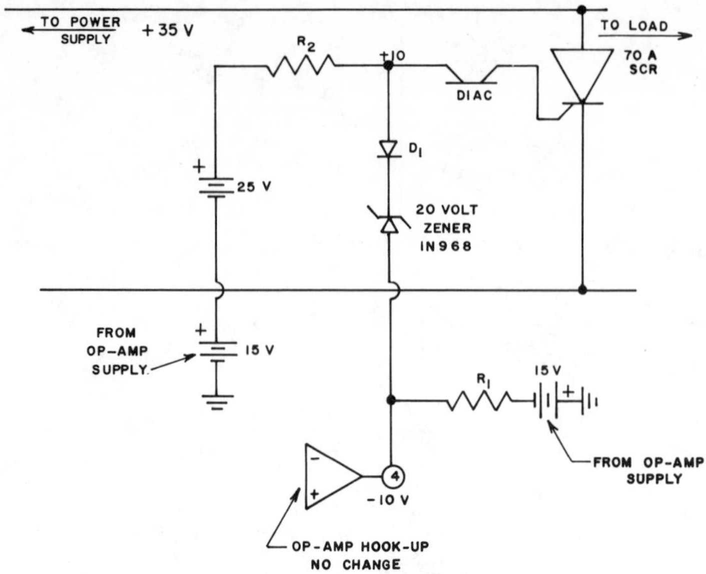

Figure 3-81. Op-amp driven crowbar circuit utilizing a DIAC.

The DIAC acts like a diode in reverse bias until a specific trigger voltage is reached. Then it passes current almost like a short circuit. This is ideal for triggering an SCR of almost any size because the DIAC provides a large current pulse when it fires. There is one small problem (isn't there always?): a DIAC needs from 28 to 32 volts to fire. To provide this, we use the system illustrated in Figure 3-81. This is a somewhat complex circuit, but it demonstrates some interesting features that are well worth discussing in detail.

When the op-amp output is -10 volts, the 20 volt drop across the zener diode puts +10 volts at the DIAC input and a drop of 35 volts across R_2. The diode D_1 is in forward bias, so we ignore its voltage drop. We pick R_2 to pass a current of 5 mA. This gives a value of $R_2 = 40$ volts/5 mA = 8 kΩ. The zener diode must be rated at 20 volts at 5 mA. When the op-amp goes to +10 volts, the top of the DIAC must be 20 volts higher (because of the zener diode again), or 30 volts. This fires the DIAC, which in turn fires the SCR. Simple, isn't it?

What about the 15 volt power supply and the resistor R_1 connected to the op-amp output (4) that are shown in the lower right of Figure 3-81? Well, it turns out that when an op-amp is in its (+) output mode, it wants to deliver current. The 15 volt battery and R_1 give the current a place to go. If the drop across R_1 is 25 volts at, say, 5 mA, R_1 must be 5 kΩ.

Power Supply Regulation

In many experiments we must have a constant output voltage from a power supply in the presence of line or load variations. If we can afford a regulated supply, that solves the problem. But suppose we can't?

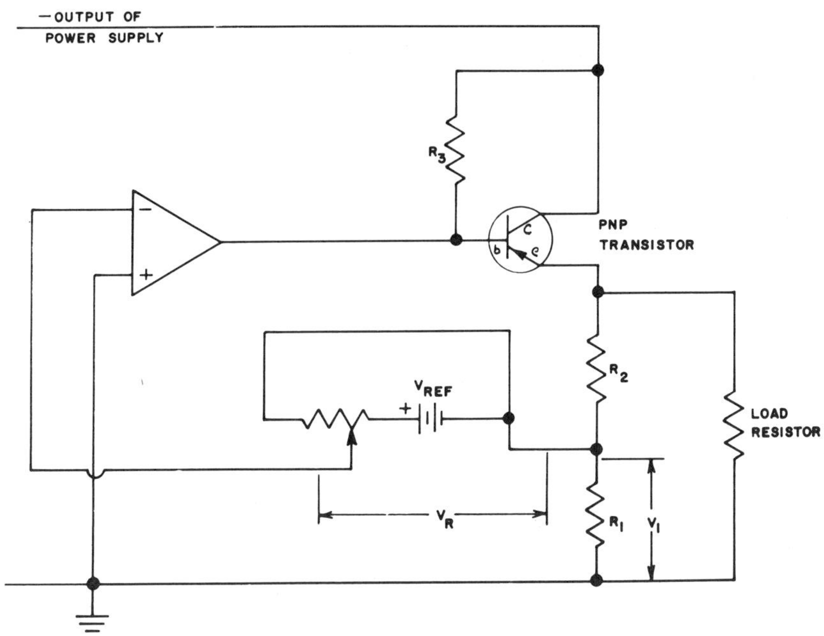

Figure 3-82. Op-amp power supply regulator.

Figure 3-82 shows a system to control the variations that might occur in the (-) output of a DC power supply. In this case the PNP transistor is a big husky beast fastened to a heat sink that carries the full output of the power supply. The op-amp provides the control current to the base of the transistor. The limit here is the 10 mA output of the op-amp. You should recall that we ran into this problem before with the hi-fi amplifier design. In the present case, we can do better because the op-amp transistor system need not be very linear: it is almost an off-on type of control circuit. The system operates by comparing the voltage drop across R_1, i.e., V_1, with whatever fraction of V_{ref} we have set with the potentiometer V_R. Essentially, V_1 and V_R are two batteries in series but opposed in polarity, as shown in Figure 3-83. If V_R is greater than V_1, current flows counterclockwise, and vice versa. To apply this to Figure 3-82, if V_R is greater than V_1, the output of the op-amp is (-); this increases the current flow through the transistor, thereby increasing V_1. If $V_1 = V_R$, it stays just where we want it.

Now, how do we pick component values for a circuit like this? How to pick transistors is always a question of great interest to electrical engineers and other people who must use electronics. Herewith we give the fastest outline in the world on how to do it. Any resemblance between this discussion and a textbook on electronics is purely coincidental. (You purist types can start reading Chapter 7 if you want the full details.)

Consider the common-emitter power transistor circuit shown in Figure 3-84. Let's assume that we want to pass a maximum of 500 mA through the load resistor, R_L. Given the 12 volt supply, the load resistor is limited to a maximum value: $R_L =$

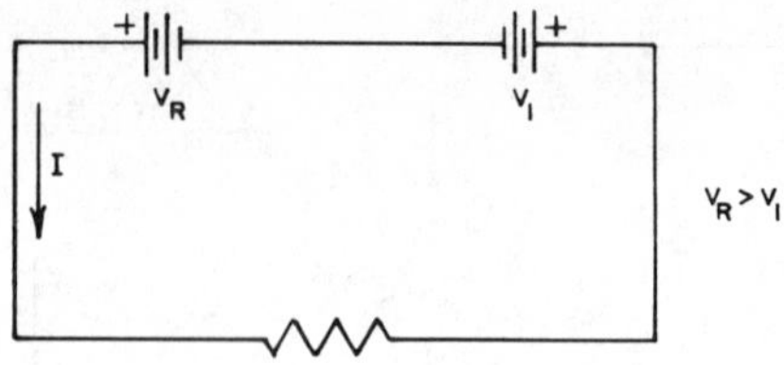

Figure 3-83. Equivalent two-battery circuit (see Figure 3-82).

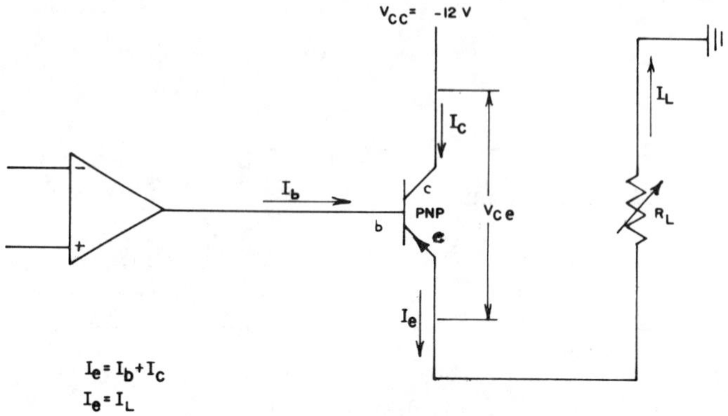

Figure 3-84. Common emitter transistor with load.

12 volts/500 mA = 24 ohms. This is the case when $V_{ce} = 0$, but when $I_c = 0$, V_{ce} must be equal to 12 volts because $V_{ce} = V_{cc} - I_L R_L$. This 12 volts is the largest reverse-bias voltage that can occur between the emitter and the collector. We have stipulated a maximum current of 500 mA; this would occur when I_b = 10 mA from the op-amp. The current gain must then be I_c/I_b = 500 mA/10 mA = 50.

Having determined these parameters, we turn to our trusty Allied catalog and look at the RCA power transistors. Why RCA? No reason except that the catalog opened to the RCA section. We pick a type 40050 germanium device. Why germanium? Well, for about 99¢ you can't lose much. The maximum values of the important parameters are listed below:

V_{ceo} = 40 volts
I_c = 5 amps
P_t = 12 watts
h_{fe} = 50 (I_c = 1 amp)

These terms tell us the following things: V_{ceo} = 40 says that the collector-to-emitter voltage must be 40 volts or less. I_c = 5 amps means that you don't exceed 5 amps. P_t = 12 watts says not to put more than 12 watts into the transistor. The term h_{fe} =

50 tells us that the current gain, I_c/I_b, is 50 at I_c = 1 amp. (Any EE who wants to quibble about our use of "h_{fe}" for direct current gain can take his trade elsewhere or wait for the detailed discussion of this concept in Chapter 7.)

For our application these values are great: h_{fe} = 50 means that 10 mA input yields 500 mA output. The maximum power input would be 12 volts × 0.5 amp = 6 watts. Actually, it could never reach more than one-quarter of this value, because maximum current and voltage conditions can't happen at the same time. Even if they did, we are still safe. Our maximum current is 500 mA and the transistor will carry 5 amps, so again we are in good shape. In fact, this design is far too conservative. The next step would be to start reducing R_L to obtain larger values of I_c. The limit to this procedure is set by (1) the transistor's current or power limit, (2) the power supply capability, or (3) the actual value of h_{fe} as I_c changes. You won't get I_c up to 5 amps, but this is a good circuit to start with. After all, what harm can you do except burn out a 99 cent transistor?

So the design is done – or is it? Suppose you really needed 5 amps instead of 500 mA. It would be back to the Darlington circuit again. Our 40050 PNP transistor serves as the first stage to drive the base of an NPN device as shown in Figure 3-85. The first transistor operates just the way it did in the previous circuit. The second transistor is rated at h_{fe} = 30, power dissipation P_t = 50 watts, maximum I_c = 10 amps, and V_{ceo} is about 80 volts. A quick calculation of the maximum power dissipation – ¼(12 volts) × (5 amps) = 15 watts – convinces us that we are okay in that area, and that's all we need to know for now.

There are, of course, many other points and details we could talk about, but this discussion must suffice at present. For further details, you should start reading things like Chapter 7 or the General Electric *Transistor Manual* and Texas Instruments' *Transistor Design Manual* (see Appendix). If you're interested in going down that path, our book has achieved its purpose. Bon voyage!

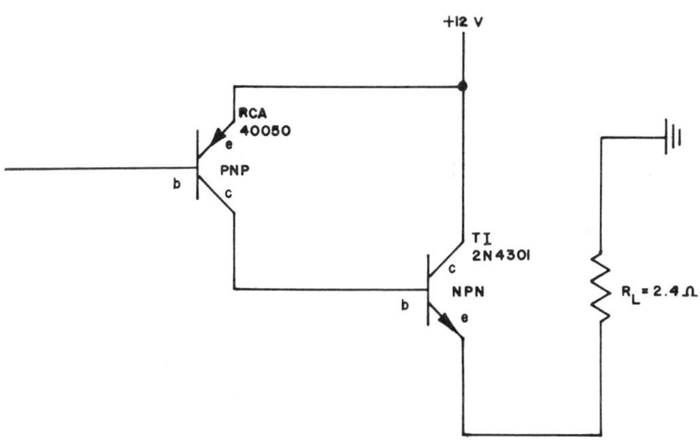

Figure 3-85. Darlington current booster.

One last point: the op-amp with feedback takes care of the biasing and thermal drift problems that plague designers of semiconductor circuits. If it weren't for the op-amp, things would be much more difficult.

Constant Voltage or Current Sources

CONSTANT VOLTAGE SUPPLY

The *constant voltage supply* shown in Figure 3-86 is typical of many instruments of this type. Constant voltage is analogous to constant current, and a typical constant current supply is shown in Figure 3-87.

In the circuit shown in Figure 3-86, V_p represents some unregulated voltage source, perhaps a storage battery or a bridge rectifier including a filter capacitor as shown in Chapter 1. V_{ref} is a stable reference voltage source. A zener diode circuit or a mercury cell could be used for this, as well as the +15 volt side of an op-amp power supply. The combination of the op-amp and the transistor has the property of balancing the bridge formed by V_p, the transistor, V_{ref}, R_1, and R_2. The transistor is a current controller that can control the output voltage by controlling the current in R_1. To see how the system works we note that the voltage at the (-) input of the op-amp is

$$(V_{out} + V_{ref})\left(\frac{R_2}{R_1 + R_2}\right)$$

The voltage at the (+) input is just V_{out}. Equating these two voltages and solving for V_{out} we obtain

$$V_{out} = V_{ref}\left(\frac{R_2}{R_1}\right)$$

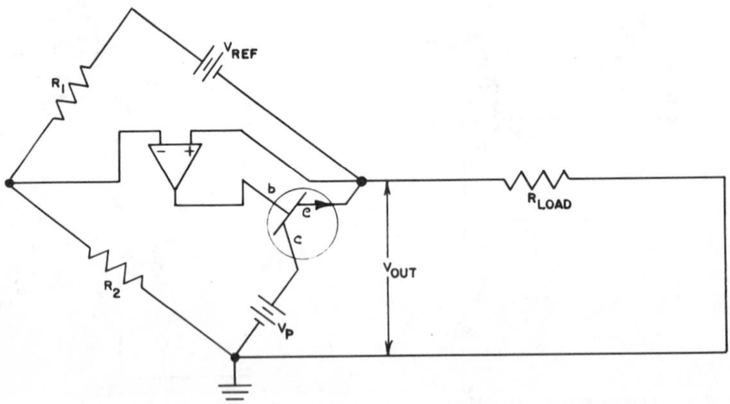

Figure 3-86. Constant voltage supply.

By making R_1 or R_2 variable resistors we can change V_{out} *but* V_{out} must always be about 2 volts less than V_p to give the transistor a voltage to operate on.

CONSTANT CURRENT SUPPLY

The constant current supply operates by sensing the voltage changes across a resistor R_s in the load circuit and adjusting the input to an op-amp so that the op-amp output will control the transistor in the current loop.

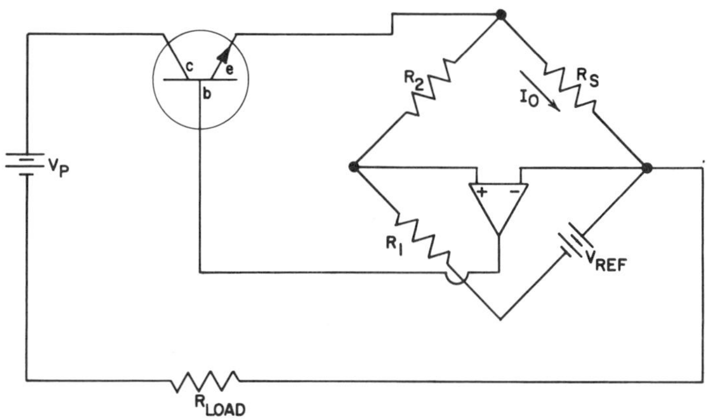

Figure 3-87. Constant current supply.

To analyze this circuit (Figure 3-87) we note again that the two op-amp input voltages must be equal. The voltage at the (-) input is just V_{ref}. The voltage at the (+) input is higher. It is

$$(V_{ref} + I_0 R_s) \frac{R_1}{R_1 + R_2}$$

where the voltage at the top of the bridge in Figure 3-87 is $V_{ref} + I_0 R_s$ and the resistors R_1, R_2 act like a voltage divider. Equating this to the voltage at the (-) input (V_{ref}) we get

$$V_{ref} = (V_{ref} + I_0 R_s) \frac{R_1}{R_1 + R_2}$$

$$V_{ref} \left(\frac{R_1}{R_1 + R_2} - 1 \right) = \frac{-I_0 R_s R_1}{R_1 + R_2}$$

$$\frac{-V_{ref} R_s}{R_1 + R_2} = \frac{-I_0 R_s R_1}{R_1 + R_2}$$

or

$$I_0 = \frac{V_{ref} R_2}{R_s R_1}$$

Again we can adjust I_0 by adjusting R_s, just as we adjusted V_{out} in the constant voltage circuit.

The point of this comparison is that in constant *voltage* operation, a *drop* in R_L causes an increase in load current. In constant *current* operation, a drop in R_L will induce an increase in voltage across the transistor to hold the current constant in spite of this drop in R_L; this is why we call it a *constant current supply.**

Power Supplies for Op-Amps

Earlier we promised that we would discuss a power supply to drive op-amps. This will require two circuits because we are also using op-amps to regulate the op-amp power supply. If that isn't lifting yourself up by your bootstraps, what is?

The circuit of Figure 3-88 begins with a transformer (a 36 volt center-tapped one) to provide ±18 volts rms with respect to ground. This is rectified to pulsating DC, which we partially smooth with two big capacitors. You might recall that the output from the bridge should be 18/0.707 = 25.4 volts peak: the 24 volts DC is an estimate that takes account of the voltage drop in the diodes when current is drawn by a load.

The rest of the power supply, the regulator circuit, is shown in Figure 3-89. The

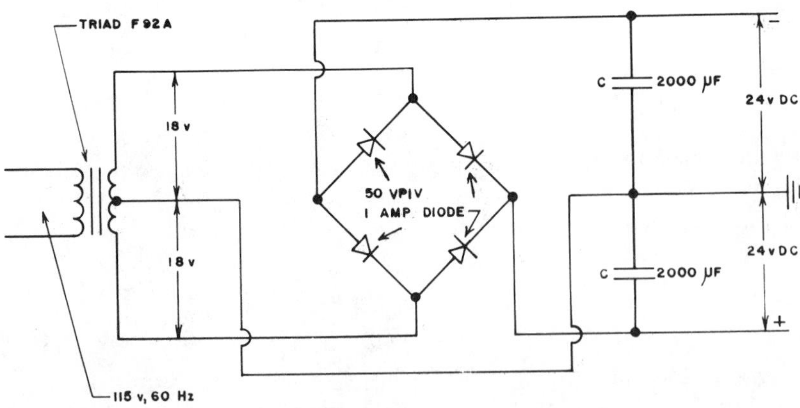

Figure 3-88. DC power source (unregulated).

*For more information on P/S, you might want to consult the *DC Power Supply Handbook,* available through your local Hewlett-Packard representative.

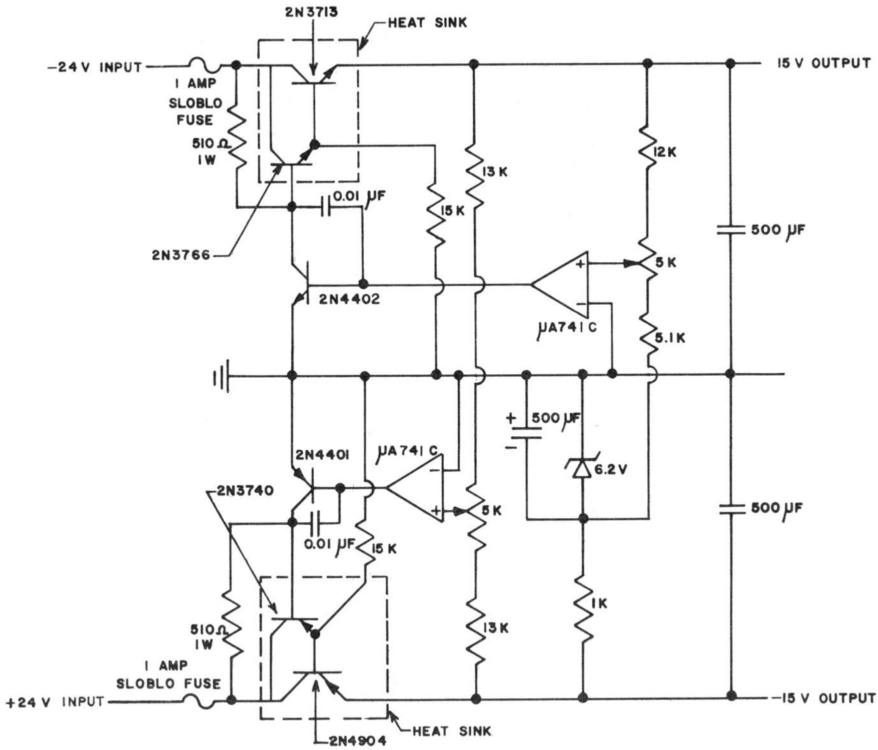

Figure 3-89. Op-amp power supply regulator circuit.

input is ± 24 volts from the unregulated DC source of Figure 3-88. In this circuit, the op-amps are Fairchild μA 741C, and all resistors are 0.5 watt, 5%, unless otherwise marked. The circuit itself is our old friend, the Darlington circuit, with the collector current from one transistor serving as the base current for the next.

The op-amps are set by voltages from the output. The 5 kΩ variable resistors should be used to balance the outputs to ±15 volts with respect to ground. Once this is done, the output voltages will remain constant despite line or load changes.

The op-amps take their supply voltage from the output of the regulator; these connections are not shown in Figure 3-89 to avoid making the drawing more cluttered. The transistors should be bolted to an aluminum heat sink if the circuit is to deliver its rated 0.5 amp output.

At this point you might wonder if you couldn't use a 30 volt battery to drive an op-amp. The answer is that you can, and the circuit is shown in Figure 3-90. The biggest problem with this circuit is what to pick for the resistances R_1. If you make $2R_1 = 30$ kΩ, the battery will last a long time, but any imbalance in your load will produce currents in the ground circuit that will upset your ±15 volt potentials.

A better system to beat this problem is shown in Figure 3-91. Here the zener

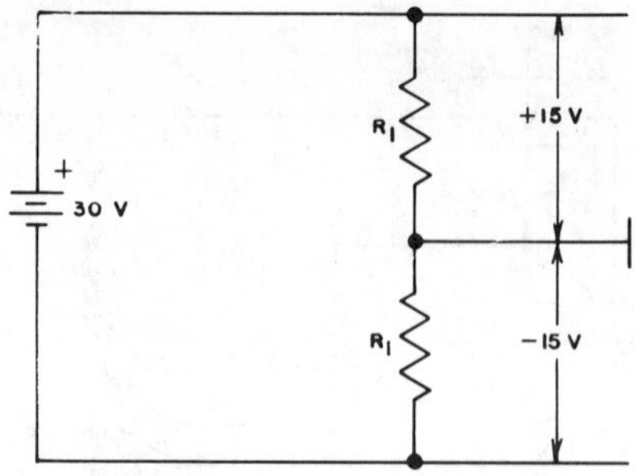

Figure 3-90. Battery power supply for an op-amp.

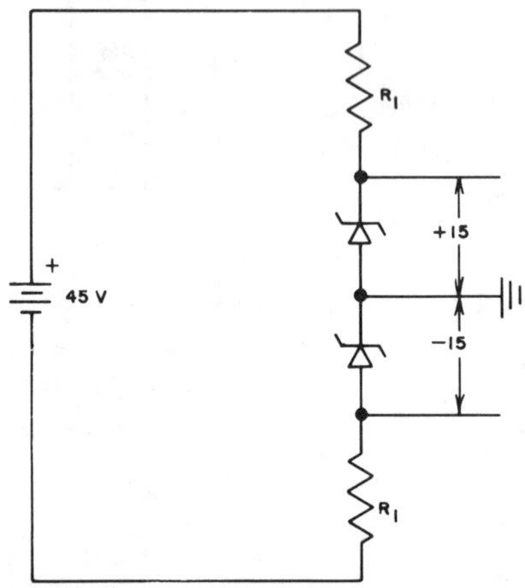

Figure 3-91. Zener-diode regulated power supply for an op-amp.

diodes are matched 15 volt types; the remaining 15 volts is dropped across the two R_1 resistances. To choose R_1, we use the following criteria:

1. The battery voltage range is 38 to 45 volts.
2. Assume the maximum load current is 5 mA.
3. The current through R_1 must be some ten times larger (to hold the output voltage constant in the presence of varying load currents) so $I_z > 50$ mA.
4. When the battery voltage is down to 38 volts (38 volts is V_{min}) we must drop 19 - 15 = 4 volts across R_1 so at 50 mA, $R_1 = 4V/50$ mA or $R_1 = 80$ ohms.
5. When the battery is up to 45 volts the current will be 22.5 - 15 = 7.5 volts divided by 80 ohms or 93 mA. The maximum I_z is 93 mA at 15 volts so the power dissipation is 1.41 watts. A 2 watt, 15 volt zener will do the trick.

If there are ground loop currents, the change in your ±15 volt potentials will be much less because the current flows through a zener diode resistance of perhaps 100 ohms rather than through a 15 kΩ resistor.

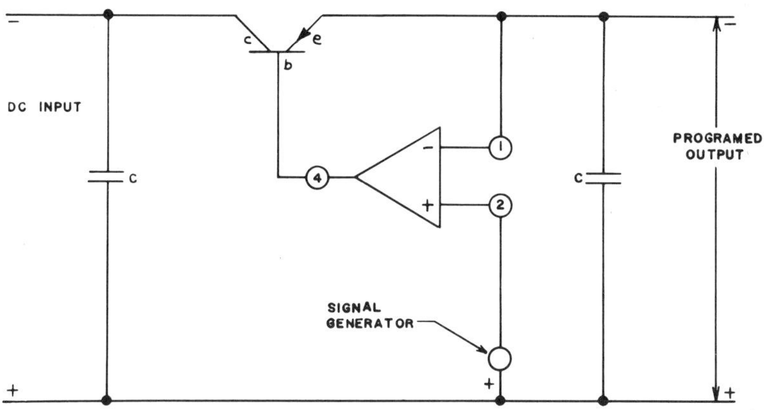

Figure 3-92. Programed voltage source.

Programed Power Supplies

Programing a power supply is another application of op-amps. Programed voltages are used in geology, for example, for induced polarization studies. In physical chemistry, voltage ramps are used for coulometric studies. Electrical pulses are often used in "electrofishing" to attract or immobilize fish. In most of these applications, considerable current is needed. If the available pulse or signal generator doesn't have enough current capability, what do you do?

The op-amp circuit for this purpose is shown in Figure 3-92. The operation of this circuit should be obvious by now. The signal generator drives the op-amp, the op-amp drives the transistor, and the transistor controls the output to the load. There is, however, no feedback control to handle things like changes in the transistor output with temperature; if you want to get that fancy, you should write to KepCo for their *Power Supply Handbook* which is listed in the Appendix.

This concludes our discussion of op-amps as they apply to power supplies. There are many other applications, but the ones we have discussed will enable you to read the literature and follow the discussions in works like the KepCo handbook.

HOW TO MAKE USE OF INTEGRATED CIRCUIT OP-AMPS WITHOUT INTERNAL COMPENSATION

Integrated circuit (IC) op-amps are inexpensive (sometimes even free) because they are easy for manufacturers to build at low cost. Unfortunately, part of their IC nature is their lack of *internal compensation,* i.e., of capacitors and resistors. This means that to use these pearls, you have to compensate them. We will tell you how to do this with a particular op-amp, the Motorola 1437P. The application to other IC op-amps is more or less the same.

The specifications for the 1437P are reproduced here by the courtesy of the Motorola Semiconductor Products, Inc., Phoenix, AZ. Some of this material needs

Table 3-2. Maximum Ratings for the Motorola 1437P Op-Amp*

Rating		Symbol	Value	Unit
Power supply voltage		V^+	+18	Vdc
		V^-	-18	Vdc
Differential input signal		V_{in}	±5.0	Volts
Common mode input swing		CMV_{in}	±8 V	Volts
Output short-circuit duration		t_S	5.0	s
Power dissipation (package limitation)		P_D		
Ceramic package	MC1437P		750	mW
Derate above T_A = +25°C	MC1437P		6.0	mW/°C
Plastic package	MC1437P		625	mW
Derate above T_A = +25°C	MC1437P		5.0	mW/°C
Operating temperature range		T_A		°C
	MC1537		-55 to +125	
	MC1437		0 to +75	
Storage temperature range		T_{stg}	-65 to +150	°C

*Reproduced courtesy of the Motorola Semiconductor Products, Inc., Phoenix, AZ.

explanation; some of it you can ignore. However, you have to know which is which, so away we go.

Figure 2-2C shows the packaged op-amp, and Table 3-2 gives the maximum ratings. Note that the output *won't* stand an indefinite short circuit, and that the differential input limit is ±5 volts. The circuit schematic (which is not shown) can be ignored, but Figure 3-93 gives the *equivalent* circuit: we *won't* ignore that. Also note that this package has two separate but equal op-amps (in spite of, but with all due respect to, the decisions of the Supreme Court).

The ±15 volt supply is attached to terminals 7 and 14 of the circuit shown in Figure 3-93. On op-amp A, the (-) input (inverting) number is 9. On op-amp B, it is number 5. The (+) inputs are 8 on op-amp A and 6 on B. The outputs are obviously 12 on op-amp A and 2 on B. Now, what about terminals 10, 11, 13, 4, 3, and 1? These are where we put the compensation. In Figures 3-94 and 3-95, we show – courtesy of Mr. Michael Pomeroy – an enlarged package drawing and the equivalent circuit with compensation added.

One term of importance that is commonly not understood by engineers is the *derating factor* (shown in Table 3-2). This lack of knowledge has caused a number of otherwise good engineering projects to fail, so we propose to explain it here. All electrical devices, motors, transformers, and op-amps are not 100% efficient and

*You will notice that on this particular op-amp there is no connection marked *ground.* In this particular circuit no special ground connection to the power supply is necessary, but you will have to watch this on each circuit. Some op-amps have a ground connection point; if there is one, you have to connect it to ground on the power supply.

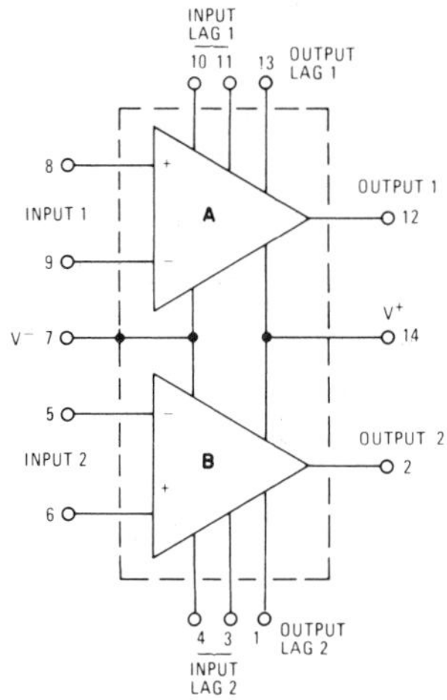

Figure 3-93. Motorola 1437P integrated-circuit op-amp equivalent circuit.* (Courtesy of the Motorola Semiconductor Products, Inc.)

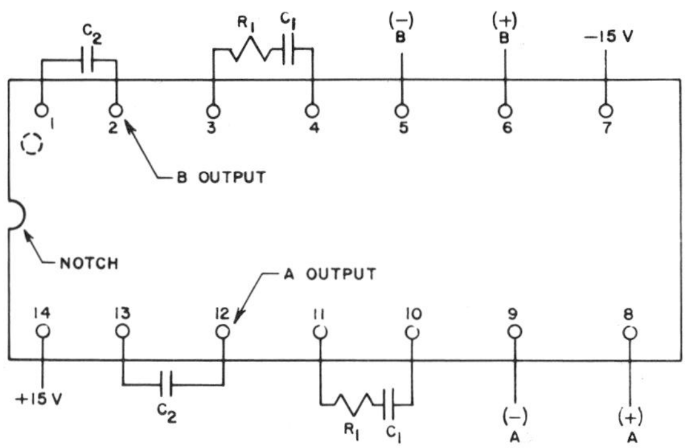

Figure 3-94. Motorola 1437P (bottom view) with compensation.* (Courtesy of Michael Pomeroy, University of Arizona.)

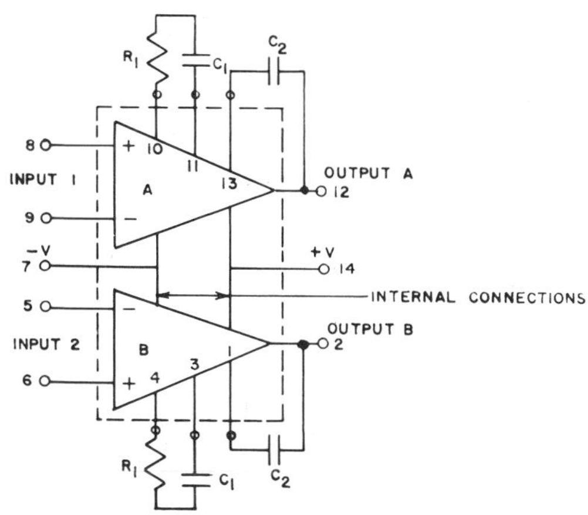

Figure 3-95. Motorola 1437P equivalent circuit with compensation added.* (Courtesy of Michael Pomeroy, University of Arizona.)

Table 3-3. Suggested Compensation Values

	R_3 (kΩ)	C_1 (pF)	C_2 (pF)
For general application	1.5	500	20
Low noise but restricted bandwidth	1.5	100	3
Higher noise but broader bandwidth	0	10	3

must dissipate the wasted energy to the environment. In most cases, the design is based on a 25C ambient with no air movement. If you operate the system at a higher ambient temperature, you have to add a fan or conductive heat sink to take away the heat. If you don't, the device will die – usually at the worst possible time.

If you must derate, do it as follows. Given that the rating is 750 mW at 25C, to operate at 50C (for a ceramic package) you must derate by (50 - 25)C × 6mW/C = 150mW; therefore, the allowed package dissipation is not 750mW but 600mW. Since the input and output currents will vary with the load, the only thing you can do is control the input voltage from the power supply. Assuming that ± 18 volts was okay for 750mW, we must reduce the supply voltage by the ratio of 600/750 to derate the unit; our new *maximum supply voltage* is ±14.4 volts.

The Motorola Corporation has a table of suggested values for R_3, C_1, and C_2. Our values are given in Table 3-3; there really isn't much difference, but these values worked best in our applications.

A typical inverting amplifier circuit is shown in Figure 3-96. Note that gain A is $-R_0/R_1$. The input impedance, R_1, is 1 kΩ and R_0 is 100 kΩ, so A = 100.

This has been easy so far, but what about the rest of Motorola's specifications that are given in Table 3-4? Let's go through them one by one:

1. Gives open loop gain (no feedback), op-amp output impedance, and input impedance. Note that such spec sheets usually give minimum, maximum, and

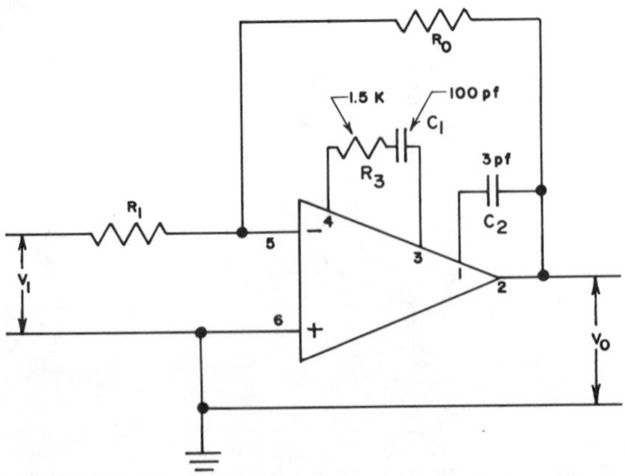

Figure 3-96. Inverting amplifier circuit using an IC op-amp with compensation.

Table 3-4. Electrical Characteristics of the Motorola 1437P Op-Amp

ELECTRICAL CHARACTERISTICS – Each Amplifier (V^+ = +15 Vdc, V^- = −15 Vdc, T_A = +25°C unless otherwise noted)

		MC1537			MC1437			
Characteristic	Symbol	Min	Typ		Min	Typ	Max	Unit
Open Loop Voltage Gain (R_L = 5.0 kΩ, V_o = ±10 V, T_A = T_{low} ① to T_{high} ②)	A_{VOL}	25,000	45,000	70,000	15,000	45,000	–	–
Output Impedance (f = 20 Hz)	Z_o	–	30	–	–	30	–	Ω
Input Impedance (f = 20 Hz)	Z_{in}	150	400	–	50	150	–	kΩ
Output Voltage Swing	V_o							V_{peak}
(R_L = 10 kΩ)		±12	±14	–	±12	±14	–	
(R_L = 2.0 kΩ)		±10	±13	–	–	–	–	
Input Common-Mode Voltage Swing	CMV_{in}	±8.0	±10	–	±8.0	±10	–	V_{peak}
Common-Mode Rejection Ratio	CM_{rej}	70	100	–	65	100	–	dB
Input Bias Current $\left(I_b = \frac{I_1 + I_2}{2}\right)$,	I_b							μA
(T_A = +25°C)		–	0.2	0.5	–	0.4	1.5	
(T_A = T_{low} ①)		–	0.5	1.5	–	–	2.0	
Input Offset Current	$\lvert I_{io}\rvert$							μA
(I_{io} = I_1 − I_2)		–	0.05	0.2	–	0.05	0.5	
(I_{io} = I_1 − I_2, T_A = T_{low} ①)		–	–	0.5	–	–	0.75	
(I_{io} = I_1 − I_2, T_A = T_{high} ②)		–	–	0.2	–	–	0.75	
Input Offset Voltage	$\lvert V_{io}\rvert$							mV
(T_A = +25°C)		–	1.0	5.0	–	1.0	7.5	
(T_A = T_{low} ① to T_{high} ②)		–	–	6.0	–	–	10	
Step Response								
Gain = 100, 5% overshoot, R_1 = 1 kΩ, R_2 = 100 kΩ, R_3 = 1.5 kΩ, C_1 = 100 pF, C_2 = 3.0 pF	t_f	–	0.8	–	–	0.8	–	μs
	t_{pd}	–	0.38	–	–	0.38	–	μs
	dV_{out}/dt ③	–	12	–	–	12	–	V/μs
Gain = 10, 10% overshoot, R_1 = 1 kΩ, R_2 = 10 kΩ, R_3 = 1.5 kΩ, C_1 = 500 pF, C_2 = 20 pF	t_f	–	0.6	–	–	0.6	–	μs
	t_{pd}	–	0.34	–	–	0.34	–	μs
	dV_{out}/dt ③	–	1.7	–	–	1.7	–	V/μs
Gain = 1, 5% overshoot, R_1 = 10 kΩ, R_2 = 10 kΩ, R_3 = 1.5 kΩ, C_1 = 5000 pF, C_2 = 200 pF	t_f	–	2.2	–	–	2.2	–	μs
	t_{pd}	–	1.3	–	–	1.3	–	μs
	dV_{out}/dt ③	–	0.25	–	–	0.25	–	V/μs
Average Temperature Coefficient of Input Offset Voltage	$\lvert TC_{Vio}\rvert$							μV/°C
(R_S = 50 Ω, T_A = T_{low} ① to T_{high} ②)		–	1.5	–	–	1.5	–	
(R_S ≦ 10 kΩ, T_A = T_{low} ① to T_{high} ②)		–	3.0	–	–	3.0	–	
Average Temperature Coefficient of Input Offset Voltage	$\lvert TC_{Iio}\rvert$							nA/°C
(T_A = T_{low} ① to +25°C)		–	0.7	–	–	0.7	–	
(T_A = +25°C to T_{high} ②)		–	0.7	–	–	0.7	–	
DC Power Dissipation (Total) (Power Supply = ±15 V, V_o = 0)	P_D	–	160	225	–	160	225	mW
Positive Supply Sensitivity (V^- constant)	S^+	–	10	150	–	10	200	μV/V
Negative Supply Sensitivity (V^+ constant)	S^-	–	10	150	–	10	200	μV/V

① T_{low} = 0°C for MC1437 = −55°C for MC1537 ② T_{high} = +75°C for MC1437 = +125°C for MC1537 ③ dV_{out}/dt = Slew Rate

MATCHING CHARACTERISTICS

Open Loop Voltage Gain	A_{VOL1}-A_{VOL2}	–	±1.0	–	–	±1.0	–	dB
Input Bias Current	I_{b1}-I_{b2}	–	±0.15	–	–	±0.15	–	μA
Input Offset Current	$\lvert I_{io1}\rvert$-$\lvert I_{io2}\rvert$	–	±0.02	–	–	±0.02	–	μA
Average Temperature Coefficient	$\lvert TC_{Iio1}\rvert$-$\lvert TC_{Iio2}\rvert$	–	±0.2	–	–	±0.2	–	nA/°C
Input Offset Voltage	$\lvert V_{io1}\rvert$-$\lvert V_{io2}\rvert$	–	±0.2	–	–	±0.2	–	mV
Average Temperature Coefficient	$\lvert TC_{Vio1}\rvert$-$\lvert TC_{Vio2}\rvert$	–	±0.5	–	–	±0.5	–	μV/°C
Channel Separation (f = 10 kHz)	$\frac{e_{out\,1}}{e_{out\,2}}$	–	90	–	–	90	–	dB

typical values. As you can see, input impedance may be as low as 50 kΩ at 20 Hz. If you want more, you have to buy a different kind of op-amp.

2. Is just output voltage swing into a 10 kΩ (or larger) resistance. These units are really intended to drive 1 kΩ to 10 kΩ loads, but you can try them for other applications. (If they are free, how can you lose?)
3. Is the allowed common mode voltage (±8 volts or less) and the common mode rejection ratio.
4. Is input bias current over a range of temperatures.
5. and 6. Are input voltage and current offset values. If your circuit is likely to be disturbed by them, you have to use the compensation techniques that we will describe in Chapter 6.
7. Is the slew rate (dE_0/dt) specification. Note that it varies with gain and compensation. For each gain, the maximum rate of change of the output is different.
8. Is the effect of ambient temperature on the offset voltage and current and the allowable power dissipation.
9. Tells you how things change if the supply voltages change. This can be important with battery-driven circuits.
10. Indicates how well the two op-amps are matched and what effect an output signal in op-amp A has on the output of op-amp B.

Figure 3-97 gives other useful data:

A. Motorola's test circuit and measured values.
B. Is large signal swing for various values of gain and compensation.
C. Is a partial Bode plot showing gain versus frequency for various values of the compensation resistors and capacitors. Note that this *is not* open loop gain; R_2 is not infinity.
D. Is open-loop gain where R_2 *is* infinity (open circuit, no feedback).
E. Is not important for most of our applications, which are low power ones.

Let's now consider Figure 3-98. Most of these data are self-explanatory and are only needed by the pros. Figure 3-98E, however, tells you how much "cross-talk" there is between op-amps A and B. This is important if you are going to use these op-amps for stereo applications.

NOISE REJECTION WITH OP-AMPS – CHOPPING AND SYNCHRONOUS DETECTION TECHNIQUES

The idea of *chopping* to improve signal-to-noise ratio – i.e., to get rid of drift – is an old one. (The term *chopping* does not refer to motorcycles. We chop signals, not motorcycle frames.) The advent of electronic chopping techniques has extended the use of chopping to the point that we predict most data systems in the post-1980 era will be chopped or digital. We will divide this discussion into three parts: (1) chopping to reduce drift effects and noise in op-amps, (2) chopping to remove detector drift, and (3) synchronous detection techniques.

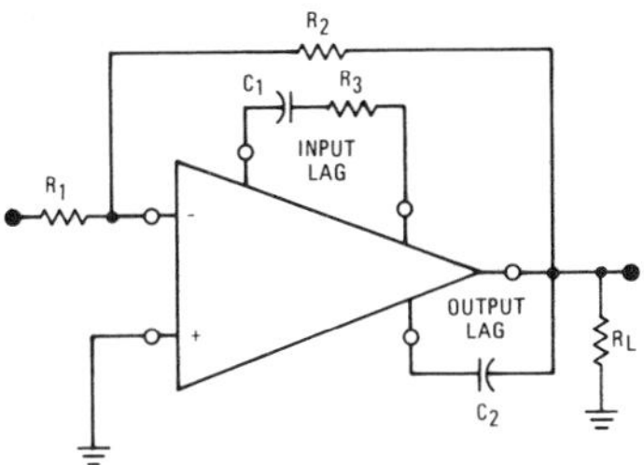

FIGURE NO.	CURVE NO.	VOLTAGE GAIN	TEST CONDITIONS $R_1(\Omega)$	$R_2(\Omega)$	$R_3(\Omega)$	C_1(pF)	C_2(pF)	OUTPUT NOISE (mV[rms])
b	1	1	10 k	10 k	1.5 k	5.0 k	200	0.10
	2	10	10 k	100 k	1.5 k	500	20	0.14
	3	100	10 k	1.0 M	1.5 k	100	3.0	0.7
	4	1000	1.0 k	1.0 M	0	10	3.0	5.2
c	1	1	10 k	10 k	1.5 k	5.0 k	200	0.10
	2	10	10 k	100 k	1.5 k	500	20	0.14
	3	100	10 k	1.0 M	1.5 k	100	3.0	0.7
	4	1000	1.0 k	1.0 M	0	10	3.0	5.2
d	1	A_{VOL}	0	∞	1.5 k	5.0 k	200	5.5
	2	A_{VOL}	0	∞	1.5 k	500	20	10.5
	3	A_{VOL}	0	∞	1.5 k	100	3.0	21.0
	4	A_{VOL}	0	∞	0	10	3.0	39.0
	5	A_{VOL}	0	∞	∞	0	3.0	—

A

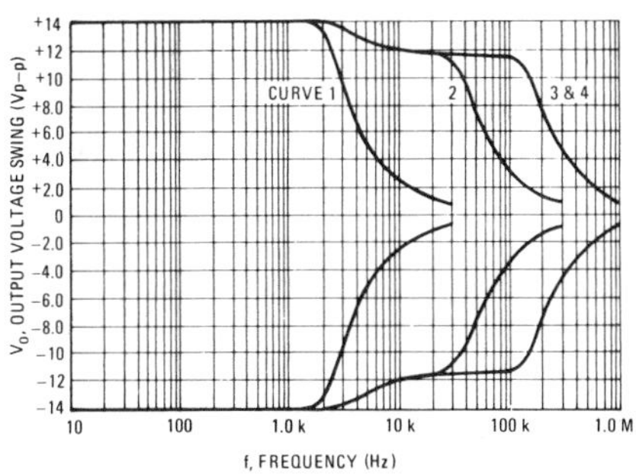

B

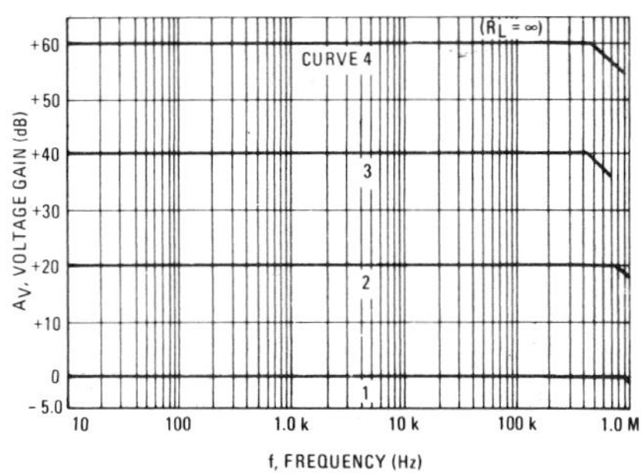

C

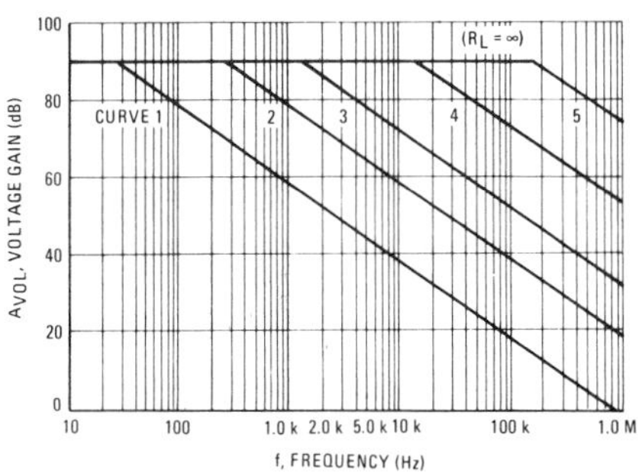

D

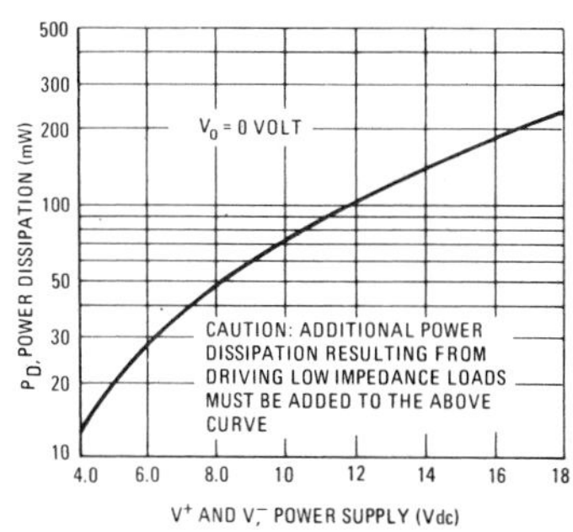

E

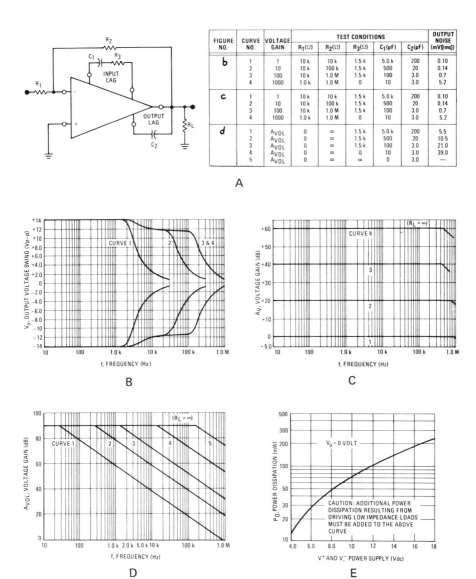

Figure 3-97. Motorola 1437P op-amp characteristics. *A.* Test circuit and test conditions. *B.* Large signal swing versus frequency. *C.* Voltage gain versus frequency. *D.* Open-loop voltage gain versus frequency. *E.* Total power dissipation versus power supply voltage. (Courtesy of the Motorola Semiconductor Products, Inc.)

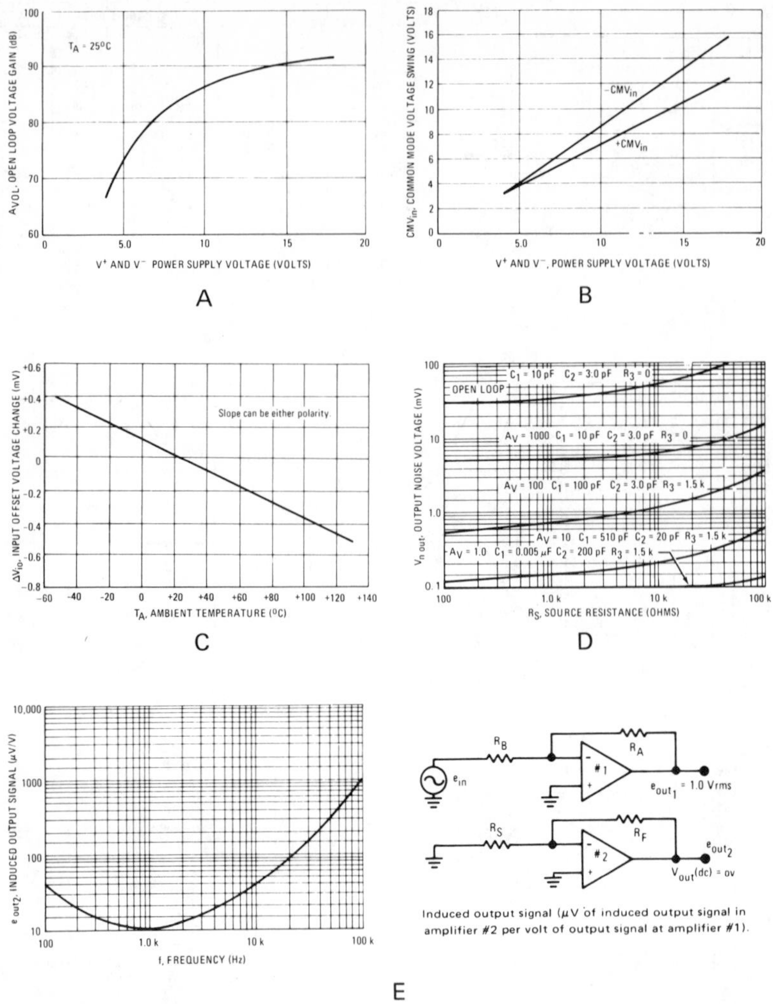

Figure 3-98. Motorola 1437P op-amp characteristics (continued). *A*. Voltage gain versus power supply voltage. *B*. Common input swing versus power supply voltage. *C*. Input offset voltage versus temperature. *D*. Output noise versus source resistance. *E*. Induced output signal (channel separation) versus frequency. (Courtesy of the Motorola Semiconductor Products, Inc.)

Op-Amp Drift Reduction

Chopper-stabilized op-amps are used to get rid of the thermal drift that occurs in *all* DC amplifiers. Drift is due primarily to the thermocouple effect that exists whenever two dissimilar semiconductors come in contact. Op-amp manufacturers try to reduce these effects, but you have to pay for it in dollars. With a simple chopper, you can make a low cost op-amp act like an expensive one.

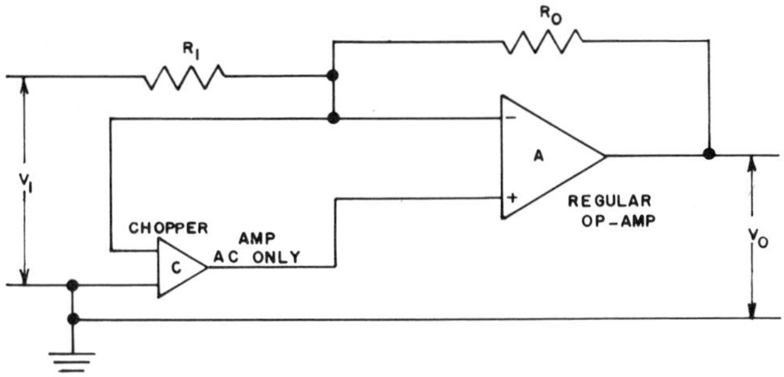

Figure 3-99. Chopper-stabilized op-amp.

A chopper-stabilized op-amp is really two op-amps in a single package. A typical system is shown in block diagram in Figure 3-99. To understand Figure 3-99, we have to study it closely. Note that op-amp C is an AC only amplifier, whereas op-amp A amplifies both AC and DC signals. Suppose that V_1 is a positive signal. This signal goes to op-amps A and C; the output of A is (-) and the output of C is also (-). However, this (-) output from C is fed into the (+) input of A, and that drives A still more (-) at its output. The upshot of this is that *any* signal that goes through both C and A is amplified much more than a signal that goes through A only. This point is critical so don't forget it.

When a signal is applied to this op-amp system, op-amp A amplifies it and adds drift because it is a DC amplifier. Op-amp C amplifies it without drift because it is an AC amplifier. Then the output of op-amp C goes to the (+) input of op-amp A and is amplified *again.* So the "drift-free" signal is larger than any drift noise introduced by op-amp A, by a factor equal to the gain of op-amp C, which might be 5000. So the signal is now at least 5000 times bigger than the noise.

Now all we have to do is show you how to get DC signals through an AC amplifier. To solve this DC-through-an-AC-amplifier problem, we introduce the synchronous chopper rectifier circuit shown in Figure 3-100. Consider two input signal frequencies passing through R_1 when the switch is in position *2.* High frequencies go through C_1, are amplified by the op-amp, and pass through C_2. They *do not,* however, appear at the output because the switch in position *2* shorts all output signals to ground. Low frequency signals are blocked completely by C_1. When the switch is in position *1,* the output is zero again, as you can see for yourself. So the output is zero when the switch is in *either* position *1* or position *2.* But what happens when the switch is in the middle? Then high frequencies go in through C_1 and out through C_2, so V_0 is not zero. Low frequencies are again stopped cold by the capacitors C_1 and C_2.

How, then, do the low frequencies get amplified? To see the answer to that question, look at Figure 3-101. This illustration shows the result of our chopping system: a series of high frequency ($f_0 = 1/\tau_0$) pulses that pass through the AC amplifier without trouble. The envelope of the pulses is the low frequency signal we want to amplify. The term *envelope* refers to a line passing through the maximum value

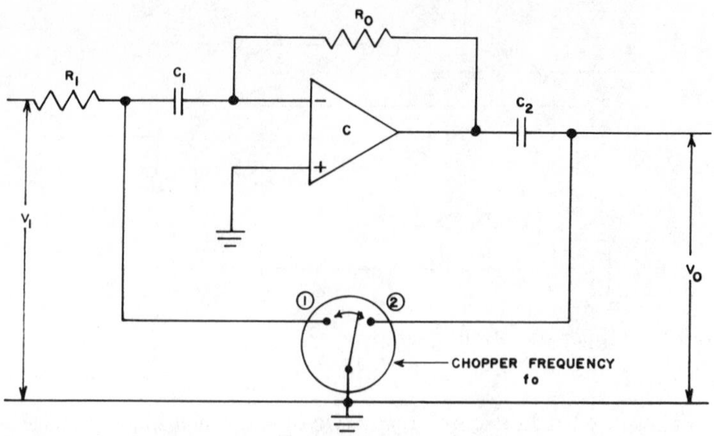

Figure 3-100. Synchronous rectification and amplification.

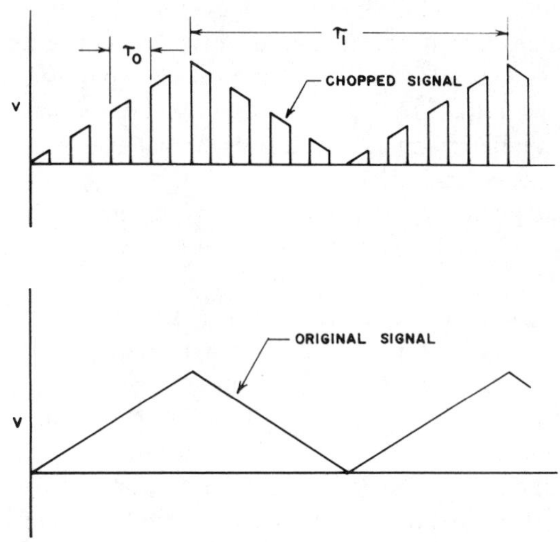

Figure 3-101. Chopping of a low frequency signal.

of each chopped segment in Figure 3-101. In effect, this line *envelopes* the maximum values. Clever, isn't it?

If you really want to understand this thoroughly, we suggest you go back and review the material on Figure 3-99. If you just want to use chopper-stabilized op-amps, you can buy them from our good friends at Burr-Brown.

We slipped in synchronous detection while you weren't looking. Look back at Figure 3-100. The only time that signals can come through the system is when the switch is in the middle position. The output is therefore *synchronized* to the input: *only when* signals *can enter* the amplifier are they allowed *out.* Any input or output noise between input signals is ignored. There is more about synchronous detection on page 198.

The important thing to remember is that this *only* works for signals with frequencies less than the chopper frequency, f_0. In fact, your maximum signal frequency, f_1, should be no higher than $1/3\ f_0$.

Detector Drift Reduction

Our next problem is the case of drift in a detector that is not an op-amp. Let's consider a saw-tooth optical signal coming into a photodetector that has *drift.* This is shown in Figure 3-102. Figure 3-102A shows the signal and the drift, and Figure 3-102B shows the combined signal plus drift. This is an example of a very common problem, and if the noise (drift) is greater than the signal, you have *trouble.*

The application of chopping to detector drift may be understood in terms of Figure 3-103. Here we used the previous example of a saw-tooth optical signal of frequency f_1. (Why saw-tooth waveforms all the time? They're easy to draw, that's why!) The saw-tooth signal is chopped by a mechanical chopper (see Figure 3-100) at a frequency f_0, where f_0 is greater than f_1, and is sent to the detector. The detector is drifting, and the output is our chopped saw-tooth signal *plus* drift in electrical form. Now we send the signal through a filter tuned to f_0. All the drift and noise that is below frequency f_0 is stopped by the filter. Our output is a chopped saw-tooth signal, and if we want to get rid of the chopped effect, we can add a filter that is tuned to stop all frequencies of f_0 and above. The final output, then, is an

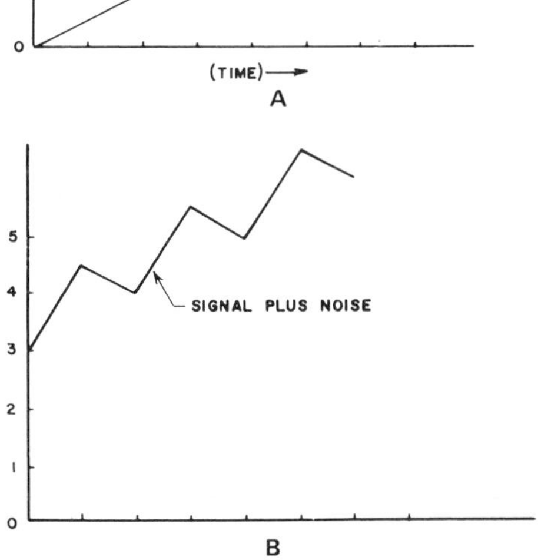

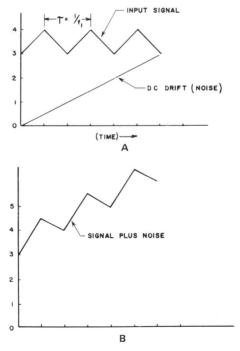

Figure 3-102. Typical DC noise and drift problem. *A.* Detector input signals. *B.* Output signal.

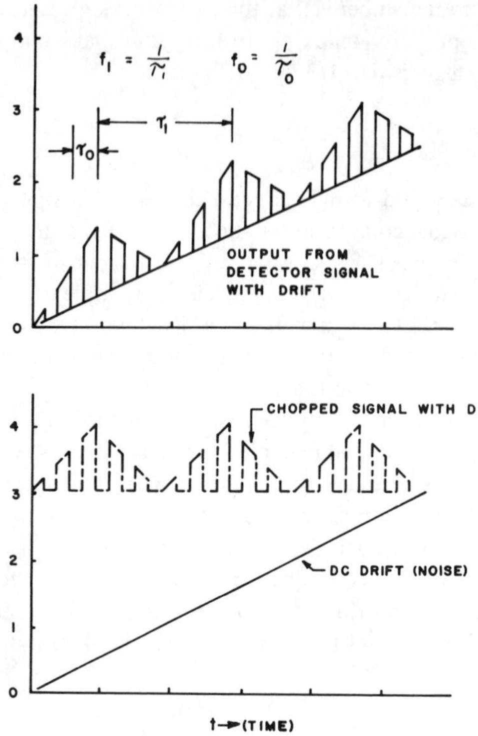

Figure 3-103. Typical chopper solution for detector drift compensation.

electrical signal that is a direct representation of the original optical input. The sequence of operations is shown as a block diagram in Figure 3-104. The wave shape (except for the final filtering) is shown in Figure 3-103 in the center of the figure as chopped signal with drift removed.

Synchronous Detection Techniques

The above methods, though helpful, are not all we can do to get rid of noise. Consider the situation shown in Figure 3-105. Here we not only have our old problem of low frequency drift, but we also have high frequency noise from the chopper, power lines, and so on. This noise, which is at frequencies *above* f_0, comes through our f_0-tuned filter. Therefore we must use another trick: the *phase-locked system,* which utilizes a second chopper driven at frequency f_0 to act as a controlled switch. The switch is so controlled that signals get through it *only* during the time when a signal from the first chopper is *due* to arrive.

It works like Maxwell's demon: if you had a box full of mixed red and white balls all bouncing about and you had a demon who opened a door when a red ball came along, you could separate the balls. Our system doesn't violate the Second Law of Thermodynamics, as Maxwell's demon would, but the idea is the same.

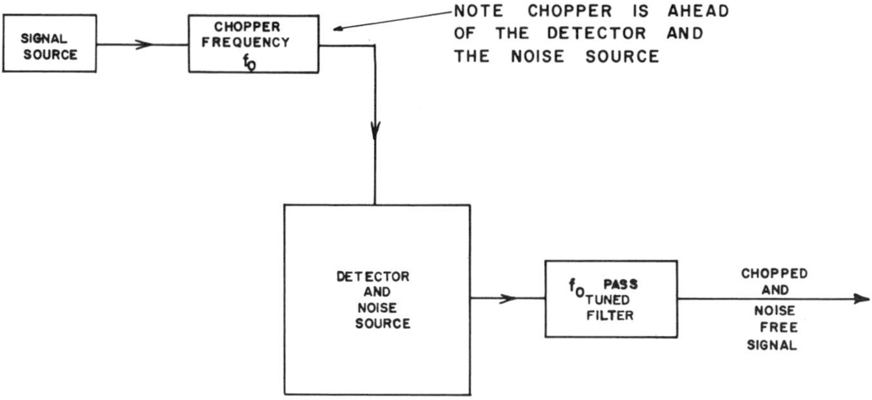

Figure 3-104. Chopper circuit for DC noise and drift removal.

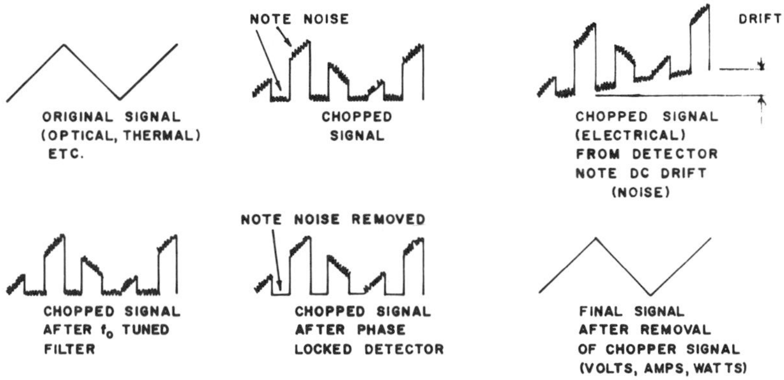

Figure 3-105. Signal sequence through phase-locked loop detector system.

The change in the signal as it progresses through the synchronous detection loop is shown in Figure 3-105. A block diagram of such a loop is shown in Figure 3-106.

At this point you might ask, what about the noise on top of the signal in Figure 3-105? Well, you can get rid of that by an *averaging* technique. Recall that for AC signals,

$$V_{\text{ave}} = \frac{1}{\tau}\int_0^{\tau} V\,dt = 0$$

So you could "integrate out" this noise, or you could "time-correlate" it out. Such techniques are beyond the scope of this book, and you would do well to consult the experts, such as Princeton Applied Research (see Appendix).

If you want to understand how signal averaging works, read on. Someday, when NSF or NIH comes through with a fat grant, you can buy some signal-averaging gear. Signal averaging (SA) is the *only* technique to use when you have a signal buried in

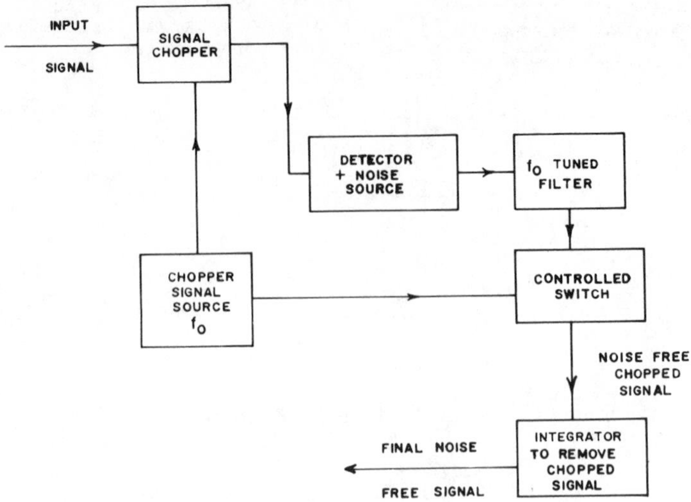

Figure 3-106. Synchronous detection circuit for noise reduction.

white noise. *White noise* is, analogously to white light, a composite of all frequencies, just as white light is composed of light of all wavelengths. White noise or shot noise has such a wide bandwidth that you can't filter it out, but you can take advantage of the fact that white noise is AC and the average of *any* AC signal is *zero.* Of course, the average has to be taken over many cycles, but that doesn't really hurt.

Let's assume that you have a repetitive signal, say a 10 Hz square wave of amplitude 10 mV that is buried in white noise 100 mV high. Obviously the square wave is invisible in the noise. But now suppose you could average the signal plus noise over some period like 10 seconds. That gives you 100 square waves and many, many cycles of white noise. The white noise being an AC signal, or really a multitude of AC signals at the same time, averages out to zero. The square wave will appear at the output almost as if the noise never existed.

Suppose now that you only have a one-shot nonrepetitive signal. Are you still able to use this technique? The answer is yes: you can use it if you are smart enough to get a magnetic tape recorder with an endless tape. Put your one-shot signal on the tape and run it through the averaging circuit as often as you like. This statement does not imply that running the data through the averaging system "N" times will improve it by some factor of "N." It is just a way of storing one-shot data on an endless tape so that you can view it and remove the noise at your leisure.

Returning to the synchronous detection system for a moment, we ought to consider two questions. First, does this work on *all* kinds of noise? Second, how do we set up a demonstration?

To answer the first question, let's consider an experiment in which a photocell converts light energy into an electrical signal. The relation between the two is called the *transfer function.* In its simplest forms, this relation is $I = K\Phi$, where I is amps, Φ is light intensity, and K (a constant) is our transfer function.

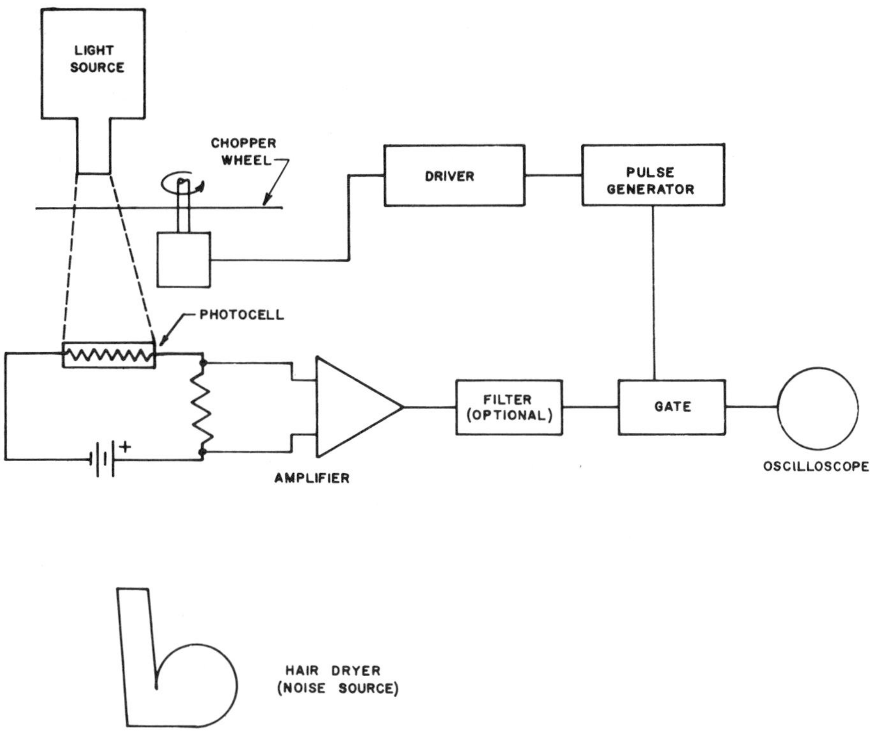

Figure 3-107. Synchronous detection demonstration system. (Courtesy of Michael Pomeroy, University of Arizona.)

This is an ideal situation; most real photocells have some current output even when Φ is zero. We might therefore write the relation as $I = K\Phi + C$, where C is another constant. However, there is yet another complication, namely, drift. This can take two distinct forms: one we can live with, the other we cannot. In the first case, if $C = At + B$, the drift is independent of Φ and is only a function of t (time). We can then write $I = K\Phi + At + B$. This case can be handled by the methods shown in Figures 3-105 and 3-106. We can chop Φ mechanically with a slotted wheel and filter out *all* effects of drift. In the second case, K is a function of time, say, $K = F(t) + G$. Thus our equation for I has the form: $I = [F(t) + G]\ \Phi + C$. Chopping will not help us because of the product of $F(t)$ and Φ; the transfer function is nonlinear and we are up the creek.

Most photocells are of the former type (thank God as well as a number of human engineers), and, using these, it isn't too difficult to build a demonstration system.* In Figure 3-107, we have indicated a cadmium sulfide photocell, a microscope light for a source, a wheel with holes in it as a chopper, and, last but not least,

*This was done by Mr. Michael Pomeroy of the University of Arizona Electrical Engineering Department.

a hair dryer to change the parameters of the photocell (this simulates the effect of detector drift).

The oscillator drives the chopping-wheel motor and a pulse generator. The latter in turn drives the second chopper, or gate. By changing the adjustment of the pulse generator, we can shut out any noise generated by the chopper. By changing the pulse width and timing, we can focus our attention on any important part of the signal. We used the hair dryer to induce photocell noise and show that this noise is easily removed by the chopping phase-lock loop circuit. If we wanted to chop at a higher rate, we would have to use a vibrating reed chopper for the light.

To chop electronic signals at high frequency, we use a field effect transistor (FET) chopper, as shown in Figure 3-108. This circuit was developed by Mr. Theodore Sammis of the University of Arizone Hydrology Department. Basically, the op-amp is an oscillator that produces alternating ±10 volt square waves. These signals are amplified and serve to drive an FET. The FET goes from the short-circuit state to the open circuit state as the amplified square waves hit the gate. This alternation in turn produces the chopped signal that is needed. In this circuit, Q_1 is a 2N3988 transistor, Q_2 is a 2N699, and Q_3 is a 2N5555. That's all there is to this circuit. By now you're probably asking, "How do we learn to design circuits with FETs?" Well, have patience; we will get to that problem in good time. There's a whole section on FETs in Chapter 7.

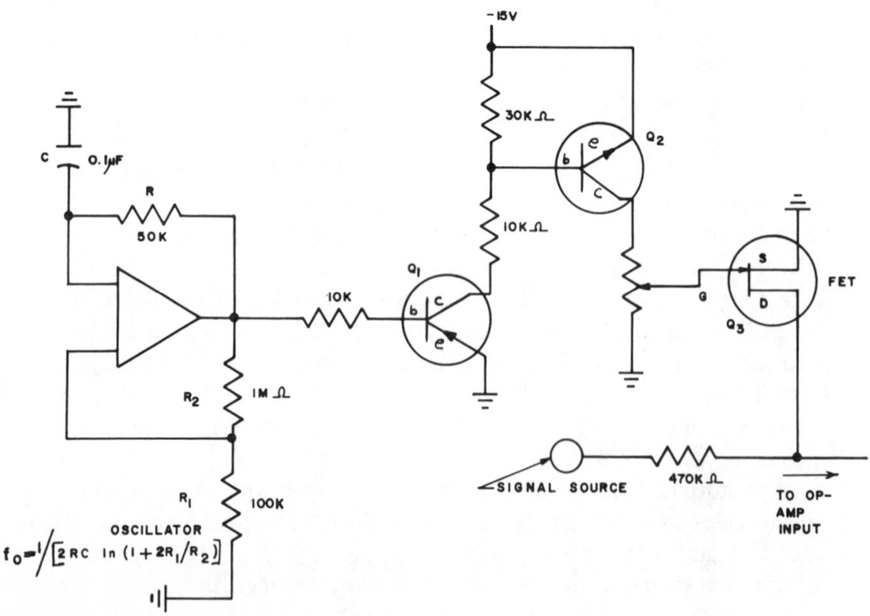

Figure 3-108. FET chopper circuit for high frequency signals. (Courtesy of Theodore Sammis, University of Arizona.)

COUNTING SINGLE PHOTONS

You often hear about this technique these days and it does have some advantages, especially in low light situations. It is a little more complex than operating a photomultiplier (PM) tube in the analog or DC mode, but you have to have manure to grow potatoes.

Each time a photon hits the cathode of your PM tube, it produces secondary electrons that are multiplied from dynode to dynode* until at the output you get a pulse of current that may be as large as 0.1 μA. This pulse is put through a good AC amplifier (100 MHz bandwidth). This allows us to ignore all the DC baseline drift and dark current that is so prevalent with DC operation. These pulses from the amplifier go to a discriminator which *you* set to ignore all pulses *below* a certain amplitude. Once again, this helps get rid of stray noise. The pulses that pass the "amplitude test" are sent to a counter; then you can count the incident photons by counting the pulses they produce.

There are a few other tricks you can play. For example, if two photons arrive in the PM tube at once, you might wish to record this fact. You can do this by using a fancy discriminator circuit which gives you pulses that are all the same height but whose duration is proportional to the amplitude of the current pulse from the PM tube. If you display the sum of the number of pulses times their duration during a given interval, the result will tell you how many photons arrived during that period *even if* they come in two or three at a time. If the pulse amplitude is A and the pulse duration is Δt, the sum S during an interval T may be written as

$$S = \sum_{1}^{N} A_x(\Delta t)_x$$

where N is the number of pulses that occur during interval T.

If all of the above hasn't bugged you too much, you might want to look at Figure 3-109, where we show a system for single-photon counting. For more details, you can look up an article by M. R. Zatzick in *Electro-Optical Systems Design* for June 1972 (page 20). His company – SSR Instruments, Santa Monica, California – will be happy to tell you all about single-photon counting, and they will even sell you the equipment for this type of work.

GROUNDING AND SHIELDING PROBLEMS AND THEIR SOLUTIONS

If you work with either low level signals or in noisy (electrically speaking) environments, noise pickup is a problem. The differential input circuit (review Figures 2-16

*A dynode is a special type of cathode used in electron multipliers; when a single electron hits the cathode, the cathode responds by emitting anywhere from 2 to 10 secondary electrons. That is why the system is called an *electron multiplier.*

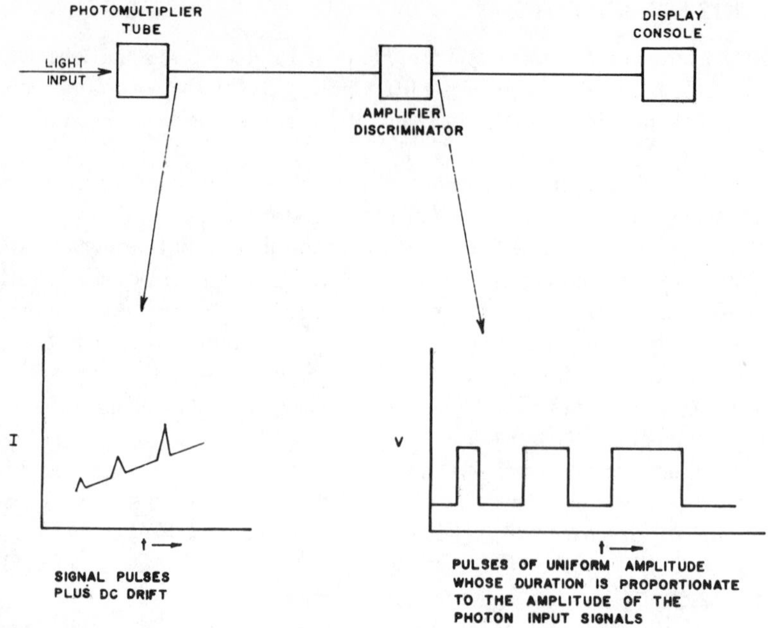

Figure 3-109. Single-photon counting system.

to 2-18) helps by reducing common mode noise, but in many cases you must use shielding techniques. A *shield* is a conducting braid or foil that stops unwanted electrical signals from getting into your system. Of course, everyone *knows* about shielding, like they do about sex and politics, but they sometimes don't know how to make proper use of it. (All right. With sex, instinct helps a lot, but have you ever heard of "grounding instinct," except in woodchucks and moles?) One of the best references on this matter (grounding, not sex) is the book by R. Morrison, *Grounding and Shielding Techniques in Instrumentation,* New York: Wiley, 1967. For those who are too cheap, or too lazy, to get Ralph Morrison's book, we give herewith a few simple rules and examples.

The first thing to remember from Chapter 1 is that electricity comes out of the "hot" side of the outlet and returns to the power plant via the "neutral" side. "Neutral" should never be called "ground"; *ground* is a water pipe or a stake buried in the earth. The third opening in a three-wire outlet is "ground," but you might want to make use of a water pipe, too (we will discuss this point in more detail in the following pages).

Let's assume that you have a grounded signal source connected by a shielded wire to an op-amp, as is shown in Figure 3-110. (The quality of the insulation between the shield and the wire is important for low level signals, e.g., 10 mV. Use Teflon insulated wire if you can afford it.) Note that we have *not* used the shield as our ground signal reference; the shield is grounded only at the source, and we have run a separate ground signal wire. An even better system is shown in Figure 3-110B.

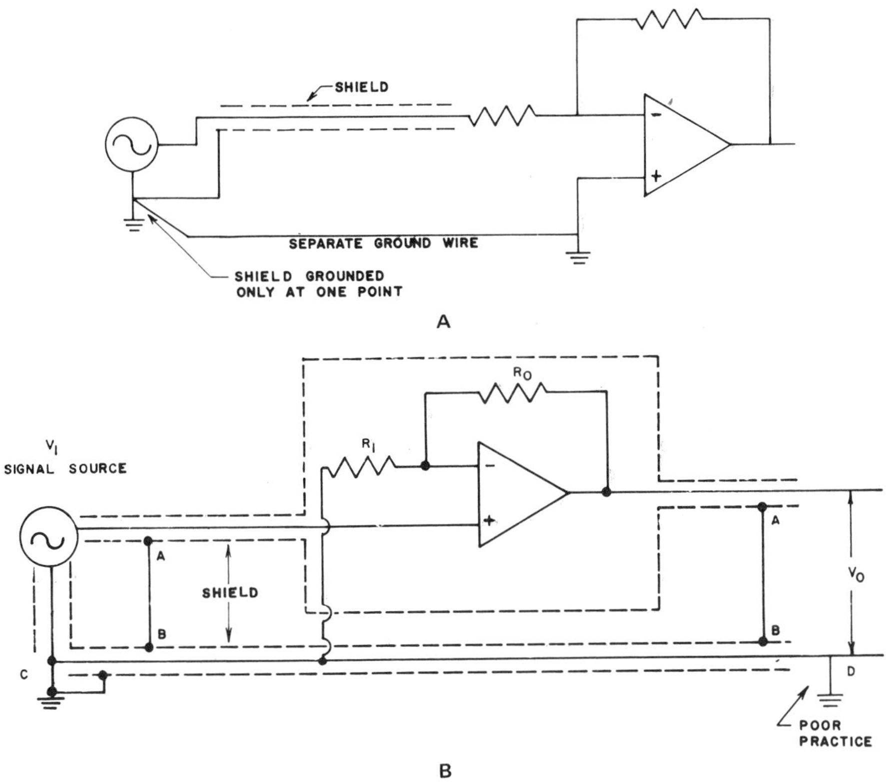

Figure 3-110. *A.* Proper use of shielded wire. *B.* Noninverted op-amp circuit with shielded cables.

Here we have the system connected up with single-lead, shielded wire. The first question is, where do we hook up the shields? Well, we had damn well better connect them together, as shown by the lines *A–B.* If both leads are inside one shield and *twisted together,* so much the better. The shield is grounded *only at the signal source,* as shown in the illustration.

You might wonder why the ground signal lead is shielded. If it weren't shielded, it might pick up noise that would be impressed on the (+) input of the op-amp via the signal source. This would show up as amplified noise at the output. Note that the ground signal lead is grounded at point *C* and at point *D.* This *isn't* the ideal way to do it, because *C* and *D* might be "grounds" at slightly different potentials. If you can't float the circuit at point *D,* you might want to run an extra heavy wire (No. 14) between *C* and *D;* this is shown as an extra line in Figure 3-111. Sometimes you might have to use this trick. Even in hospitals, the grounds are often only connected by metal conduit that is pinched (not screwed) together.

You might also wonder what to use for a metal can to put the op-amp in. You can buy metal boxes, but if they are aluminum, soldering to them is difficult. We

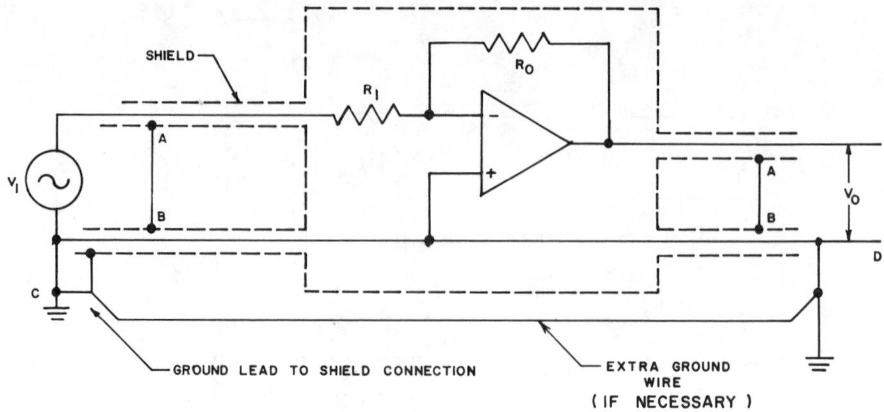

Figure 3-111. Shielded inverting op-amp circuit.

always use No. 10 tin cans. They are easy to solder, and you can cover the open top with copper window screen.

The application to an inverting op-amp circuit is very similar to the previous case. The ground lead is connected to the op-amp shield, but only at the signal source itself. This is shown in Figure 3-111. Note that in this circuit, the (-) op-amp input is *not* connected in any way to the shield, whereas the (+) input is tied to the shield *only* at the signal source. Connecting the shield to ground at more than one point is all right in this case because of the extra ground wire, but it should still be avoided if possible.

One or two points on recorders and we can go on to other things; in most cases you will be lucky enough to have a modern recorder with a three-wire cord and a three-input terminal connection. In this case, the outputs of the op-amp go to the two inputs, and the ground wire from point *C* is connected "naturally enough" to the ground terminal. This suggests that three-wire, three-input terminal recorders are "no problem." The next question is, what about two-wire, two-input terminal recorders?

This is a tough one, and there are no simple answers, since all of those "old-style" recorders are different. In most cases, one terminal, usually the black one, is marked with a "ground" symbol to show that it is connected to the chassis of the unit. NOTE, IT IS NOT, REPEAT *NOT,* CONNECTED TO POWER GROUND BECAUSE YOU HAVE ONLY A TWO-WIRE CORD. IN THIS CASE CHASSIS GROUND IS NOT POWER GROUND. If you meet this situation you have to be a little sneaky by first connecting the two op-amp inputs to the recorder input terminals and leaving the "ground" wire floating loose. If that gives you a lot of noise on the trace you will have to try connecting a wire from C to first one and then the other of the input terminals. The point at which you get the best signal and the least noise is the place to stop. If you do get a lot of 60 Hz pickup, you may have to build the 60 Hz noise filter circuit we discussed earlier in Chapter 1 and put that in the line. Unfortunately, with old recorders, there is no simple solution.

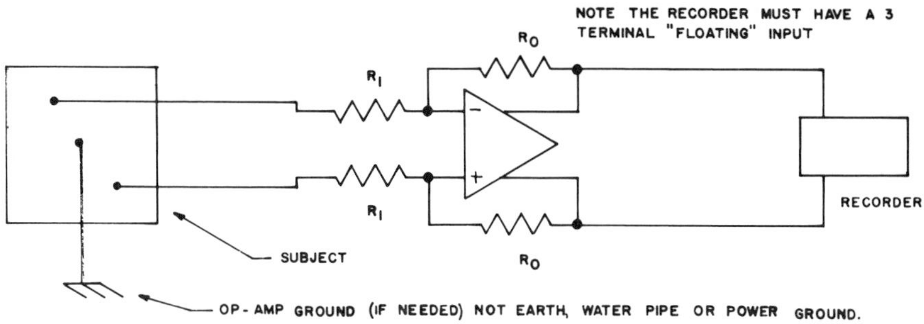

Figure 3-112. Measuring body potentials.

One last point to remember. If you are working with low signal levels and long (3 meters or more) distances between the sensor and the recorder you have to *make sure* that the ground at sensor *C* is at the same potential as that at the recorder, point *D.* Even in new buildings with three-wire outlets, the neutrals are at different potentials and the grounds are connected only by dirty or corroded metal conduit. To be safe, connect all grounds together with No. 14 wire and hook that wire to a water pipe!

That's about all we care to say on this point except for a suggestion about measuring human body potentials (more will be said about biometrics in Chapter 4). The body is a good antenna, and one is tempted to ground off pickup voltages to a water pipe.* Don't do it! Usually it makes things worse, and it may create a danger to the subject. The best thing might be to place the electrodes symmetrically about a center point and use it as an op-amp *circuit ground,* not an earth ground. This is shown in Figure 3-112. In this case we assume you have a typical "medical" op-amp with two outputs. If you don't have one, just use the circuit of Figure 2-17, taking care not to injure your subject.

CONVERTING SINGLE-ENDED SIGNALS TO DIFFERENTIAL OUTPUT

Sometimes you may have a signal from a so-called *single-ended* or *grounded* device. This just means that your signal is referenced to ground. The problem with such signals is noise pickup during transmission (we discussed that in Chapter 2). The way to fix this is to change the signal into a differential form with an op-amp. A typical circuit for doing this is shown in Figure 3-113. Of course, not all op-amps have two outputs, but they *are* available, for example, from Burr-Brown (see Appendix).

Note in the circuit of Figure 3-113 that now we have an output that is *not* referenced to ground. You might wonder why the ground symbol is between V_{01} and V_{02}. Well, V_{01} is just the V_0 that you would get with respect to ground. V_{02}

*The important point here is that working with live subjects, particularly human beings, is a special situation in which the usual rules of grounding do not apply. When working with people or animals it is best not to try to ground off any stray signals that appear on the subject.

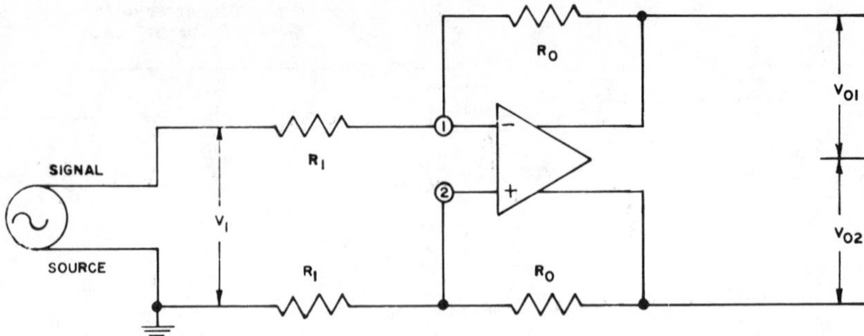

Figure 3-113. Circuit for converting single-ended signals to differential signals.

is equal to V_{01} but opposite in sign, so the voltage between the two op-amp outputs is $2V_0 = -2\,(R_0/R_1)\,V_1$.

The important point is that with a differential output, you can transmit your signal over long wires and then cancel out noise pickup with a differential input op-amp. In case you are interested, the input impedance of this circuit is $2R_1$.

APPLICATIONS OF OP-AMPS IN COMMUNICATIONS SYSTEMS

When we say *communications systems,* we don't mean AT & T or citizens band transmission (that's not in our line in this book). Let's suppose, for example, that you would like to monitor an animal's temperature as it walks about. Ideally, you would like to do this without using wires so the animal can be free of restraint. This leads us to the use of *telemetry* via radio.

Transmitting Our Signal

Let's assume that the important variable to be measured – temperature, blood pressure, or whatever – can be obtained as a voltage and that we want to transmit this voltage signal through the air. The signal must be AC because we don't know of any way to transmit DC by radio.

Suppose that we want to transmit a signal of, say, 1000 Hz. The wavelength, λ, will be determined by $\lambda f = c$ where c is the speed of light, or 3×10^8 meters per second. On our assumption that $f = 10^3$ Hz, then $\lambda = 3 \times 10^5$ meters. This presents our first problem. It turns out that an efficient antenna must have a length $L = \lambda/4$, which in this case would be 75 kilometers or about 50 miles. Obviously this is not the way to go! So let's raise our sights to a carrier wave of 3000 kHz or 3 MHz. Then $\lambda = 100$ meters, and the length of our antenna, L, will be 25 meters, or roughly 75 feet. This is possible with a few tricks, but suppose we go to 30 MHz. This brings the length of our antenna down to about 7 feet, which we can handle easily, but

"Your FCC license, please."

now we run into another problem: high frequency signals travel in straight lines and are severely attenuated by things like mountains. This problem is negligible at 3 MHz or below, but at 30 MHz, it can be troublesome. For this reason, we will stick to below 3 MHz which unfortunately includes the standard Am radio broadcast band. If you are an EE, it is obvious that the AM radio broadcast band is that frequency range for which the antenna length isn't too long and the radio waves aren't too directional. After all, the important point is to "broadcast" the sponsor's message to *all* nearby consumers as easily as possible. If we are careful, the FCC will stay away from our door. We will discuss the FCC problem in more detail later (p. 213).

We pick a *radio carrier frequency* (RF) of exactly 1 MHz. How do we get this carrier wave of 1 MHz to carry our audio signal? We do it by mixing the audio and RF signals to get a *modulated carrier* (Figure 3-114). The *mixer* is a device that takes the 1 MHz carrier and modulates its amplitude with the low frequency audio signal. Thus the audio signal is the envelope (look at the definition in the lowest part of Figure 3-115); and the carrier signal oscillates inside the envelope. Simple diode mixer circuits will be discussed later (p. 213); for now, let's just assume that mixing can be done. Figure 3-115 shows the 1 MHz RF carrier wave, the 1 kHz audio signal, and the modulated carrier wave that we feed to an antenna.

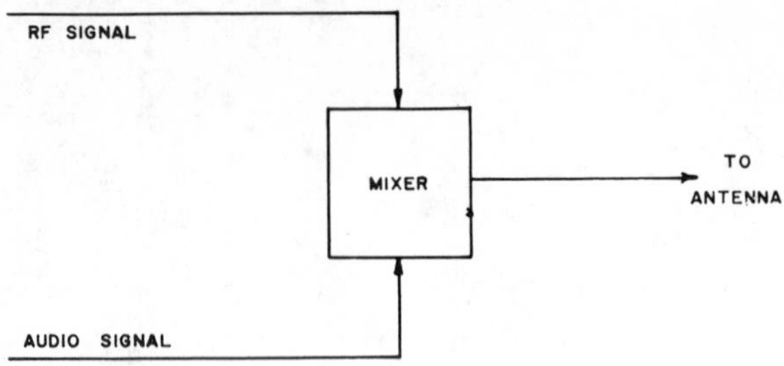

Figure 3-114. AM modulation system.

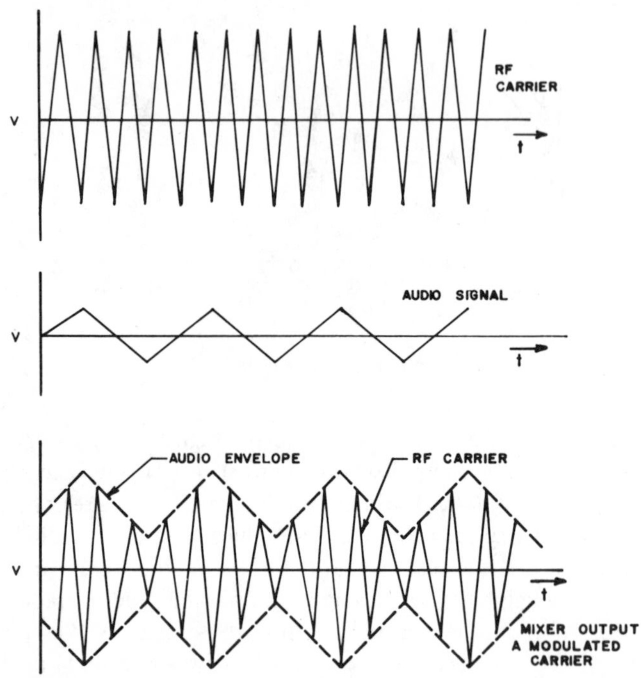

Figure 3-115. AM modulation signals.

Receiving Our Signal

Having modulated our signal one way or the other, what do we do next? Well, we can send our signal to an antenna. From the antenna the signal is broadcast to a breathlessly waiting world (do you feel like Marconi?). Now we want to *receive* the signal. First we must detect the carrier and then recover the audio signal. We must realize, however, that we are *not* alone in the use of carriers with AM or pulse-width modulation; there may be other carriers around. Assume that we are operating at a carrier frequency of 1000 kHz and other people are using 800 kHz, 900 kHz, and so forth. In our receiver, we need a way to pick up *our* 1000 kHz carrier. We do it with a *resonant circuit.*

RESONANT CIRCUITS

What, you ask, is a resonant circuit? Rather than just saying that it is a circuit having capacitance and inductance, and then referring you to Chapter 1, let's use an analogy. Consider a flat spring that is vibrating. The spring has *two* ways to "store" energy. At the end points of its arc, the energy is *all potential;* the spring is fully deflected, and, for an infinitesimal instant, is stationary. The velocity at that point is zero, so its kinetic energy must be zero. At the center of its arc, the deflection is zero. The potential energy stored in deflection of the spring must also be zero. The energy then is *all kinetic.* The point is that the energy is *transformed back and forth* from kinetic to potential as the spring vibrates. The *rate* of vibration is set by the condition that this energy balance is achieved.

An *inductance-capacitance* (*LC*) circuit stores energy in an inductor (L) as current and in a capacitor (C) as charge. The oscillation frequency is *just* that required for all the charge in the capacitor to be converted into current in the inductor, and vice versa. The energy would oscillate back and forth indefinitely if there were no losses due to resistance.

Figure 3-116 shows an *LC* resonant circuit that might be used in a radio receiver to pick up a 1 MHz carrier signal. The oscillator's natural frequency is $f_0 = 0.159/\sqrt{LC}$, L and C are in henrys (H) and farads (F), respectively. If we choose $f_0 = 1$ MHz, then

$$LC = \frac{(0.159)^2}{10^{12}} = 2.5 \times 10^{-14}$$

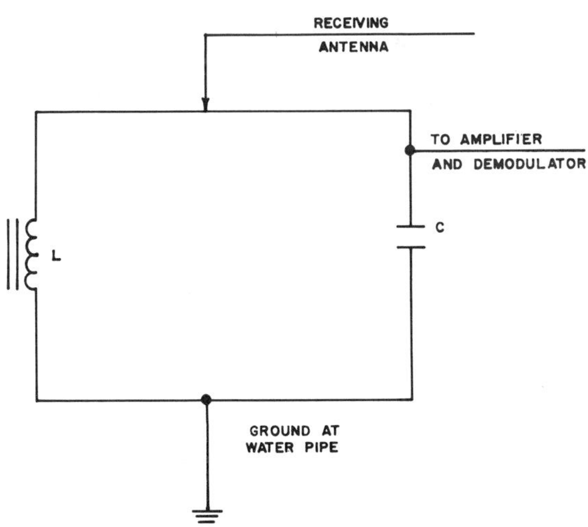

Figure 3-116. Resonant circuit.

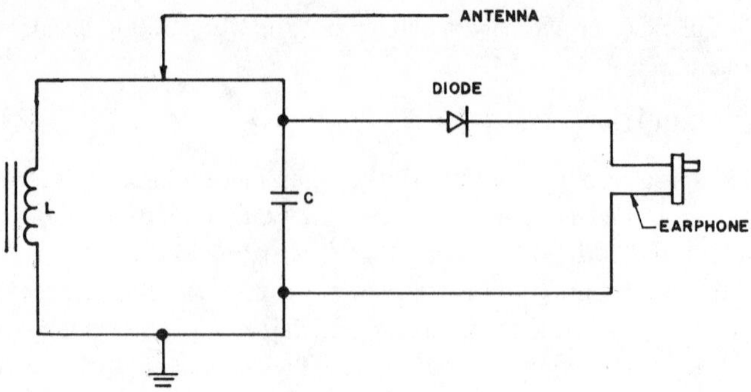

Figure 3-117. *LC* filter receiver with a diode detector.

If we choose $C = 100$ pF, then L comes out to be 2.5×10^{-4} H, or 0.25 mH. The point to remember here is that, when we set up this circuit for a resonant frequency of 1 MHz, we are arranging it to "stop" that frequency so that the 1 MHz signals will go to the amplifier and demodulator. Other frequencies will see the resonant circuit as a very low resistance and be short circuited to ground.

Now let's put in a diode as shown in Figure 3-117. The diode rectifies the RF by chopping off the bottom half of the waveform. The earphone still can't respond to the RF, but there is a net audio variation in the RF envelope to which it can respond. The result is a signal that we can hear, and hopefully it is similar to one we started with.

What if the signal isn't strong enough? Then we use an op-amp amplifier; you know *all* about that already. You might wonder what the op-amp does with a modulated 1 MHz signal. Few op-amps have much gain at 1 MHz. However, after detection with the diode, we could amplify the audio signal and that is what you want to hear anyway.

MODULATION AND SIDE-BANDS

Now that we have gone over the descriptive part, we can settle down to the nitty-gritty of *how to do it!* Before discussing modulators, we should make sure that we know what *modulation* is and clarify the electrical engineering jargon in this area. Consider first a diode having a curve as shown back in Figure 1-21. The *I-V* relation is $I = k\,V^2$ in the forward direction, and we shall assume that reverse-bias leakage is zero. A circuit including such a diode is shown in Figure 3-118. Note that p and q are symbols for $2\pi f_1$ and $2\pi f_2$, where f_1 and f_2 are the carrier and audio frequencies involved. The current I is given by the square of the sum of all three voltages: $I = k\,V^2 = k\,(V_0 + V_1 \cos pt + V_2 \cos qt)^2$. If we expand this formula and use identities, we get the *(UGH!)* result

$$I = k\left[V_0^2 + \frac{V_1^2}{2} + \frac{V_1^2}{2}\cos 2pt + \frac{V_2^2}{2} + \frac{V_2^2}{2}\cos 2qt + 2V_0V_1\cos pt + 2V_0V_1\cos qt + V_1V_2\cos(p+q)t + V_1V_2\cos(p-q)t\right]$$

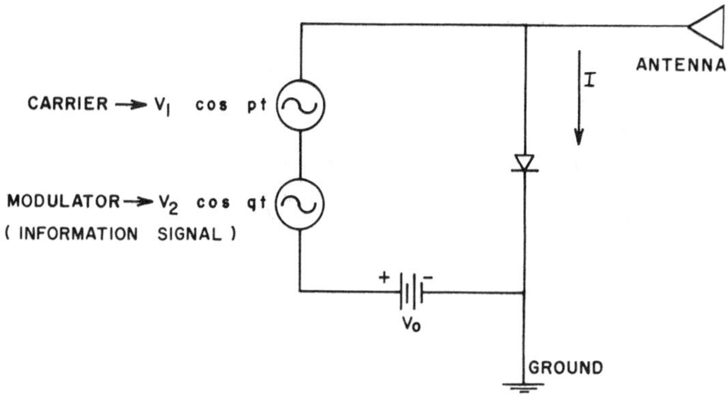

Figure 3-118. Diode mixer circuit.

This indicates that the modulated current I has components of the form cos 2 pt and cos 2 qt, as well as side-band frequencies cos $(p + q)\ t$ and cos $(p - q)\ t$. In this diode modulator we have created signals of frequency (p - q) and (p + q); since they are either higher or lower in frequency than the original carrier frequency (p) we speak of upper and lower side-bands. Note that the side-bands contain all the information, the carrier (p) and the modulator (q).

Let's introduce a filter that removes all the signals except those having the frequencies cos qt, cos $(p - q)\ t$, and cos $(p + q)t$. The remaining signal is our filtered, modulated carrier that we can put into an antenna. You should note that we have *three* frequencies: q, $p - q$, and $p + q$. These are the fundamental frequency and the upper and lower side-band frequencies. For some applications, we need only to transmit one of the side-bands – say, cos $(p + q)\ t$ – and filter out the other frequency. This is called *single side-band* operation. Each side band has the q frequency, which is *all* we want to transmit as far as information is concerned.

Applied Telemetry

Having waded through all this stuff about radio, now how do we do telemetry? We begin by admitting that we won't use the above techniques directly; it would be too complex and costly. Why, then, waste time on it? The answer is that the above ideas and notations are used in telemetry. You have to know the rules to play the game!

The first rule is an important one. The Federal Communications Commission (Washington, DC 20554) controls the use of all radio signals in the United States. Upon application, they will send you their bewildering list of regulations. The list is enough to discourage Marconi himself, but as usual there are ways to squirm out from under the thumb of "Big Brother."

The first and "best" way is to operate with those frequencies and signal strengths authorized by the FCC. The problem with this solution is cost, both of time and money. So the next best way is to stay with low strength signals so that "The Man" doesn't know you exist. The problem here is that your range is limited because of the weak signal strength, and there isn't any easy way to beat that without getting in trouble with the FCC. The last and hardest way is to get smart: if you can operate at a frequency that the FCC doesn't monitor, you can do what you like. The problem here is that many electrical engineers (as well as other people with perhaps less honorable intentions) are looking for that solution, too. We suggest, therefore, that you stick with the first two solutions suggested above.

The second rule is to get Xerox copies of C. L. Strong's article, "Little Radio Transmitters for Short Range Telemetry," in *Scientific American* for March 1968. That article recommends a book called *Biometrical Telemetry* by Stuart McKay that you should also obtain if you are further interested. His circuits use transistors; in some cases, these transistors can be replaced by op-amps to great advantage (we will discuss this below).

Having set forth some rules, we can begin the game. The first thing one does in telemetry is *tracking.* This means that in our example of monitoring an animal, we attach a transmitter to the critter that allows us to follow and locate it. The signals should be high frequency waves (provided the animal isn't allowed to stray too far) so we can use small antennas. There is no need to transmit continuously; a series of pulses will do and will conserve battery power.

The first circuit we show for doing this is a *neon-bulb pulse generator* (Figure 3-119). The operation of this circuit is simple. The capacitor charges through the resistor until the firing voltage of the neon bulb is reached. When the neon bulb fires, the capacitor discharges through the bulb until the capacitor voltage is too low to keep the neon bulb in the conducting state. At this point, the bulb goes out and the capacitor begins to charge again. The time between flashes is given by $T = RC$. If R is a thermistor that changes its resistance as a function of temperature, the rate of flashing will be a function of temperature.

We can modify this circuit to produce a burst of RF energy with each flash. A circuit for this purpose is shown in Figure 3-120. In this case, the capacitors and the inductor form a resonant circuit when the neon bulb fires. The resonant frequency, f, is given by

$$f = \frac{1}{2\pi\sqrt{L\,C_T}}$$

where

$$\frac{1}{C_T} = \frac{1}{C_1} + \frac{1}{C_2}$$

Each time the bulb fires, the circuit will oscillate a few times and then stop until the capacitor C_1 charges up again. The signals that are radiated by the inductor L in Figure 3-120 can be picked up by the cheapest hand-held, FM radio. This is strictly a short range circuit – 50 meters or so. If you want to transmit farther than that,

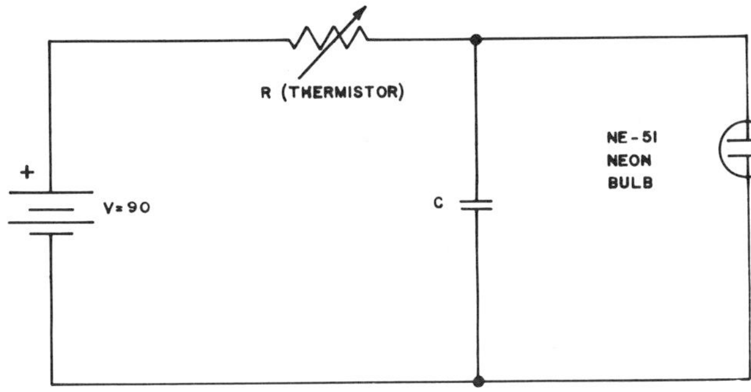

Figure 3-119. Neon-bulb pulse generator.

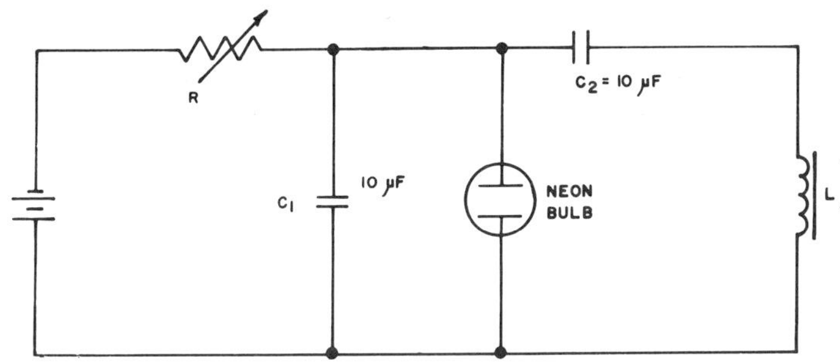

Figure 3-120. Neon pulse RF circuit.

you will have to read the section on AM radio. Each pulse is a burst of static. By counting pulses and turning the radio (FM antennas are directional) you can measure an animal's temperature and keep track of its location; you can't beat that for $12.00.

The only problem with this circuit is that a 90 volt battery is required. Since 90 volt batteries aren't easy to put on an animal, we can use a *transistor oscillator circuit,* as shown in Figure 3-121. In this circuit, note that the base of the transistor is forward biased through the thermistor. This allows current to flow from the collector through the emitter to the inductor. The current is delivered to the coil asymmetrically because the ratio of turns in the coil is 100 to 150, and this causes the resonant circuit to oscillate. During these oscillations, the emitter coil tap goes negative in part of the cycle. This in turn puts a forward bias on the emitter-base junction. The 10 μF capacitor then charges negatively, which stops the collector-to-emitter current. The oscillations then die away until the 10 μF capacitor charges up again through the thermistor. (The cheap FM portable radio should work here, too, as a detector. Try it!) The time τ between the pulse packages (Figure 3-122) is determined by the thermistor and the 10 μF capacitor.

The problem with the transistor circuit of Figure 3-121 is that the transistor parameters change with temperature. The best way to avoid this is to use an *op-amp*

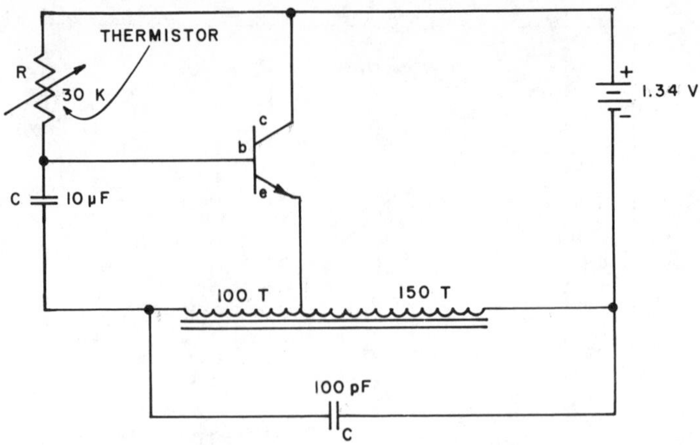

Figure 3-121. Hartley oscillator circuit.

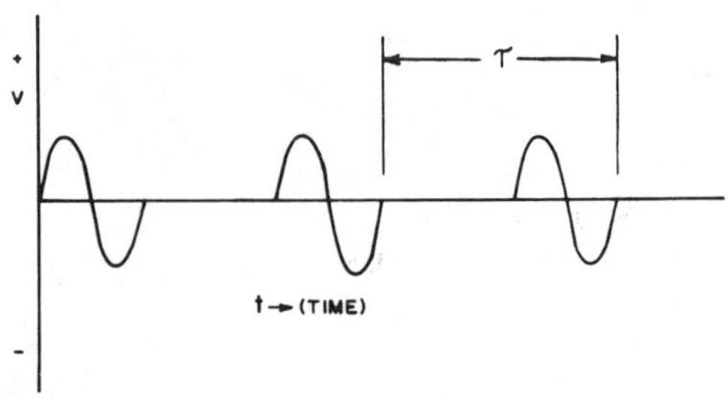

Figure 3-122. Oscillator output signal.

oscillator, as shown in Figure 3-123. This is a good circuit to set up and play with. We don't guarantee that it will work for every problem, but you should be able to set it up and find the correct values for the components.

The operation goes as follows: The R_1 C_1 circuit charges up and drives the (+) input of the op-amp. The output (V_0) comes through Z_2 and R_2 C_2 until Q_1 turns on and short circuits the charge on C_1 to zero. This brings V_0 to zero and effectively pulses the oscillator circuit on the right. It oscillates until C_3 charges up and stops the current.

The disadvantage of this circuit is the need for a ±15 volt supply,* the advantage is that the time between pulses is controlled by R_1. Temperature changes may affect Q_1 and Q_2, but these changes will not affect the time between pulses, which is the important effect you want to transmit.

You might wonder about Z_1. This is a pair of zener diodes, back-to-back, to keep the op-amp from doing what is called *latch-up.* Latching up can occur when

*Potential users of this circuit will be happy to know that, as of 1980, there are 3 volt op-amps available from our friends at Burr-Brown.

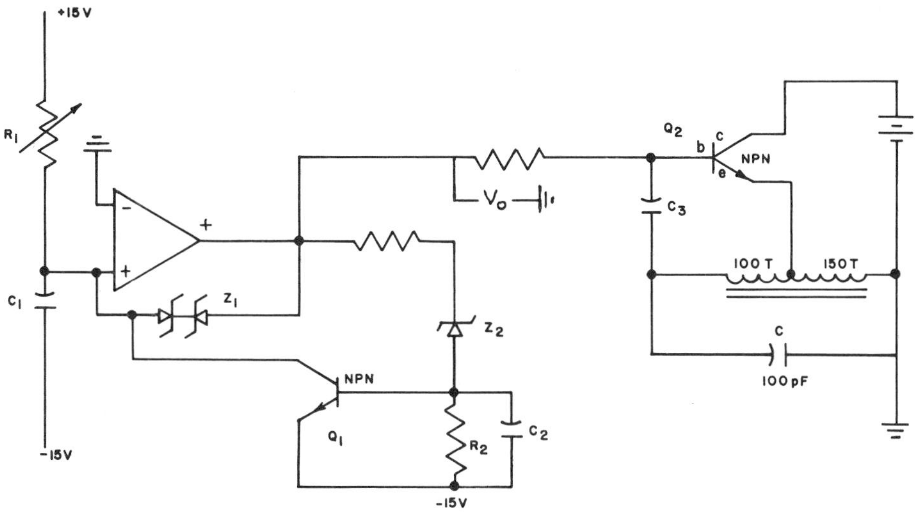

Figure 3-123. Op-amp pulsed oscillator circuit.

there is positive feedback. It forces the op-amp into saturation and holds it there; that is bad for op-amps.

This is about all we care to say about telemetry. If you are rich, you can buy ready-made telemetry gear. Details of many experiments are given by MacKay in his book. He uses transistor oscillators of all sorts; naturally, we suggest op-amps for these applications.

Heterodyning

MacKay mentions heterodyning in his book, and that concept is worth a page or two. Suppose you have an electrical signal containing three separate frequencies, f_1, f_2, and f_3, of different amplitudes, A_1, A_2, and A_3. Let's assume that $f_1 = 10$ kHz, $f_2 = 11$ kHz, and $f_3 = 12$ kHz and that we are dealing only with sine or cosine signals. (This doesn't mean that you can't also handle square or saw-tooth waves. We will describe the case for sine or cosine waves because the math for these is simpler.) Our heterodyne test system consists of an op-amp slot filter designed to pass a narrow band of frequencies centered at, say, 500 Hz. To this filter, we add a frequency generator and a diode mixer (which was discussed earlier). The incoming signal is assumed to be of the form $A_1 \cos 2\pi f_1 t$; the frequency generator output is of the form $A_0 \cos 2\pi f_0 t$. The analyzer system is shown in Figure 3-124.

For operation, the signal generator is set at $f_0 = 9.5$ kHz. The mixer output contains frequencies $f_1 - f_0, f_1 + f_0, 2f_0, 2f_1, f_0$, and f_1. Notice that all of these frequencies are much higher than $f_1 - f_0$ and are therefore rejected by the slot filter, which is set for 500 Hz. We measure the output signal, $A_1 A_0 \cos 2\pi (f_1 - f_0)$.

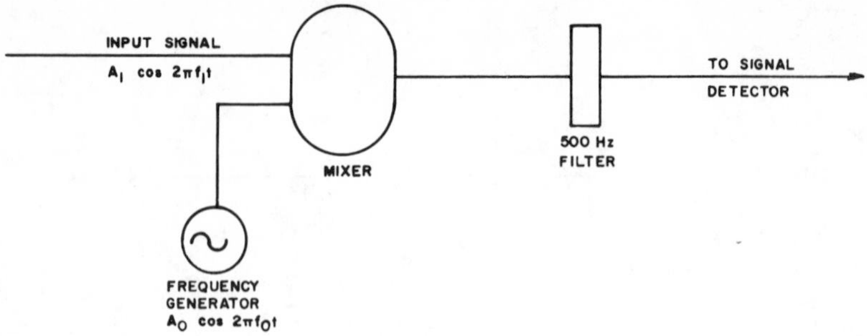

Figure 3-124. Heterodyne frequency analyzer.

Then we change the signal generator to f_0 = 10.5 kHz and measure the amplitude of f_2, and so on. (Note that A_0 is known, so we can find A_1 quite easily.)

How do we know what frequency to pick? The answer is that in practice, we don't! We scan the generator over the frequency range and observe the filter output. If we see an increase in output, we know that there must be a signal frequency 500 Hz above or below the generator frequency. This is how the frequency analyzers work, and it justifies our detailed discussion of sine wave generators.

There are two other important applications of heterodyning. One arises when we want to amplify a high frequency signal, say, 10 MHz. Our op-amp won't handle it, but if we mix the 10 MHz signal with a 9.98 MHz signal and take $f_1 - f_0$, the result is a 20 kHz signal (this is an old trick in AM radio). The 20 kHz signal is called an *intermediate frequency* (IF), and our op-amp amplifier works just fine.

Another application of heterodyning is in the reception of telemetry signals that involve a change in frequency. Suppose you have a pulse rate of 10 pulses per second (pps) that changes by one pulse per second when something happens. A change of 1 pulse in 10 is hard to detect, but if you mix the 10 pps signal with a 4 pps signal, the output changes from 6 pps to 5 pps as the signal changes from 10 pps to 9 pps. We all know that 1 in 6 is much easier to detect than 1 in 10.

You might ask, "how do we get accurate oscillators for all these applications?" One way is to build them and then calibrate them against the frequencies broadcast by radio station WWV. WWV can be picked up on any short-wave receiver. It broadcasts at standard frequencies of 440 Hz and 2.5, 5, 10, 15, 20, and 25 MHz. This station also gives you the correct time (the nation's tax dollars are well spent by the Bureau of Standards).

Magnetic Tape Recording for Those Poor Scientists Who Don't Have a Grant – Or Japan to the Rescue

Suppose you are a geology or hydrology student and you want to record data in the field without a big and expensive commercial system. Your data are available as electrical signals of 1 or 2 volts and the signal frequency is somewhere between 0 and 500 Hz. (This doesn't mean only AC sine wave signals can be recorded; we refer

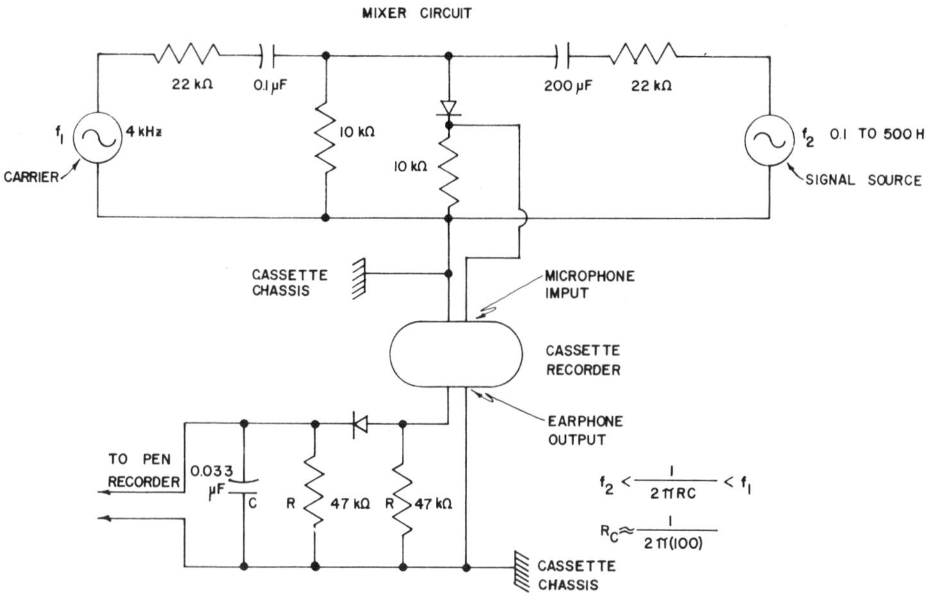

$$f_2 < \frac{1}{2\pi RC} < f_1$$

$$R_C \approx \frac{1}{2\pi(100)}$$

Figure 3-125. Cassette recorder input and output circuits.

only to the rate at which the signal varies with time.) Having the signal available, you go to your nearest discount house and buy a cheap cassette recorder. Don't buy one for "voice recording only"; what you want is a regular rock-and-roll "teenage" type – the inexpensive Japanese-made variety is adequate for the purpose. The recorder won't accept 10 Hz signals, so we must build an AM modulation system.

To begin with, you beg, borrow, build, or otherwise obtain a 4 kHz sine wave oscillator. A single op-amp oscillator will do the job, and you can reserve the begging or borrowing for later. The next thing you need is a cheap diode; anything with a PIV of 5 volts or less will do fine. The circuit is shown in Figure 3-125. The similarity to our AM modulation circuit of Figure 3-118 is *no* accident.

What happens here is simple enough. The signal you want to record is mixed with and modulates the 4 kHz carrier. The modulated carrier is fed into the recorder via the microphone jack. You turn on the recorder, set the level (turn off the automatic level control if you have one), and record away.

Now you traipse in from the field, pick the cactus spines or the burdocks out, and decide to put your data through a pen recorder or a computer. If you have a computer, you are out of our league: go find your own analog-to-digital converter. If you are still in the pen recorder or "peasant stage," we can help you.

Connect the output jack (it is marked "earphone") to the demodulation circuit and the recorder; then play back your data. If the output isn't strong enough to drive your recorder, put in an op-amp amplifier, and away you go. How else can you record so much data for so little money?

OP-AMP MODULATORS

One last item might be about the application of op-amps to modulation. Well, suppose we take our audio signal source, $V_2 \cos qt$, and our carrier source, $V_1 \cos pt$, and use them to drive an op-amp as shown in Figure 3-126. Assuming for the moment that $R_1 = R_0$ we can see that whenever the curves cross, $V_0 = 0$ because of common mode rejection. Conversely, when the two signals have the maximum values at the same time (and opposite signs), the output will be at its maximum value. What we have is an *adder circuit* that gives us $V_1 \cos pt - V_2 \cos qt$, rather than $V_1 V_2 \cos (p + q)\, t$, as will be the case with the diode modulator shown in Figure 3-118.

This circuit has some advantages – including having an easy mathematical analysis – which we can see by adding some zener diodes to prevent saturation and making $R_0 = 1000\, R_1$. Now the output goes to the voltage set by the diodes (5 volts) whenever either of the input signals is above about 0.05 volt. This makes the output exquisitely sensitive to the *difference* between the two voltages. The output becomes a sequence of square waves whose duration is controlled by the modulation signal, $V_2 \cos qt$. Figure 3-127 shows the input signals and the output signal. The pulse length is modulated by the $V_2 \cos qt$ signal. We have built a *pulse-width modu-*

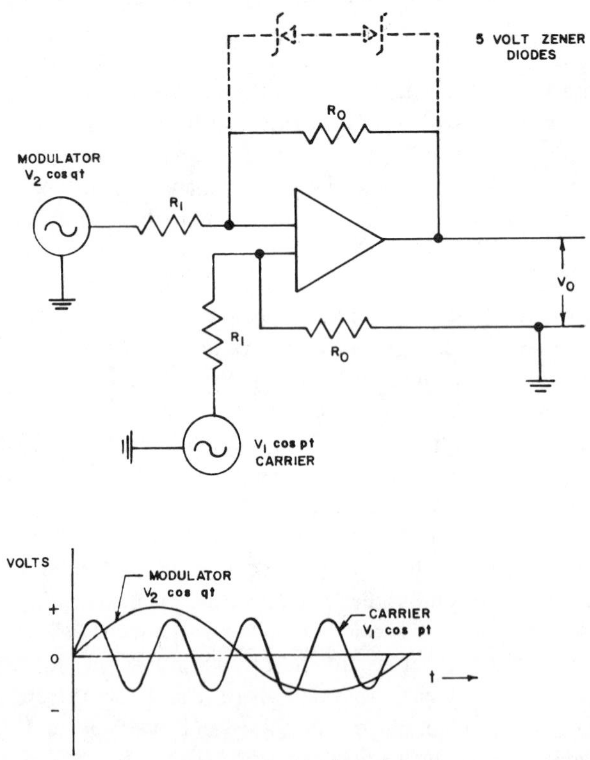

Figure 3-126. Op-amp modulator circuit and input signals.

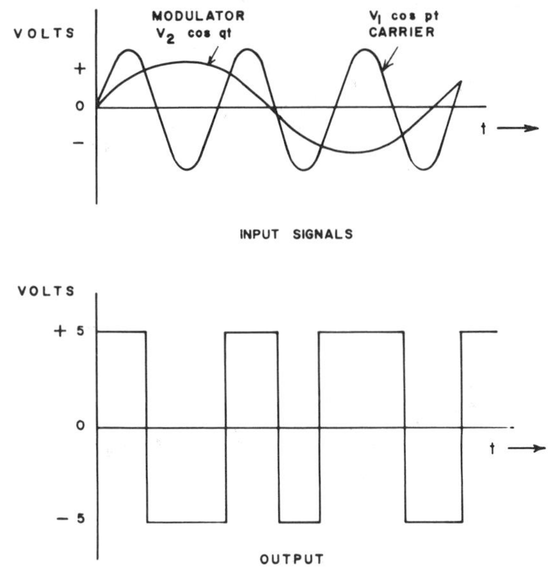

Figure 3-127. Op-amp modulator input and output signals.

lator. When you get interested in digital circuits or advanced telemetry, you will want to build circuits of this type. We do REPEAT that this is a pulse-width signal generator, not an AM modulator. To transmit data by this technique you are going to have to take a course in digital data transmission. Sorry, we can't teach it all.

POSTSCRIPT

In this chapter, we have provided you with a multitude of op-amp circuits, some of which, we hope, you will find valuable. The next chapter deals with some specific applications of op-amp circuits – namely, those in the biomedical field. However, even if you're a geologist, you should read the next chapter; you might pick up some ideas that can be applied in your own field. (That's often how great ideas are born!) The circuitry for measuring EEG or ECG signals, for example, might be of interest to a seismologist. We supply the tools; you supply the imagination – as they say on Madison Avenue, "Think Big."

TEST YOURSELF PROBLEMS AND SUGGESTED EXPERIMENTS

Test Yourself Problems

1. Discuss the use of transistors to boost the current from an op-amp. Why do we have to use PNP, NPN transistors to accomplish this?
2. Discuss the differences and similarities between zener diodes and regular diodes.

Draw voltage versus current curves for both kinds of diode and show the regions of operation.

3. How are zener diodes used as voltage regulators? Can you design a circuit to provide a regulated 12 volts from a voltage source that swings from 12.7 to 15 volts? How do you provide for a change of temperature and its effect on the zener voltage?
4. Design a gate circuit that switches from 0 output to (+) output when the input signal goes to (-) 3 volts.
5. Why do we limit the output voltage of an op-amp to some level below the normal 10 volt maximum?
6. What do we mean by saying that a real diode in an op-amp feedback loop can (if we are clever) act like an "ideal" diode?
7. Draw up some simple op-amp circuits that might be used for addition and subtraction of electrical signals.
8. Draw an op-amp integration circuit and show what the output would be for a: sine wave, steady DC signal, DC signal plus noise.
9. Design an op-amp differentiator circuit and show what the output looks like for: a sine wave, a cosine wave, and a square wave input.
10. Discuss the application of the "leaky integrator" circuit to the suppression of noise that might come along with a low-frequency signal. How and why is the noise suppressed?
11. What are the advantages of active (op-amp) filters over the simple RC filters discussed earlier? Make sure you mention at least two advantages.
12. What is a twin-tee filter used for?
13. Draw a phase-shift oscillator circuit (copy from the text if necessary) and discuss *why* the system oscillates but does not saturate.
14. Repeat question 13 for the Wein bridge oscillator.
15. What are square waves used for? How can they test both the AC and the DC response of a system? How are they used for calibration?
16. We say that a programmable ramp generator can be built using diodes. Explain how this is done.
17. Discuss how square waves are used to calibrate and test apparatus. What do we mean when we say that a square wave tests the AC and the DC characteristics of a system, i.e., recorder or amplifier?
18. Given a multivibrator circuit, discuss how and why it oscillates.
19. Repeat question 18 for the monostable and bistable.
20. Given a Schmidt trigger circuit, can you explain how and why it switches at a given voltage? Think before you say "yes"; it isn't so simple.
21. Define (in your own words) the terms *ripple, regulation, short-circuit protection, current limiting, crowbar protection,* and *outputs* with and without respect to ground.
22. What is an SCR? Can you draw a voltage current diagram for an SCR? How do you turn off an SCR? What happens if you put the wrong bias on the "gate"? How can you prevent that from happening?

23. Discuss how an op-amp might be used in conjunction with a transistor to regulate an unregulated power supply.
24. For effective transistor design we need four parameters: V_{ceo}, I_c, P_t, h_{fe}. Can you define what these symbols stand for?
25. Define the terms *constant current* and *constant voltage supply.* How can a constant current supply be dangerous if it does not have a voltage limiter? If someone offered to sell you a constant current, constant voltage supply (both at the same time), what would your response be?
26. What are the advantages and disadvantages of integrated circuits versus discrete op-amps? Which would you use for: (a) high frequency, (b) great accuracy, (c) resistance to electrical damage, (d) thermal stability?
27. With integrated circuit op-amps, we must supply external compensation. Why is this and what does this compensation do?
28. Chopper op-amps are used to reduce thermal drift. Draw a typical chopper op-amp circuit and explain how it works. Here again you have to think. It is not simple.
29. Chopping of signals is a common noise reduction technique. Can you explain why it works and how? Are all types of noise subject to reduction by chopping? If not, which are and which are not?
30. Synchronous detection is another noise rejection technique. Can you draw a block diagram of such a circuit and explain how it works?

Suggested Experiments

A. GATES

Construct the limiting, or negative gate, circuit of Figure 3-128 (p. 131). Build the circuit *without* the optional adjustment and draw the circuit with $R_1 = 10\ \text{k}\Omega$, $R_0 = 22\ \text{k}\Omega$, and $D_1 = D_2$.

1. Using the function generator as an input, sketch the input and output waveforms on Figure 3-128.

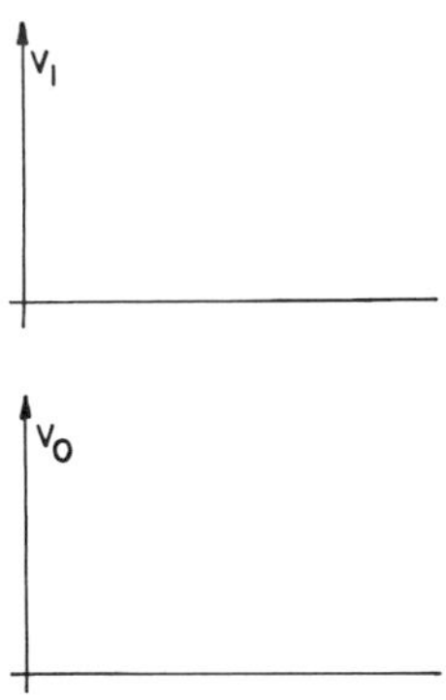

Figure 3-128.

2. Mark the part of the input that is being passed, undistorted, to the output.
   a. Is the amplifier inverting the input in addition to clipping it?
   b. What is the clipping level?
3. Add the optional adjustment with R_G = 22 kΩ. Start with V_G at zero volts and explain what happens as V_G is varied.
4. For V_G = 1 volt and V_1 = 2 volt peak, sketch the input and output waveforms on Figure 3-129.

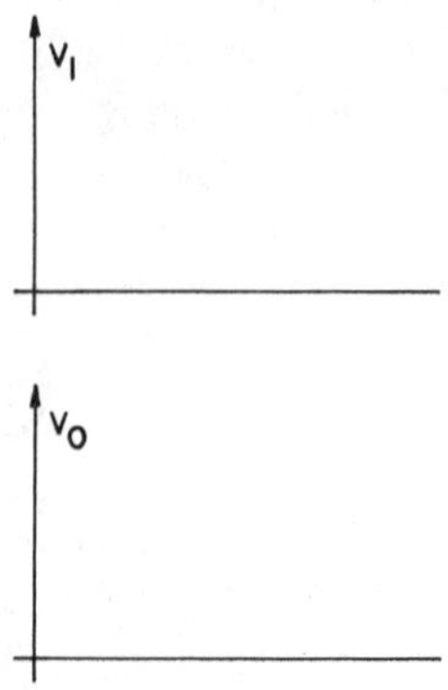

Figure 3-129.

Could this circuit be used to prevent large, unwanted voltages from overdriving sensitive circuitry?

B. FILTERS

Assemble the filters of Figure 3-130. Using both traces on the scope, plot the 5 points indicated on Figure 3-131. V_1 should be held at a constant peak voltage, as monitored by one channel of the scope.

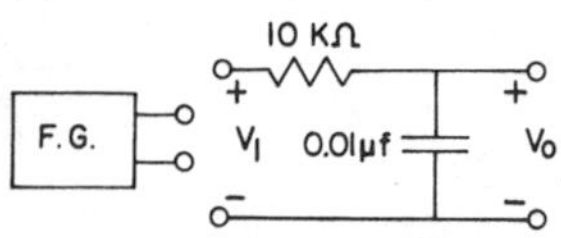

Figure 3-130.

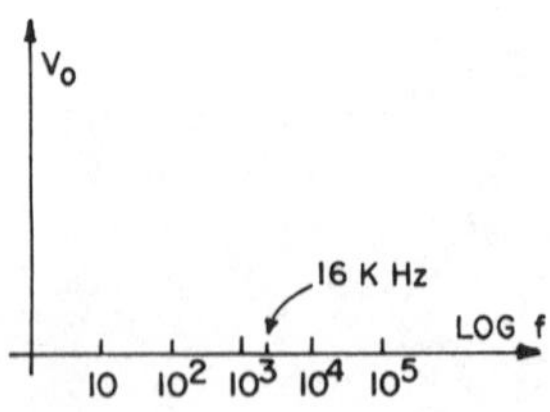

Figure 3-131.

Build the circuit of Figure 3-132. Using Figure 3-131 again, plot the values of V_0 at the 5 frequencies indicated.

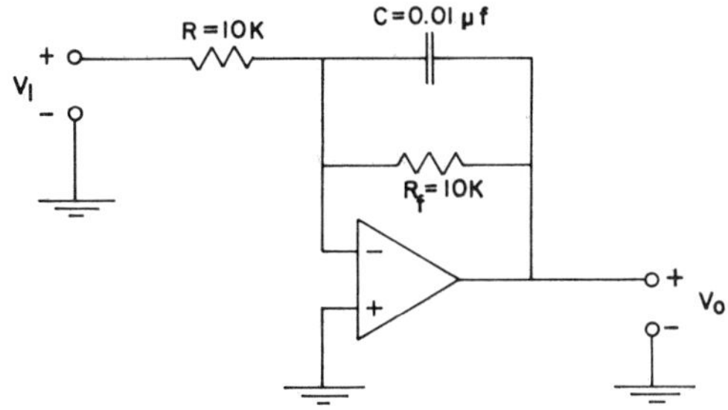

Figure 3-132.

1. How do the two plots compare?
2. What is the advantage of Figure 3-130 over Figure 3-132?
3. What is the advantage of Figure 3-132 over Figure 3-130?
4. What is the input and output impedance of Figure 3-130?
5. What is the input and output impedance of Figure 3-132?

C. OP-AMP OSCILLATORS

Whip up the circuit of Figure 3-61, and draw the schematic.

1. What frequency does the circuit oscillate at?
2. How could you change the frequency of oscillation?

D. SCRs

1. Draw the current versus voltage curve for an SCR (Figure 3-133). Indicate which part of the curve represents the forward- and reverse-biased regions.

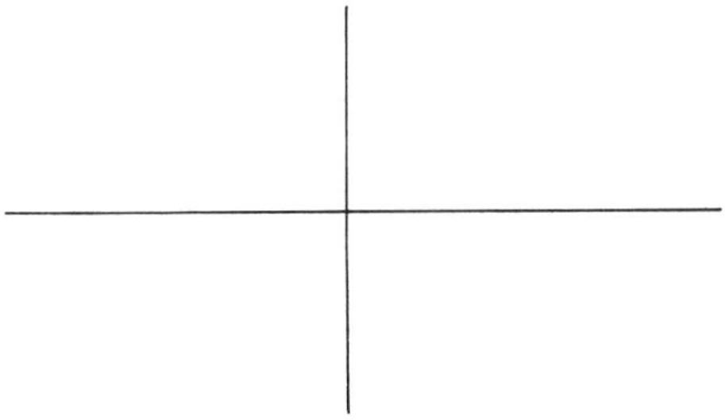

Figure 3-133.

2. For the diagrams of Figure 3-134, draw an electrical model (of each situation) in the boxes. Assume that the SCRs are in a circuit with the voltage polarity, across the SCRs, being given by the + and – signs. (Hint: you should be interested in the resistance between points *A* and *B*.)

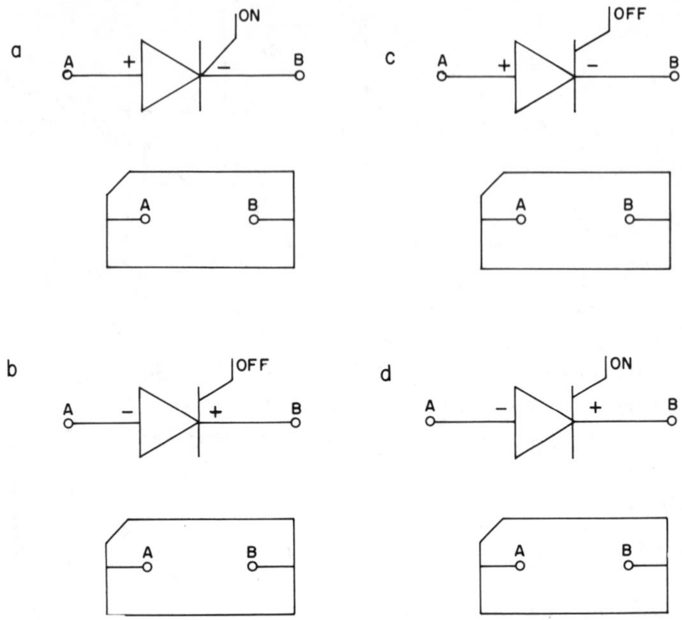

Figure 3-134.

3. Construct the circuit of Figure 3-135.

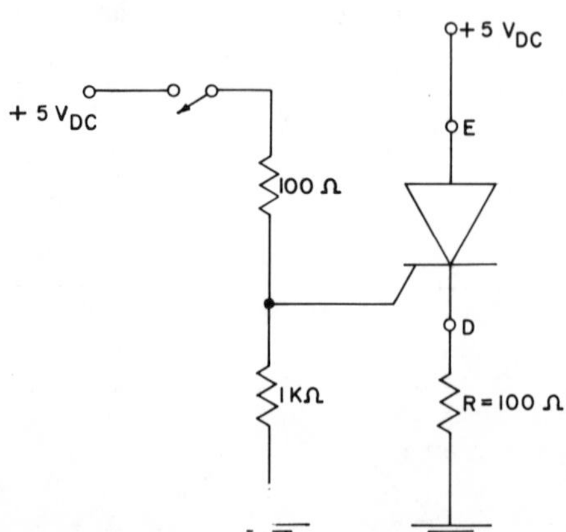

Figure 3-135.

Monitor the voltage from point *E* to ground with one channel and point *D* to ground with the other channel of the scope.

With the switch off:

a. What is the current through *R?*
b. What is the resistance between points *D* and *E?* (Hint: $R_{SCR} = V_{SCR}/I$)
c. Which model, in Figure 3-134, does this situation apply to?

With the switch on:

a. What is the current through *R?*
b. What is the resistance between points *D* and *E?*
c. Which model, in Figure 3-134, does this situation apply to?

E. WIRING OP-AMPS FROM SCRATCH

Using the book, pages 84 and 187, construct a unity gain follower for general applications. Draw a *complete* schematic.
What does compensation do for you?
Check your circuit to make sure it works.

F. RISE TIME, OVERSHOOT, SETTLING TIME

Using the same circuit you made in Experiment E (a unity gain follower), input a 1 kHz, 4 volt peak to peak, square wave signal. Adjust your scope until you can observe the rise time, etc. Draw a picture of the input and output waveforms. Approximately: What is the rise time? What is the overshoot? What is the settling time?
Indicate how these were measured.

G. SHIELDING

Obtain shielding and insert a wire, with banana plug terminators, through the shielding. The wire should be about the same length as the shield.

1. Plug the wire into a scope input, without grounding the shield, and observe the noise picked up by the wire. The wire is acting like an antenna.
2. Turn on a signal generator near the wire with the amplitude on maximum (make no connection to the signal generator). You should be able to induce the signal onto the wire (try observing this on the scope).
3. Now connect the shielding to the scope ground. Did you notice a reduction in the noise level?

H. INTERFACING

In the magazine *Machine Design* (*49,* (21), 1977, p. 22), a corrosion detector is discussed. This unit (consider it a black box) has an output voltage of 0.2 volt when the coolant rust inhibitors are working and a voltage of about 0.75 volt when the inhibitor is exhausted.

1. Design and construct the interface between the corrosion detector and an alarm buzzer (or light).
   a. Draw a model of the corrosion detector using a battery, one fixed resistor, and a variable resistor (rheostat).
   b. Draw a model of the buzzer (or light).
2. Since we really don't have any corrosion detectors in the lab, use a DC power supply (0-0.8 volt) to simulate the detector and just use a resistor to simulate the buzzer. When the voltage across the resistor is high, the buzzer is ON; when the voltage is approximately 0 volt, the buzzer is OFF.
   a. Draw a block diagram of your implementation.
   b. Draw a circuit diagram.

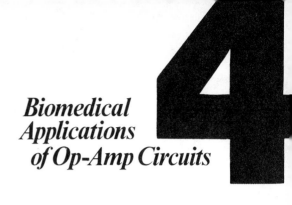

4 *Biomedical Applications of Op-Amp Circuits*

If you are interested in the biomedical area and you have read this far in our book, you should have all sorts of experiments in mind already. All we can do is provide a few hints and suggestions based on our own limited experience and the advice of various University of Arizona staff members.

We are going to assume that you have some background in this field. If however, you are as green in this area as we were a few years ago, you should write to Tektronix Incorporated (see the Appendix), and order a copy of their book, *Biophysical Measurements.* The price will be about $10.00, which is money well spent. If you live in a city that has a Tektronix representative (check the Yellow Pages), you can contact him. He might have copies to give away to potential customers. If he won't give you a copy, pay for it. (There is a limit to how cheap you can be.) Another good book in this area is that by L. Geddes and L. Baker, *Principles of Applied Biomedical Instrumentation,* New York: Wiley, 1968. You really should have both these books. The book by Geddes and Baker is useful as a survey of the field. The Tektronix book is more up-to-date in the electronics area.

Assuming that you have read or will read a book on this topic, we will first consider a common application of op-amps: getting electrical signals through more or less undamaged skin (either human or animal). Skin may have all sorts of nice properties, such as rejecting germs and protecting the animal inside, but electrically it is *very* bad. Skin acts like a resistor whose value changes with voltage, current, and temperature. There are also capacitive and inductive effects, as well as changes due to patient irritation, anger, and restraint.

The techniques for attaching electrodes to skin, for restraining animals, and so on are discussed in various books on biomedical instrumentation; our interest is what you do with the signals that you get. Typical signals might be related to skin resistance or potential level. Other signals might be due to changes in these levels when some stress is applied or motion is initiated.

A typical example of this type of study is the change in skin resistance due to some external stress. Here we can use a constant voltage or constant current to observe the change. Let's assume that the constant current generator discussed earlier (p. 82) is to be used between the two hands of a subject to generate a voltage proportionate to skin resistance (Figure 4-1).

The *first* consideration is that the current be limited to avoid shock to the patient. This means the *maximum* current must be 10 mA or less. You *must* make

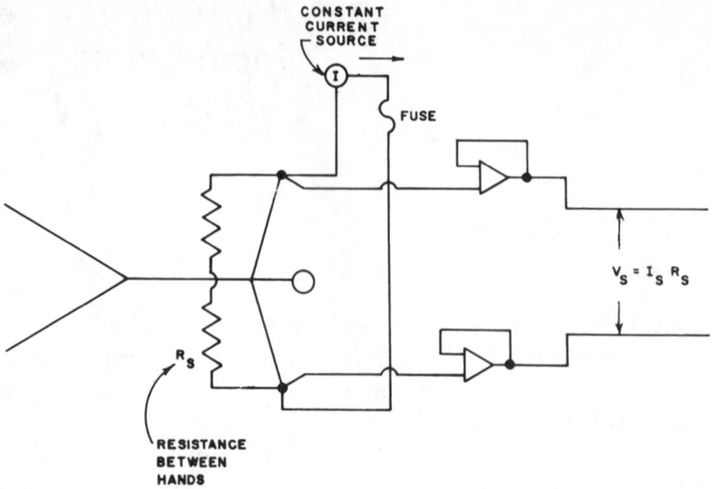

Figure 4-1. Measurement of skin resistance, left hand to right hand.

sure this level is not exceeded, even if something *or everything* goes wrong. The simplest way to make sure of this is to insert fuses. There are 10 mA fuses; there are even 2 mA fuses. However, all fuses require a finite time to interrupt a current. You may therefore want to use more exotic devices, such as *field effect diodes* or *current-limiting modules.* In essence, they pass signals at the 1 μA level and stop them at the 300 μA level. Since their availability is limited, we will go on to current-limiting modules. These are rather expensive packages that limit the voltage that will pass to the output. As the input voltage increases above about 30 volts, the output voltage is held at 30 volts. This limits the current through a 10 MΩ load to a safe 3 μA. At the time of this writing, most workers in the field are using *isolation amplifiers,* and we will discuss them in detail later (see page 239). For the moment, keep in mind that in all experiments, *subject safety comes first!*

Returning to our first experiment – measurement of skin resistance – you have to make a big decision: "Shall we use AC or DC current for this measurement?" The problem is that skin has both capacitive and inductive properties as well as polarization effects. We shall assume that you know all about these phenomena (if you don't, we suggest you read the book by Geddes and Baker) and have decided what frequency to use. Most experimenters to date seem to use DC, which may be why they see so many polarization effects. In any case, you use a signal source or stable DC voltage (remember the zener diode) to power a constant current generator. The output goes through a 5 mA fuse and then to the body. To measure the resistance, you can calculate the voltage from Ohm's law; this gives you $R_s + \Delta R_s$, where R_s is the skin resistance and ΔR_s is the change due to some external effect (Figure 4-1). Notice in Figure 4-1 that we have used two op-amps as buffers: this allows us to detect V_s *without* drawing any current from the system. This also *pro-*

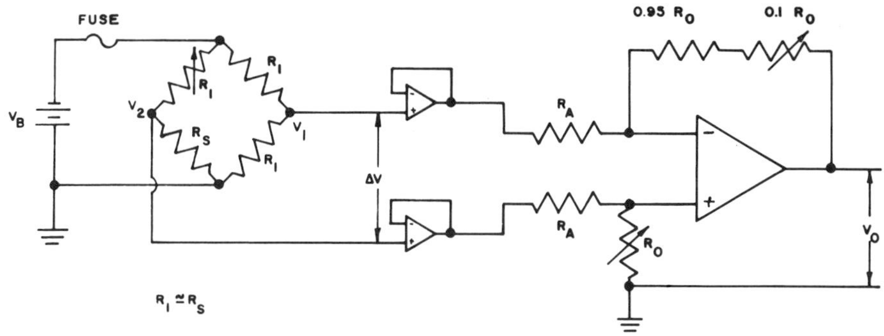

Figure 4-2. Balanced circuit to measure changes in skin resistance.

tects the subject from any electrical surges due to defects or mistakes in the recorder circuit.

There is one problem with this simple circuit. The output V_s is related to R_s, the total skin resistance. Because of this, a small change in R_s (i.e., ΔR_s) will be hard to detect in the presence of the larger R_s. In this event we must go to a *balanced circuit,* such as that shown in Figure 4-2. In this circuit, we balance our R_s by adjusting the potentiometer R_1 until $\Delta V = 0$. This removes any signal due to the normal skin resistance, R_s. Then any changes in R_s will cause ΔV to vary according to the formula:

$$\Delta V = V_2 - V_1 = \frac{V_B}{4}\left(\frac{\Delta R_s}{R_s}\right)$$

To see how this was done we need only note that

$$V_1 = V_B\left(\frac{R_1}{2R_1}\right)$$

$$V_2 = \frac{(R_s + \Delta R_s)\,V_B}{R_1 + R_s + \Delta R_s}$$

and

$$\Delta V = V_2 - V_1 = V_B\left(\frac{R_s + \Delta R_s}{R_1 + R_s + \Delta R_s} - \frac{R_1}{2R_1}\right)$$

If we simplify this formula by writing $R_1 + R_s = R = 2R_1$, we have

$$\Delta V = V_B\left(\frac{R_s + \Delta R_s}{R + \Delta R_s} - \frac{R_1}{R}\right)$$

and if we take R out of the parentheses, we have

$$\Delta V = \frac{V_B}{R}\left(\frac{R_s + \Delta R_s}{1 + \Delta R_s/R} - R_1\right)$$

This can be written as

$$\Delta V = \frac{V_B}{R}\left[\frac{R_s + \Delta R_s - R_1 - \left(\frac{R_1}{R}\right)\Delta R_s}{1 + \Delta R_s/R}\right]$$

Since when the potentiometer is balanced, $R_s = R_1$, then $R_s - R_1 = 0$. The above equation thus simplifies to

$$\Delta V = \frac{V_B}{R}\left[\frac{\Delta R_s(1 - R_1/R)}{1 + \Delta R_s/R}\right]$$

If now we expand $(1 + \Delta R_s/R)^{-1}$ as an infinite series (see your calculus textbook) and discard all but the first term, the final result (recalling that $R = 2R_1$) is

$$\Delta V \approx \left(\frac{V_B}{4}\right)\left(\frac{\Delta R_s}{R_1}\right)$$

Of course, this formula assumes that $\Delta V = 0$ when the bridge is balanced, i.e., when $R_1 = R_s$. However, your op-amp output won't be zero if there is any error in the differential amplifiers. To ensure that the op-amp output is zero when it should be, place a jumper between the inputs to the two buffers and then adjust R_0 until $V_0 = 0$. This adjustment will be critical for high values of R_0.

This is the best circuit for most skin resistance studies. While we are on the subject, you may want to try out your skin experiments on a piece of chamois skin (a "shammy" at Checker Auto Parts). Wet it with salt water and it is a fair substitute for the real thing. It doesn't wiggle around, either, which helps while you're learning.

Four-Point Probes

Another idea you may want to try is the *four-point probe*. Four-point probes are based on the concept that it is better to use two heavy electrodes to carry current and two separate, light electrodes to measure potential. It yields superior results in terms of freedom from noise and contact problems. To build one, you take four phonograph needles (the old-fashioned steel kind) and mount them in a row on a Plexiglas strip. A spacing of 1 cm center-to-center is about right. Now hook a constant current generator on the outer electrodes, press the pins on the shammy, and measure the potential between the inner pins (using op-amps of course).

In books on this subject – for example, see R. Plonsey, *Bioelectric Phenomena,* New York: McGraw-Hill, 1969 – the details of four-point probes are discussed. Consider first an infinite area that is *thick* compared with the pin or probe spacing (b). In this case, the *skin conductivity* σ is $\sigma = (I/2\pi\, b\, V)$ ohm-cm. Here I is the current between the two outer probes, V is the voltage measured at the two inner probes, and b is the probe spacing in centimeters. Conversely, if the tissue thickness t is *very thin* compared with b, the formula becomes $\sigma\, t = (I \ln 2)/V$. Here I and V have the same meaning as before. The product $\sigma\, t$ is a *surface resistance* having the units of ohms, where t is in centimeters. It is often called *ohms per square,* and it is interesting to note that it doesn't matter if your unit of area is centimeters, feet, or miles; the resistance of a square is the same.

Of course, all of these examples apply strictly to large skin areas that are uniform in their electrical properties. You have to realize that when you apply this to a living creature smaller than an elephant, the formulas are not quite accurate. Nevertheless, four-point probes have been used for a number of experiments in cardiology, plethysmography, and electroencephalography. If you go from a DC current source to an AC source, you can control the depth to which the electrical signals penetrate the body (this is called the "skin effect" by EEs). If you go to a driving frequency of from 10 kHz to 40 kHz, most of the signal will pass through the skin. This allows you to bypass the blood and muscle tissue that might well affect a DC skin-resistance measurement.

Another application of this technique exists in geological explorations for minerals. In mineral studies the skin effect can be exploited to look for minerals at depth or near the surface depending upon what the geologist is looking for. We repeat that the "depth" to which the current flows is a function of the distance between the outer electrodes.

OP-AMPS IN ECG AND EEG WORK

Other applications of op-amps exist in electrocardiography (ECG) and electroencephalography (EEG). In both cases, small electric potentials appear on the chest or head due to heart or brain action. The details of how and where to put on contacts is up to you. The interpretation of ECG and EEG signals is a specialty in itself, and here again we must cop out. However, there are some electrical details we can discuss that may be helpful.

In working with EEG and ECG signals, you have a series of repetitive pulses that can be thought of as a combination of several AC signals of different amplitude and frequency. The signal levels are quite low, and amplification is necessary for viewing and recording them. To demonstrate the application of op-amps, we choose a typical ECG voltage-versus-time signal as it might appear "at the body" without electrode or amplifier problems (Figure 4-3).

One of the easy things to do is measure the frequency, f, of these signals ($f = 1/\tau$). For that purpose, you do not need to amplify everything: clip off all signals below some level (recall clipper circuits from Figures 3-26 and 3-27) and

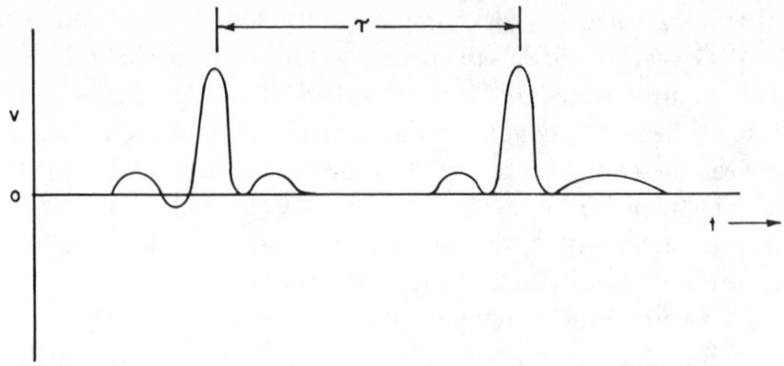

Figure 4-3. Typical ECG signal.

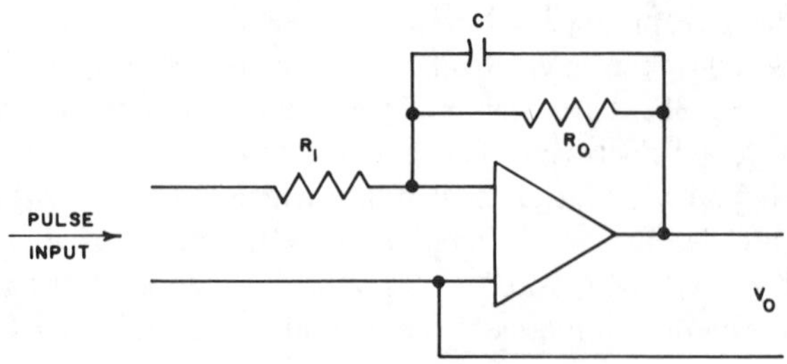

Figure 4-4. Pulse-rate-to-voltage converter.

count pulses. If you want a voltage output that is proportional to the pulse rate, you use a leaky integrating circuit as shown in Figure 4-4. The output (if you choose the capacitor large enough) will be a DC voltage (with a pulsating component) that is proportional to the rate of pulse input.

This is an easy circuit to build, but you may need to play with component values. The signals are stored by the capacitor, C, and dissipated by the resistor, R_0. If the resistance R_0 is too high, the op-amp saturates, and if R_0 is too small, the output is quite low. All pulses must be of the same amplitude and duration. If your pulses aren't all the same, use them as the input to a Schmidt trigger circuit (Figures 3-72 and 3-73) and count the output of that circuit.

Of course, counting isn't all you may want to do with the signal shown in Figure 4-3. An experienced reader of ECG curves can detect disease and malfunctions by looking at changes in the form of the total signal. Obviously, this physician's talents and experience are wasted if the record is distorted by defects in the electronic amplifying and recording systems. It will be *your* responsibility to see that this does *not* occur.

To see how this works in actual practice, let's consider an almost trivial case in which we have attached our EEG or ECG electrodes. Now we put in buffer op-amps to limit current drain from the patient and add amplifiers as shown in Figure 4-5. The next question is, what does this system do to the signals besides amplify and record them? To find out, you get a low-frequency square-wave generator and check

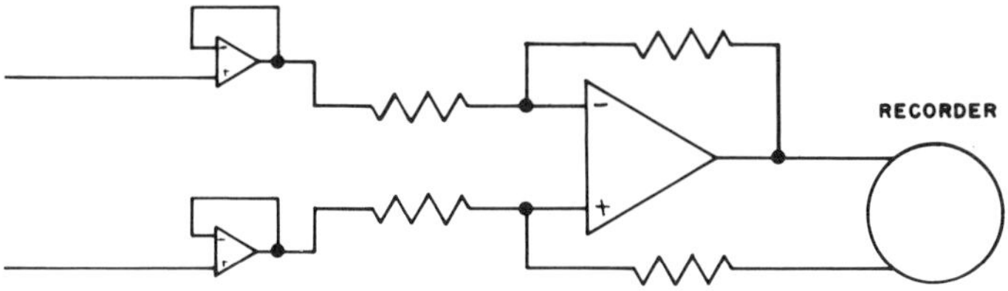

Figure 4-5. Typical simplified EEG or ECG measurement system.

its output with a good oscilloscope. Then use the generator to run your buffer, amplifier, and recorder system. Do a square-wave test, just as we did before (see Figure 3-63), with a frequency halfway between the lowest that might appear in your real signal (1 Hz) to the highest (50 Hz). Make sure your system reproduces them properly.

Figure 3-63 shows the sort of problems you can have. If your output looks like "Brand X" (Figure 3-63B or C), the odds are the trouble is the recorder. Check this by trying the oscilloscope at the amplifier output. Look at the instruction manual first, it may suggest some "fix" for your problem. If it doesn't, you can try a capacitor across the input terminals to cure overshoot or oscillation problems.

If the problem is in the op-amps, you will have to try some compensation techniques, i.e., a capacitor across R_0. If that doesn't work, you either can get better op-amps from someone like Burr-Brown or Analog Devices, or you can learn more about compensation techniques. You can begin with Chapter 6 of this book and work up to the book sponsored by Burr-Brown entitled *Operational Amplifiers,* by J. Graeme et al., New York: McGraw-Hill, 1971.

Assuming that your system has passed the square-wave test, you are *almost* done. It remains only to check the system for *frequency-dependent phase shift.* If this phase shift exists, it might shift your wave shapes around and that is a *no-no* in this business. The trick is to use a technique employing Lissajous figures (recall or review Chapter 1, p. 56). Use a signal generator to put a signal into the system, and put the output of the system into the Y input of an oscilloscope. The direct output of the signal generator is used as the X (horizontal) drive on the oscilloscope, and the trace should be a 45-degree down-sloping line to the Y and X axes. Why? Well, the total fixed phase shift through a single op-amp is 180°, so when the X signal is at its maximum positive value, the Y signal is at its maximum negative value. The result is a trace at 45° if both the X and the Y signals have the same magnitude. If they don't you will still have a straight line but it won't be at 45°. If you have two inverting op-amps in series or no inverting op-amps at all, the fixed phase shift will be either 360° or 0°. In that case, the trace will be an up-sloping line, again at 45°, provided the X and Y signals have the same magnitude. The *important* thing is to have a straight-line trace that *does not* change to an ellipse as you vary the driving frequency from the signal generator. The point is that fixed phase shift itself is harmless; it is the phase shift that varies with frequency that hurts you.

If at some frequency your straight line turns into an ellipse or a circle, you have the bad kind of phase shift. The first thing to try is a small capacitor (100 pF) across the R_0 resistor. If that doesn't work, start reading the Burr-Brown book on op-amps again. The main thing is not to try amplifying any biomedical signal with an amplifier or recording system that has frequency-dependent phase shift.

ELECTRICAL SAFETY AND PATIENT RESISTANCE

Up to this point, there have been no real differences beteen the experimental techniques used for biomedical studies and the methods that would be used in any other type of work. There are, however, two very significant differences that you must now take into account. The first is electrical safety; you will be reading a lot about that. The second is more insidious and always bites you when you least expect it. We can warn you, but only experience will qualify you to deal with the problems.

The first thing to appreciate is that all, repeat all, the body tissues except the skin, are *good conductors.* This includes the mucous tissue of the mouth and any skin that happens to be wet or has been treated with some electrical contact paste of the type used for ECG studies. It follows then that dry skin is our *only* protection against electricity. This also makes it difficult to get electrical signals in or out of the body.

The next "fact" is that low-frequency DC currents are most dangerous in terms of skin burns and electrocution. However, the conductivity of human skin has its lowest value at low DC frequencies. As the signal frequency increases the conductivity of human skin goes up. As a result, we find that the danger of electrocution and skin burns is greatest at frequencies up to 60 Hz, while for experimental studies, i.e., skin resistance, it is more convenient to use higher frequencies, i.e., 1000 Hz.

Another factor in work with humans is that the body seems to act like an antenna to pick up any stray signal, i.e., 60 Hz from the power lines, that is floating in the air. The thing *not* to do is to try grounding the body to a water pipe in the hope of getting rid of the pickup signals. It won't work, and it might be dangerous for the subject.

We will discuss some solutions for these problems in the next section, but we warn you that working with living subjects is not easy.

Signal Sources and Shielding

In the biomedical area, you are likely to run across references to all sorts of grounded, floating, and balanced signal sources. By the time you have worked through all these concepts, you might be ready to float off yourself. The best thing to do is write to the Brush Instruments Company (see the Appendix), and ask for their booklet entitled *Signal Conditioning.* It is well written and there is no need for us to paraphrase it, but some of the terms might not be familiar to you and are therefore worth discussion.

Why go to this much trouble? Usually you do it in cases where the noise level is so high that the simpler circuits just do not work. Many times the biomedical sources you have to work with are driven off ground by AC pickup, and you have to know what the source is like to pick the signal out of the noise.

To begin with, recall that the usual signal source is either grounded or floating (see Chapter 1, Figure 1-35). If you want a mental picture of a signal source, think of a dry-cell battery. We can connect either, or neither, terminal to ground before we connect it to the amplifier. The next case is a little more difficult: we have a second battery between one terminal and ground. This is a *single-ended* signal source *driven off ground;* the circuit is shown in Figure 4-6A. With two batteries, we can also use a connection to their common point; this is called a *balanced* source. If we ground the common point, it is *balanced to ground.* If we don't, it is a *balanced-floating* source. The balanced-to-ground circuit is shown in Figure 4-6B. Last but not least, is the *balanced, driven off ground* system in which a third battery has been put between the common point and ground (Figure 4-6C). If the use of batteries as signal sources bothers you, just replace the batteries with signal generators; the definitions remain exactly the same.

You might ask, "How do I know what kind of a source I have?" If it is a commercial device, the manual will tell you. If not, the Brush manual has a series of tests that might help. However, the best thing, especially with biomedical situations, is to

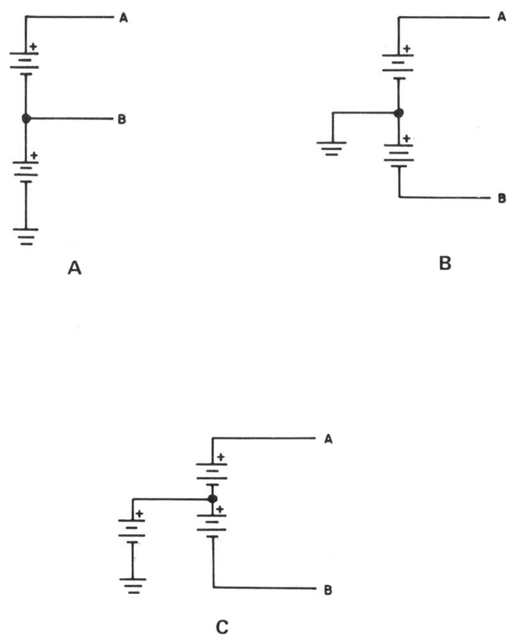

Figure 4-6. Various types of signal sources. *A*. Single-ended, driven off ground. *B*. Balanced-to-ground. *C*. Balanced, driven off ground.*

*For AC signals, you can think of the batteries as signal generators.

assume that the source is either single-ended or balanced and always driven off ground.

The next question is how to hook up the source to your op-amp. We will assume that you are smart enough to use a good grade of shielded wire: two twisted, insulated wires inside a braided shield. If you can afford Teflon insulation, so much the better. If not, buy a good grade of microphone cable (you can always use the short pieces for hooking up your hi-fi). The how-to-do-it with simple grounded or floating sources was covered earlier (Figures 3-110 to 3-112). Above all, don't connect the shield to ground at more than one point. With a *grounded* source, the shield is grounded at the source. With a *floating* source the shield is grounded at the op-amp.

For the case of a single-ended source that is driven off ground (Figure 4-6A), you connect the shield to output *B* (the one closest to ground potential). Output *A* is connected to one op-amp input, output *B* to the other input, and the shield is at the same potential as *B*. You don't ground the shield at the op-amp at all.

For the case of the balanced-to-ground system (Figure 4-6B), you hook output *A* to one input, output *B* to the other input, and the shield to ground at the signal source only. In principle, the shield is at AC ground and you could connect it to ground at both ends. In practice, however, this is almost never the case: every AC power ground seems to be at a slightly different voltage. Grounding the shield at one point avoids something called *ground loops.* They are bad news, because signals flowing in the ground shield will be impressed upon your op-amp inputs.

The last and most common biomedical situation is the balanced source driven off ground (Figure 4-6C). Here the *A* and *B* outputs go to the op-amp as before, and the shield is connected to either *A* or *B* since they are both at the same potential above ground. The problem here is that the shield is above ground, and that may represent a dangerous situation in a hospital. One precaution to take is to purchase shielded wire that has a rubber or plastic sheath over the shield. However, the metal jacks and plugs on your cable will still be at the shield potential, so the shield cover idea is not a perfect solution: *you just have to be careful.* There are other facets to this problem. Since your shield is "floating" at voltage V_A or V_B, you cannot connect it to power ground at the op-amp. This may induce some noise pickup at the op-amp, and, if it does, you will have to obtain a *guarded* or *internally shielded* op-amp. This internal shield is not connected to the op-amp power ground, so with any of the various types of signal sources, you can connect the cable shield to the internal shield. This will reduce noise pickup at the op-amp. Believe it or not, sometimes these tricks are useful!

All the above suggestions are useful, but we still need a feedback loop that connects the output of the op-amp to its input: this feedback loop plus various internal connections can, in some cases, pass high external voltages to or from your patient. This is, of course, a hazard, and to eliminate it we pull another rabbit out of the hat, the *isolation amplifier* (IA). In using the IA, we will rely upon your knowledge and understanding of the signal source concepts discussed above. If you get confused, go back and read the section again.

Isolation Op-Amps and Their Applications

The subject of electrical safety in hospitals is very important these days. Hospitals are getting more electrical gadgets all the time, which in turn leads to a higher probability of breakdown and defective units. The increased use of open heart surgical procedures produces more opportunities for patient injury by minute electrical currents. Normally, it requires a current of some 200 mA to 500 mA through the unbroken skin to injure a human. However, if the electrodes are applied to the exposed heart, only 20 μA may be deadly. An electrical defect that is unnoticeable on a skin electrode may be deadly once the high resistance of the skin has been penetrated.

In view of this problem, it is common to consider three levels or degrees of hazard prevention in the use of electronic equipment.

1. *Laboratory use,* where only about 500 mA will cause serious injury and users *can be expected* to report malfunctions.
2. *Patient use with unbroken skin.* Here, the current hazard level is about 200 mA, but the subject *cannot be expected* to report malfunctions. The law considers your subject to be a helpless child, and if anything goes wrong, it will be your fault.
3. *Patient use when the skin is penetrated and the heart may be exposed.* The legal situation above applies even more strongly, and the danger level may be as low as 20 μA. Here again, it is *you* who must protect the subject against accidental injury.

Don't think that you can ignore this problem by saying, "I'll only work with animals; no one cares what you do with them." You had better be aware that there is a federal law in this area that forbids unnecessary injury or cruelty to animals. A researcher who is careless with animal subjects will also be careless with human subjects and is fully deserving of the utmost penalty of the law. The mere thought of losing a valuable experimental animal in the middle of a program should be enough to convince you even if nothing else does.

When you must take every precaution to protect your experimental subjects from any chance of injury, there is nothing like the *isolation amplifier.* An isolation amplifier is a special type of op-amp that transfers a signal from the input to the output, but it stops any stray signals that might be on the line, e.g., 110 volts AC. As you can imagine, the op-amp industry has not been idle in this field. There are all sorts of isolation op-amps, and many of them are quite expensive. At the moment the major applications are in the biomedical area, but as the price comes down, the number of users goes up (this is sometimes called *Ford's rule,* after the famous Henry).

Isolation can be achieved in various ways. The best way is probably the *optical system,* in which the initial signal drives a light-emitting device like a light-emitting diode (LED). Its light output is a function of the input current. This light output is

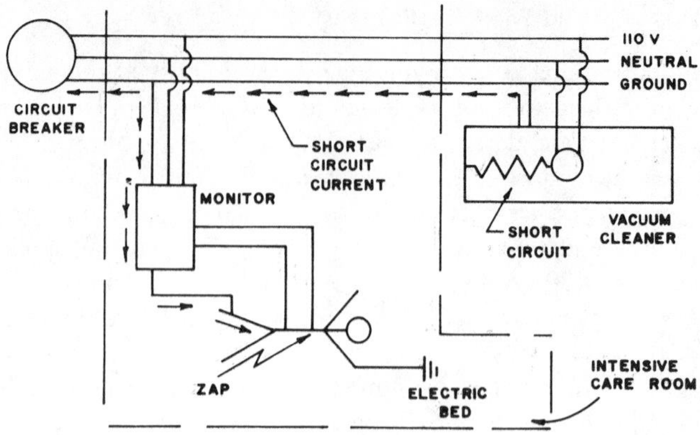

Figure 4-7. How the janitor injured the patient.

then picked up by something like a phototransistor, whose output is then amplified to provide the desired final signal. Other isolation amplifiers use *transformer coupling,* which is quite satisfactory if properly designed. We suggest that isolation amplifiers are not the place to save money. Buy from a reputable outfit and pay their price.

We have said that the isolation amplifier is a device that isolates its inputs from its outputs. This is true even at high differential or common-mode voltages (e.g., 5 kV). It also has high common-mode rejection even in the presence of unbalanced inputs. Best of all, it can have a floating internal shield so that even the *shield* on the input cable is *not* at earth ground. The advantages of these expensive goodies will become apparent as we go along. Remember that your major objective is to ensure that *even if* something goes wrong, your system should neither put a potential on the patient nor provide a path to ground so that someone else's potential can cause injury.

A common hazard in hospitals that install new intensive care (IC) facilities is the problem of grounding in old buildings. Suppose an IC room has a monitor unit that requires a grounded leg on the patient. This ground is hooked to the frame of the monitor unit and that, in turn, is grounded to the third-wire ground wire in the wall (Figure 4-7). All is well until the janitor hooks up his vacuum cleaner in the hall outside. It has a three-wire cord (if you don't remember why, go back and read about it in Chapter 1). When the janitor picks up a hairpin in the vacuum cleaner, the motor short circuits to ground. The third wire protects the janitor from getting a shock (just as it is supposed to do), but before the 15 amp circuit breaker blows, a current greater than 15 amps must pass down the ground wire.* On its way, there is a voltage drop because the ground wire doesn't have zero resistance. If our patient

*It is interesting to note the current levels needed to open a standard circuit breaker in some given period of time. At a short-circuit current of 100 amps the circuit breaker opens in 1 second, at 50 amps it takes 5 seconds, and at 35 amps some 20 seconds are required. (See *Biomedical Safety and Standards,* 6(3), March 19, 1977, p. 38.)

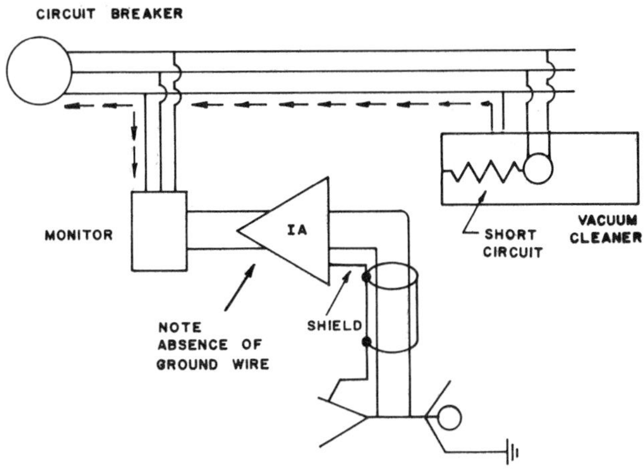

Figure 4-8. How the isolation op-amp prevents patient injury.

is in an electric bed with one hand on the grounded bed rail, and if the bed is plugged into a different branch circuit from the monitor, the patient might be injured.

In Figure 4-8, we show how to apply an isolation op-amp to prevent injury to the patient in the case discussed in Figure 4-7. The important thing to appreciate about this setup is the *complete* isolation of the patient from ground. Even if the normal "ground circuit" was somehow raised to a potential of 110 volts, the patient would still be safe. Since the signals that are read on the monitor are not referenced to ground, the absence of a ground wire is no problem.

In a crude sense, the isolation amplifier is like an ideal transformer that transmits AC or DC signals from the input to the output without a hard wire connection. Notice that there is no external feedback loop, because if there were, this would destroy the isolation. Due to this lack of feedback, the IA has unity gain. If you need more gain, you can put regular op-amps on the output.

Another interesting application of the isolation amplifier involves a patient with a battery-driven pacemaker. The pacemaker is isolated from ground, i.e., it is a single-ended floating source, but suppose you want to look at the voltage output with an oscilloscope. You could get a battery-driven oscilloscope, but that is the expensive way to go. A better way is to use an IA, as shown in Figure 4-9. Even with the worst of oscilloscopes you can't hurt the patient, but the operator has to watch out for himself: old oscilloscopes can give you a 5 kV zap!

While we are on the topic of avoiding getting zapped, we should emphasize again the importance of grounded three-wire cords and receptacles in hospitals. As an example of "why," consider the case of an old oscilloscope: the 5 kV acceleration voltage is produced by a power supply that uses capacitor-filtering to smooth the output. The capacitors carry the filtered-out signal to the frame and to ground via the third wire. However, if that third wire is broken or nonexistent, the frame and outside case of the oscilloscope are raised to some voltage above that of the

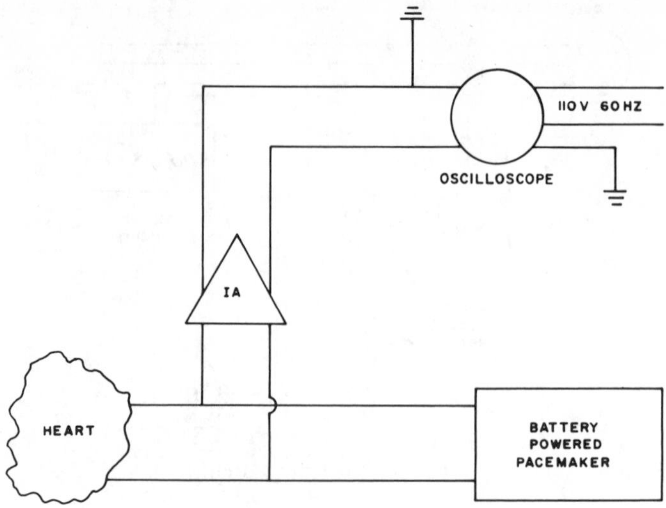

Figure 4-9. Use of an isolation amplifier (IA) with a pacemaker.

nearby sink or radiator. Then with one hand on the sink and one on the oscilloscope – *Zap!* The point is, the older an instrument is, the more likely that leakage exists. Resistors and capacitors age, insulation dries out, "moth and rust doth corrupt," and so forth. Insist on equipment with three-wire cords, and check the third wire at least once a year. If you move into a new facility with three-wire receptacles, check every one of them with a VOM. Our experience indicates that 14% of new receptacles are improperly installed by electrical contractors. *Remember: when in doubt, use isolation amplifiers.*

The isolation amplifier can also be used when you want to apply an electrical stimulus to an animal and then observe the change in its respiration or heart rate. There are two problems in doing this kind of experiment: first, the stimulus signal can drive your monitor into saturation; second, the application of the stimulus current can change the skin or contact resistance, which in turn can affect the actual stimulus signal applied to the animal. For example, suppose you applied a staircase voltage signal* to a resistive load whose resistance went up exponentially as a function of I^2R, or *joulean heating.* The current in this case wouldn't be a staircase function; it might even go down as the voltage goes up! A situation like that could make your experimental results a little hard to understand.

To solve the first problem – i.e., the stimulus overdriving the monitor – we must recognize that the effect is due to the stimulus signal arriving at the monitor in a manner that precludes common mode rejection. The worst case you could have would involve a stimulator with a grounded output and a monitor with a grounded input. Assuming that you are too smart to fall into that trap, we will consider a

*A staircase generator puts out a voltage-time signal that rises from one value to the next at regular intervals; the output looks like a flight of stairs.

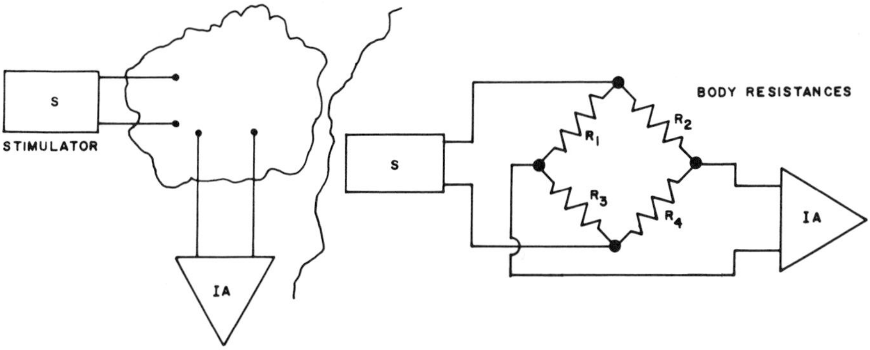

Figure 4-10. Use of an isolation amplifier (IA) for stimulus studies.

"real life" situation where here we have a floating stimulator and an isolation amplifier between the monitor and the subject, as shown in Figure 4-10. The signal from the stimulator arrives at the detector by a variety of paths through the body. This makes it very difficult to get good common mode noise rejection because of the imbalance in the signal source.

As you might have guessed, this application of the IA gets rid of all overloading and unbalance problems due to the stimulus pulse. The IA is a big help here because it has high CMR even with an unbalanced signal. Ideally, you would try to balance all the body resistances until $R_2R_3 = R_1R_4$, and if you did, you could use regular op-amps (except, of course, for the question of patient safety) that have good CMR for balanced signals. In general, you can't do this; the various resistances are just *not* equal, so you go to an IA. This will help even if you are poor and have to use a stimulator with a grounded output. If you are stuck in a situation like that (it's hell to be poor), try putting an audio transformer between the stimulator and the subject. It will give you a floating signal, and shouldn't mess up your pulse shape too much.

The solution to the second problem – i.e., the change in resistance due to the stimulus current – does not require an IA, but you might use one (or more) as part of the circuit anyway. The objective here is to deliver a known signal to a load whose impedance may be varying due to the input signal itself. The answer, of course, is to use a constant current source rather than a constant voltage source. A constant current source raises its output voltage to whatever value that may be necessary in order to push the required current through the load. You might recall that we discussed these devices previously (see Figure 3-87). You can buy them or build them, but in stimulus experiments, you had better plan to use them.

To display another facet of the isolation amplifier, let's consider a case in which you want to pick up a fetal heartbeat in the presence of noise (Figure 4-11). Here we are considering a "worst case" situation. The mother's body has a 50 mV potential, above ground, due to 60 Hz AC pickup: let's call that signal V_C. The mother's heart signal is another voltage of, say, 2 mV, which we will call V_M. The fetal heart signal, which might be 50 μV, we will call V_F. The objective is to observe V_F with-

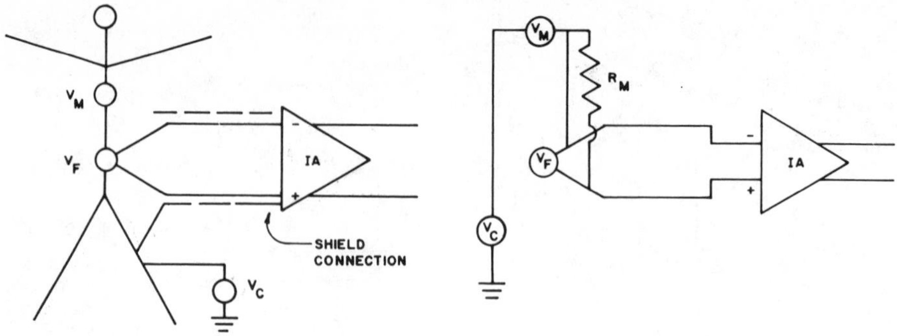

Figure 4-11. Detection of fetal heart sounds using an isolation amplifier.

out interference from V_C or V_M. V_F is a *single-ended* signal source driven off ground by 60 Hz noise (V_C) and by the mother's heartbeat signal (V_M).

We begin by putting two electrodes on the *mother's abdomen* to pick up V_F and present it to the IA as a difference signal that will be amplified. The mother's heart signal, V_M, is impressed as a common mode signal on both IA inputs and is rejected. The pickup signal is put on the shield, which gets rid of it, but the shield does *not* carry it to AC power ground. We can thus detect the baby's heartbeat without danger of injury to the mother.

This is not the whole story, because if you look at Figure 4-11 again, you will see that the signal V_M is *not* really presented to the IA in the usual common mode fashion. This is unavoidable, especially in medical studies, because the two pickups cannot be on exactly the same spot on the body. This means that the signal presented to the (+) input of the op-amp will have come through a slightly different resistance (R_M) than the V_M signal presented to the (-) input. This in turns means that we have an unbalanced input signal, but we still want common mode rejection of that signal. So we must require that our isolation op-amp have a high CMR value and that it maintain this characteristic even in the presence of, say, 5 kΩ of source unbalance. That is part of the reason why isolation op-amps are so expensive.

If you look back at our earlier discussion of common mode rejection (Chapter 2, p. 85 ff.), it was always used with a balanced input, and this made it relatively easy to get good common mode rejection. When there is an unbalanced input, the op-amp must have a very high input impedance and a somewhat special design if it is to retain its CMR characteristics. ("You pays your money and you takes your choice.")

While we are on the subject of money, we might suggest that in times of financial stress, you can use conventional op-amps for isolation *if you power them with dry-cell batteries.* Build a battery box of ½ inch thick Plexiglas, bring out insulated wires, and hook up your op-amps. Of course, the recorder or oscilloscope on the op-amp output should be battery driven, too, but in emergencies a setup something like that shown in Figure 4-12 will do. Here a grounded subject has a thermistor in one nostril. The thermistor is driven by a constant current from the battery to a temperature of about 1°C above that of the local tissue. As the lung contents ebb and

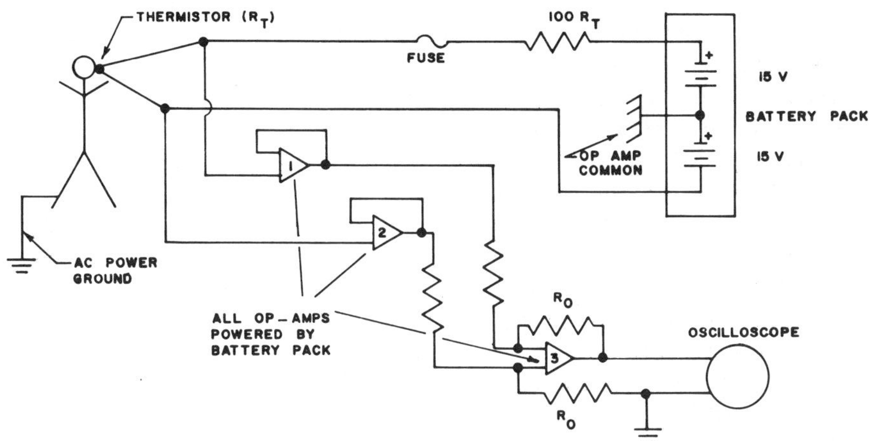

Figure 4-12. Low-cost isolation system using conventional op-amps.

flow, the thermistor will be cooled and its internal resistance will change. The voltage (derived directly from Ohm's law) across the thermistor will then be a measure of the fluid velocity.

The important point in Figure 4-12 is the isolation of the subject from the line-powered oscilloscope by three battery-powered op-amps. Unless two op-amps fail at the same time, the situation is "reasonably" safe. Of course, you must take full legal and moral responsibility for what you do. Try to keep the value of R_0, the feedback resistor, above 1 MΩ. This will reduce the danger of high voltages getting back to the input side of the op-amp.

These are only some of the many situations where isolation techniques can get you out of a sticky problem. Obviously, the whole question of hospital electrical safety is a vast area in which problems and solutions are developing rapidly. In this book, we can only hope to make you aware of prudent electrical practice and provide some examples of what can be done.

If you are directly involved in hospital electrical safety or patient electronic equipment, you should write to the ECRI and get on their subscription list for *Health Devices* (Emergency Care Research Institute, 5200 Butler Pike, Plymouth Meeting, PA 19462). Another good organization to get acquainted with is the NFPA (National Fire Protection Association, 470 Atlantic Avenue, Boston, MA 02110); their booklet, *Safe Use of Electricity In Hospitals,* is well worth reading, but it won't tell you everything.

Skin Contact Problems

A recently discovered hazard in biomedical work is low voltage DC. You wouldn't think that 3 volts DC could hurt the unbroken skin, and if you put two fingers across a 12 volt automobile battery, you would be sure of it. Here again, common sense is *wrong* when you have conductive grease around. This is the gunk that is put on the ground-return plates when cautery work is to be done. It is often used to make contact to the skin during ECG studies. In fact, in hospitals the stuff is everywhere. It turns out that when this grease becomes a conductor for DC, the sodium in it is released at the cathode and forms sodium hydroxide. This in turn can cause chemical burns on the skin and unhappiness for all concerned.

It would be easy to say, "don't pass DC of over 10 μA to any living creature." This would, of course, solve the problem. The difficulty is that we can't always use fuses and we don't always know when low voltage DC signals are present. Cautery machines, for example, often have some DC in their output. They deliver far more than 10 μA, as well, so you can't use fuses with them. There is no easy solution to this problem. You have to get a good VOM with a 0–1 volt DC scale and go around looking for small DC voltages.

Sometimes a skin-electrode contact will act like a diode and turn AC into pulsating DC; EEs call this a *non-ohmic contact.* This type of problem can sneak up on you when you least expect it. Since we have raised the question of non-ohmic contacts, we should define an *ohmic contact:* in the biomedical area, this is a contact that passes signals either in or out of the body without interference or distortion. (We did say you were on your own in this area, but we decided to include a few words on the subject.)

The skin contact problem is one of the most vexing areas in biomedical engineering. There are various books that discuss the techniques of contact attachment and the problems you can have. The best of these is by L. Geddes, *Electrodes and the Measurement of Bioelectric Events,* New York: Wiley, 1972. He suggests that the *skin-drilling technique* is the one most likely to yield a "good contact." He does not, however, provide a general technique for testing contacts, and this is where we can make a contribution. You should realize that an ideal contact would act like a pure resistance: if you apply a DC voltage, you get a current; if you reverse the polarity of the applied voltage, the current should be *the same.* If you apply a square wave pulse or a variety of sine wave signals, the contact should display no inductive or capacitive characteristics.

Of course, no real contact is this good. To find out just where you stand, we suggest that you make one contact by the skin-drilling method and check the other contacts against it. The DC test circuit for doing this is shown in Figure 4-13A. You apply a voltage, measure the current, and then reverse the polarity to see if the current direction affects the contact resistance. By doing this at various voltages, you can obtain the DC current-voltage characteristic.

To investigate frequency-dependent effects, it might be possible to use an AC Wheatstone bridge, called an *impedance* or *RLC bridge* (recall p. 44), between the skin-drilled contact and the contact being tested. A better way would involve the use of something like the four-point probe that we discussed earlier (p. 232). In this case, the two outer probes would be drilled skin contacts and driven by a pulse or signal generator. The two inner contacts would consist of one skin-drilled reference

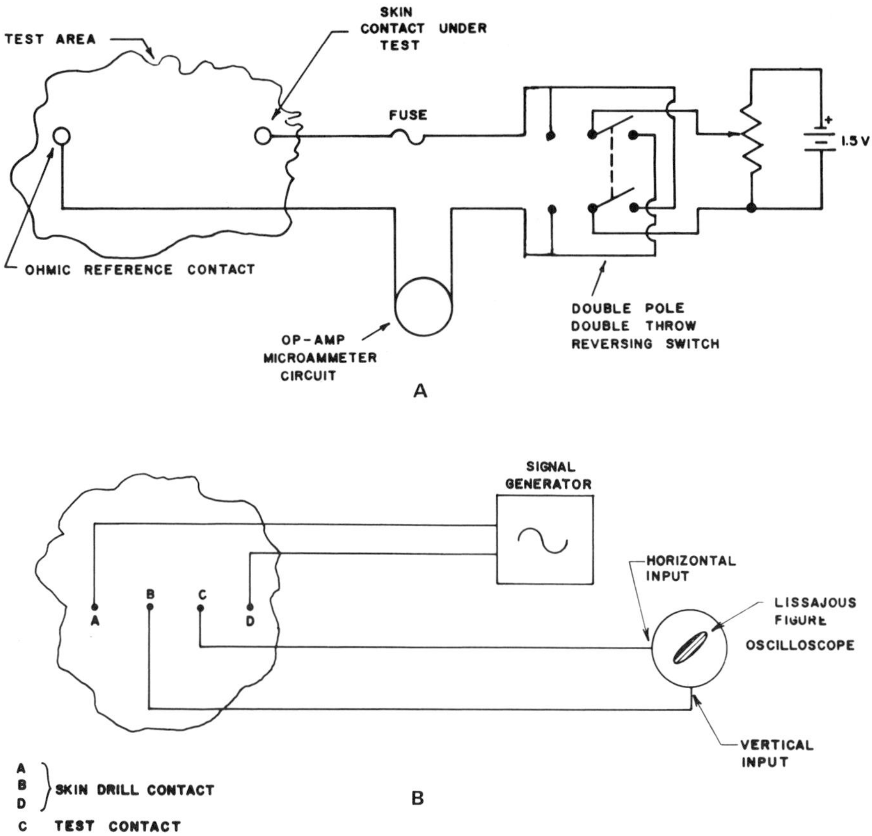

Figure 4-13. Skin-contact test system. *A*. DC test circuit. *B*. Four-point probe contact-testing system.

contact and the contact that you are testing. The phase shift, if any, between the two inner contacts would be a measure of capacitive and inductive effects.

The set-up is shown in Figure 4-13B. Ideally, you would check the test contact with a variety of frequencies, pulse durations, and voltage amplitudes.

POSTSCRIPT

The time has come to end this chapter on biomedical op-amp applications, not because we have run out of applications, but because even a textbook must be finite in size. We hope that you are ready to design and build your own op-amp circuits, and that you will stay with us for Chapter 5, where we will talk about new applications of the computer for nonelectrical engineers.

5 The Computer and Its Applications (Or Space War for Fun and Profit)

You might be wondering what a chapter on digital computers is doing in a book on operational amplifiers (which are definitely analog type devices). The answer is that in the first edition of this book we had a chapter on analog computers, but they have now "gone out of date." We had to fill the gap with something and decided to discuss how computers can be used for engineering and scientific applications, by people who are *not computer experts.* In that sense, the material in this chapter fits in with the rest of the book, which is devoted to electronic applications for people who are not electronics experts.

There seem to be only two kinds of opinion about computers – the "gee-whiz" and the "aw-nuts." The gee-whiz types think that computers can solve any problem – "human beings are obsolete." The aw-nuts types maintain that computers are a passing fad, that "there is no substitute for the slide rule."

If this isn't bad enough, we can think about "the computer is a monster" group. This sect sees the computer as an arm of "big brother," or even the actual big brother in the awful metallic flesh. Every move will be watched, and our fabled way of life (whatever that may mean) will be destroyed.*

As usual, the truth lies somewhere in between these radical ideas. Computers are useful devices, but they are only tools. What we do with them is our decision, not theirs. One thing to keep in mind is that computer data output is only as good as the information we feed into it. At Massachusetts Institute of Technology, they say *GIGO* (garbage in-garbage out). Sometimes there is even some "gain," and you get more garbage than you put in.

Having given all of this introduction, the next question is, "What shall be said about computers in a single chapter that is part of a book about operational amplifiers?" Well, the first thing *not* to do is to try to teach you how computers operate or how to use them. Operational and how-to-use-it computer books are now as familiar as the common cold; the world does not need another chapter on "that" topic. Instead, what we propose to do is discuss how the computer might be used

*This irrational fear puts me in mind of the furor raised in England about 1830 as Sir Robert Peel was organizing the first uniformed, professional police force. "It will take away our freedoms," was the cry; now no one would want to live in a city without an effective police force. Citizens who are prepared to defend their rights have nothing to fear from either the police or a computer. The founding fathers did their best to insure that by writing the Second Amendment to the *Constitution* (the right to keep and bear arms).

for various scientific, engineering, and industrial applications. We shall assume that, if you want to learn how to use a computer, you will get one of the many good books available. (A list is provided at the end of this chapter.) Also, if you want to learn to use the computer you have to learn how to program it in some sort of language. Here there is room for discussion, but we suggest that BASIC (Beginners All-Purpose Symbolic Instruction Code) is the way to go. Users of other languages, such as COBOL or FORTRAN may not agree with me – let them write their own book. The facts indicate that BASIC is becoming the almost universal language among computer users and that more and more programs and software are available in BASIC. In a sense, it is like English as a world language. English is very difficult to learn and hard to use for communication, but all over the world people seem to think that you have to know it. Even Radio Moscow offers listeners English language lessons; that tells you something.

DEFINITIONS

Before we get to some computer applications, we have to have a few definitions so that you can understand the literature. If you get stuck, there are usually Byte Shops* or one of the competing home computer companies in town that will be happy to answer questions and let you play with the equipment, and so on. Of course they don't do this just for fun; they are trying to sell you equipment. (Profit and education go hand in hand.) Radio Shack also sells lots of computers.

1. *Digital computer.* An electronic system that does mathematics in terms of digital rather than decimal numbers. For example: $10 = 2^3 + 2$. If we write

$$A2^5 + B2^4 + C2^3 + D2^2 + E2^1 + F2^0$$

where A, B, C, D . . . can equal 1 or 0, we can write any number up to 32 + 16 + 8 + 4 + 2 + 1 = 63. In fact, 63 is written in binary as 1 1 1 1 1 1. The number 30 would be written 0 1 1 1 1 0 = 30. There are systems based on 8 and on 6 as well as on 2. You don't need to learn them, so we won't mention them again.

2. *Minicomputer.* A small computer usually dedicated to a specific individual or purpose. The opposite would be a multi-user general computer.

3. *Microprocessor or Microcomputer.* A single "chip" or slab of semiconductor material that performs some computing function.

4. *Language.* The way you talk to computers. There are several languages; BASIC and FORTRAN are the most popular.

5. *Assembler.* A black box that translates BASIC, COBOL, FORTRAN, etc., into the digital language the machine can understand.

6. *Processor.* The black box that actually does the addition, subtraction, etc.

7. *Register.* A place where the computer stores data.

*Byte Shops is a franchised chain of home computer stores.

8. *Program or Flow Chart.* This is the way you tell the computer what to do. Sometimes you do a flow chart first to decide what to do, then you write the program to "do it." In many cases no flow chart is necessary.

9. *Input/Output Device.* You talk to the computer via an input device, usually a typewriter. It answers via voice, a cathode ray screen, or a printer. Sometimes the input/output functions are in the same cabinet.

10. *Bit.* A "bit" is the smallest unit of data that the computer will deal with. If you look back at our definition of a digital computer and the equation $A2^5 + B2^4$, and so on, where A, B, C, etc., could be 0 or 1, the information as to whether A, B, C, etc., are 0 or 1 is contained in a "single bit."

11. *Byte.* A byte (pronounced bite) is a word that is 8 bits long. Some computer storage systems are rated in bytes.

That completes our list of definitions, but we might mention one more distinction of interest: that between a hand calculator and a minicomputer. You can find some excellent hand calculators that can be programmed to do all sorts of useful things (see the Hewlett-Packard or Texas Instruments people for data about that). The hand calculator has one limitation in that while you can set up a program to calculate something, say the first 1000 common factors of a number, you do have to enter the data, numbers, etc., into the calculator. Another thing calculators can't usually do is *go get* some information when it is needed. If you want to use a calculator to control the gas flow to a furnace to hold the house at a constant temperature, you have to actually "key in" the data on the actual temperature from moment to moment. With a computer, you can say "go get it"; with a calculator, you have to provide it by hand.

At this point, some "calculator freak" will rise up and scream, "My calculator can be connected to a remote source to get that kind of data." I reply, "You don't have a calculator any more, you have a minicomputer."

Having gone through all of that, we shall assume that if you want to use a computer, you will get a book on BASIC and learn at least the beginnings of the language. In the remainder of this chapter we will discuss some applications other than the usual "solving the binomial equation" and "how many coconuts did the monkey steal" problems. We might warn you that some of the applications we discuss may not be readily available with the computer you have on hand. Nevertheless, we feel that they will be available over the next few years. If customers want something, they usually get it. (Remember: this book has to be more or less current for five years or so. If we wrote it to today's technology, it would be out of date before the print was dry.)

COMPUTER APPLICATIONS

Identification and Verification

Identification and verification are important applications if we wish to keep thieves from stealing via the computer. For example, at a hospital only certain people can use the computer to order drugs. At the same time, other workers, i.e., the janitor, should be able to order things like toilet paper. To begin, the user inserts a key and presses the START button. The computer responds with its ID check sequence and prints *what is your ID number?*

When you type in your ID number, and the computer verifies that it is the number that corresponds with your key, it asks, *your order?* You type in the name, quantity, etc. If this procedure agrees with what you are authorized to order, all is well. If not, the computer can say, *bad order* or it can sound an alarm if appropriate.

In this area, another major application of these systems is control of access to restricted areas. Sometimes very elaborate identification techniques, i.e., hand patterns or fingerprint readers, might be used. In other cases a special type of credit card or code word might be enough. One example occurs in a big city hotel; all sorts of problems exist because criminals may get master keys to guest rooms and steal things. The cleaning help must be able to get into the rooms and, since these workers are not very well paid, there is a high turnover and loss of keys. One solution, *already in use,* involves giving each guest a coded key card when registering.* The room lock is coded to the key; since there are thousands of possible combinations no two cards are alike. The code is changed for every guest, so the problem of lost keys has been solved. The cleaning staff is issued master cards as appropriate, but the master codes change so often that there is little danger of a disgruntled staff member carrying off a key and selling it to a hotel burglar, since the key would be useless in a day or two. This system does require that the hotel have back-up power for the computer and the electric room door locks. A loss of power would lead to a lot of unhappy guests.

Computer Memory Applications

There is a host of applications here and we will comment on only a few that might be of general interest. Obviously the field is almost unlimited.

EDUCATIONAL TRENDS BY COMPUTER

One thing colleges and universities worry about is "grade inflation," a fancy word for the process whereby a department seems to give more As and fewer Ds year by year. It may be that the students are getting smarter, but in many cases, it is the faculty getting lazier, being afraid of lawsuits or charges of prejudice. Sometimes this process is hard to see over the short time, and in some years there can be a run of good students. The computer solution is easy; you just ask for data on the fractional number of As, Bs, Cs, etc., over a period of years and look for trends. If you want a graphical plot the computer can do that too; you just have to buy the appropriate hardware. The point is that no one can argue with a graph that says, "Professor, you are getting too easy."

*One interesting device (Micro-Circuits Co., Inc., New Buffalo MI 49117) is a throwaway electronic timer in the form of a ticket. This system could be used to have the electronic room key "go blank" after the guest checks out.

We might note that this and all further applications of the computer memory do require that the data be "entered" or read into the computer. This can be a tedious process with great potential for *human* error. We have no solution for this except to hope that humans will become more accustomed to using computer materials for entering information. Most universities use this system already, so there is hope for the future. Another idea might involve building computers that can recognize handwriting and human speech. Some discussion of the latter idea follows.

DATA REDUCTION BY COMPUTER

Another application of the memory capability of the computer involves mass spectrometric data or gas chromatographic results. In either case, you get a series of peaks. In the mass spectrograph they are amplitude versus m/e ratio (m/e refers to the ratio of mass "m" to charge "e"); in the gas chromatograph they are amplitude versus time. In either case, it helps in identification if each peak is labeled with the possible formula of the material that it might correspond to. For example, a peak at m/e = 44 might be CO_2 or ethylene oxide $(CH_2)_2O$, peak 28 might be CO or N_2, and things become even more complex as the value of m/e increases.

The point is that graphical data can be entered into the computer and then printed out with the appropriate chemical formulas at each mass number. If you saw a peak at m/e = 28, you would know it might be N_2, CO, or C_2H_4. The computer doesn't solve the problem for you, but it does make the solution easier to find.

Another possibility would involve the computer's displaying some typical spectra for comparison purposes. This is very helpful in identifying new or unknown materials. You compare the unknown with the known and look for similarities.

Another application of the curve drawing system (which might well make use of a TV type screen) is the ability to compare several supposedly similar things. One example might be a medical test in which a patient is asked to draw in a deep breath and blow it out as fast as possible, to obtain a vital capacity measurement. There might be some question that the patient did not do as well as he could have on a single test, so the easy thing is to try it three times and then ask the computer to plot all the measurements at the same time on the TV. If the curves differ widely, you may have to run a few more tests. If they agree, you know the patient has done as well as he or she can.

CONNECTING THE INSTRUMENT TO THE COMPUTER

Setting up the system to allow a flow of data and instructions from the instrument to the computer and back again has always been a problem. Here we have no easy solutions; every problem has to be a special case, at least for the first time. There is some help available in this problem area, and one place you might try is a division of IBM – the Instrument Systems Division, 1000 Westchester Avenue, White Plains, NY 10604. A good book on this subject is J. Finkel's *Computer-Aided Experimentation,* New York: Wiley, 1975.

Another interesting facet of modern instruments is that they have available binary coded decimal (bdc) outputs suitable for connection to a computer. In some cases, you can even program and command the instrument by means of the same system. This bcd is convenient, and it didn't take very long before some manufacturer arranged for a system that allows this instrument-to-computer connection for a wide variety of instruments, i.e., balances, photometers, strain gauge systems, and the like. For details, contact your local Hewlett-Packard representative or Hewlett-Packard, 1501 Page Mill Road, Palo Alto, CA 94304.

TRACING APPLICATIONS

This is one of the most useful applications of the computer memory, and, as more and more libraries put their holdings into the computer, it will be easier for users to write programs and save time in finding what they want.

As an example, let us assume that you are interested in the topic of fly ash. *Fly ash* is a vague term for the lightweight residue that forms when any solid or liquid fuel is burned. Fly ash is also formed in smelters and cement plants as part of their operations. As you can see, asking for a large library's holdings on fly ash might drown you with data that you don't need.

Let us suppose you are naive and just ask for the topic fly ash. The TV screen comes back with 500 references. You don't want to look at all of them; so you ask for more specific topics like power plant fly ash control, until you have narrowed the information to the proper area. You might think that this is a trivial application, but if you think of the time that can be spent searching in libraries, it could be very important.

Another important point is that many organizations, such as National Technical Information Service (NTIS) of Springfield, VA, offer computer-generated bibliographies. Most users find that they suffer from the problem of receiving a flood of unwanted material that cannot be eliminated because the programmer does not really understand what the user is looking for. The best user of the computer is the person who has the problem. Eliminating the middleman (the programmer) is almost always an improvement.

The Computer as a Troubleshooting and Repair Manual

Science fiction stories often talk about the self-repairing computer, but anyone who knows computer repair realizes that this is almost impossible. What is needed is a computer that will help a human "repair-person" or technician find and fix the problem.

To begin with, we must realize that the typical large system (radar, power turbine, radio station) comes with instruction and repair manuals in one to six volumes. Every volume is composed of about ten sections, each of which is written by the people who built that section of the system. The difficulty with this setup is that there is no guide to say, "Given trouble X, look in Volume 4, page 16." Even if such a guide were written, it would be very difficult to use. The worst problem might be that Volume 4 would be missing just when most needed.

To "solve" this problem, think of an auxiliary "repair computer." When told, "The system doesn't work," it replies, "Have you plugged it in?" More serious problems would receive replies like "Check test point 31. If the voltage is 3.8 volts ± 10%, go to point 32. If this voltage is not present, replace circuit card 42Z."

In more complex situations, the computer could provide instructions on how to get to the defective component, how to replace it, what adjustments to make and what test requirement will be needed. Such dangers as high voltage, radiation, or solvents can be spelled out to prevent industrial injury and speed up the repair.

If you think this is too far out, we suggest you look up *IEEE Spectrum, 15*(2), February 1978, p. 2. It might surprise you.

Counting and Classifying Operations

One of the most tedious applications of the human eye and brain involve visual observation and classification (counting) of blood cells, dust particles, etc. One good example involves looking through a microscope at a dust sample and counting the number of particles with a given length-to-diameter ratio (this may be a measure of the presence of asbestos). The point is that all of this nonsense can be eliminated by coupling a microscope to an image-analyzing computer. (For details get in touch with Image Analyzing Computers, Inc., 40 Robert Pitt Drive, Monsey, NY 19052.) The equipment involved may be expensive initially, but the programming is simple, and the ultimate saving of time, eyesight, etc., is enormous. Systems of this type for counting blood cells are available from the Perkin-Elmer Corporation, Norwalk, CT 06856, among others.

We might note that these operations do not preclude the necessity for human judgment. In such cases, as in the identification of cancer cells, the skills of the pathologist are absolutely necessary if a correct diagnosis is to be obtained. Theoretically it might be possible to program all the pathologist knows into the computer, but the time and trouble might not be worth it. In some cases you just can't top a human being.

Other, rather simple, applications of the computer involve counting and classifying on a larger scale, for example, packaging of sugar in the small containers used by restaurants. The price of sugar is subject to wide variations, and the manufacturer (or the company that fills the packages) will take this into account by adding more or less sugar to each package, depending upon the price of sugar. Each package can be printed with the actual net weight, so there is no deception involved.

The point is that this is an ideal application for the computer. Programming would involve a decision on the dollar value of sugar to be put in each package. The data on the cost of sugar on that particular day would allow the computer to determine the amount of sugar per package. These data would then control the measuring-packaging machine and of course, the printer that prints "net weight in grams."

Computer Controlled Assembly and Shipping Systems

The first task in this area is getting rid of the idea that great masses of robots are going to lead to most humans sitting about collecting welfare or starving in the streets. A good discussion of the question is given in an article by J. L. Nevins and D. E. Whitney, *Scientific American, 238* (2), 1978, p. 62. Robots are excellent devices for certain types of highly repetitious or dangerous types of assembly work, but they will not replace the human hand for some time to come.

The first requirement for such a machine is that it be programmable to perform a variety of operations. This is relatively easy to do with a computer, providing that we account for the variability in all mass-produced parts – even a machine will change slightly as it repeats a series of operations. Normally, the human assembler will compensate for the variations in a production line but a machine cannot. The point of all this is that, for many operations, the best thing to do is use the automated system and the computer as aids to the human assembler rather than as replacements.

Examples of this type of system are discussed in the article by O. Firscein and associates in *IEEE Spectrum, 11* (7), July 1974, p. 41. In one case, letters and packages in the post office go by a human reader who has 1 second to read the address and press the appropriate buttons that direct the item onto the proper conveyor belt. There is some interesting software* here in that a letter may be "read" in the New York post office and directed to Hong Kong. The first question is, "surface mail or air mail?" If it is surface mail, the question of "what ships are available" will arise. (Ships sail to Hong Kong from a variety of ports.) The letter will have to be directed to the appropriate city, and to do this the computer needs to know what ships are sailing when.

Other relevant information for the computer may be that a ship from Houston will have to go through the Panama Canal and may take longer to get to Hong Kong than a ship from Los Angeles. The existence of strikes or excess aircraft capacity (so that surface mail can go by air) just adds to the fun. The point is that this is the sort of thing computers were intended for – not to replace, but to help, human beings.

Computer Controlled Checkout Systems

The application of the computer in store checkout is just the tip of an iceberg that will change the way merchandise is sold in the United States. To appreciate this, consider that most, perhaps 80%, of a retail store's expense is employee related. If the store is to hold down costs and meet competition, the available personnel must be used with utmost efficiency. "You can't have any more cats than will catch mice." This is one aspect of the problem; the other is that in many cases it is hard to keep up with what is selling and when to stock up again. All of this is greatly improved by the computer, as we shall see.

As an example, we shall consider a department store that is selling a particular item: women's sweaters. Management would like to know which models are selling best so that they can reorder and increase advertising. At the same time, the buyer

**Software* simply refers to the instructions that make the computer do its thing. The transistors, etc., that do the work are called *hardware*.

in the sweater department will have to keep an eye on his total sales so that he can schedule sales and cancel orders.

In the bad old days, all of this depended on an army of clerks who had to read the handwritten sales slips and then enter the data in the file. Even after this, it was hard to look for trends, and stores often got stuck with things that didn't sell or ran out of things that did sell better than expected.

This problem has been reduced, if not eliminated, by the use of product codes or cash register systems that can read the digits on the price tags. In actual practice, the clerk passes an optical reader over the price tag. This allows the computer to read the product code or the price tag (the product code is just another way of identifying the product). In either case the computer knows what was sold, including size, price, and color, and who it was sold to (for billing purposes). The billing is another matter; the question here is what is done with the sales data.

The sales information can be used in many ways. It can give the management a daily readout of what is selling and how the rate compares with predictions. A simple subtraction program will provide data on the remaining stock, including the supplier's information on how long it will take to get the next shipment. This will allow for quick decisions in the ordering, advertising, and shipping departments. If an item is selling well, there will be more clerks needed to restock that department, which may require transfer of personnel, overtime for workers, or other changes. The computer can provide data on the cost of various alternatives.

The same principles can be applied all over the store. If an item is selling below the rate established by experience over the past several years, it is time to ask why. Is someone in the neighborhood cutting prices, has store product quality declined, is something new on the market? This kind of information can show up the weak spots early enough to allow trouble to be stopped before it starts.

One item that many stores look at is the rate of returns and complaints as a function of department, items, supplier, etc. Each complaint that comes in may mean five more dissatisfied customers who say nothing but go elsewhere the next time. It is not unknown for an unethical department manager to order a poor line of goods and establish a great selling record by letting them go at very low prices. It looks good until the returns come back; if he isn't caught early, he may continue long enough to leave for a better job, with the store holding the bag.

The application of automatic price reading equipment is so important that we might give another example, the familiar supermarket. Here automatic checking with price codes on every item has demonstrated the following:

1. Faster checkout with less waiting in line, which makes for happier customers.
2. A significant reduction in errors, especially those due to "sweetheart sales," where a clerk deliberately undercharges a friend.
3. Better accounting for complex sales taxes and food stamps. Some items are taxed at one level and some at another level, others not at all; food stamps can be used for some things and not others – the computer takes care of all this.
4. Coupons are a popular promotion item by manufacturers, but a headache to the store. Here again the computer has demonstrated its ability to improve things.

5. If the customer wishes to pay by check, the computer can run a quick verification with the local bank to deter "paperhangers." There is the danger that if credit is refused without good reason the customer may sue. Each store has to decide how it wants to play this game.
6. The customer receives a clear and itemized sales slip for later reference or income tax deductions (for medicines or prescription vitamins, health foods, etc.).

The Computer as an Aid to Municipal Services

Here again the applications are so numerous that we can mention only a few of them, for example, a fire department in a large city. The computer would contain data on special hazards, i.e., chemical plants or flammables or toxic materials in storage, for every section of the city. If a fire occurred in a hazard area, the appropriate equipment and personnel would be dispatched automatically.

Another application might inform the fire department when a fire area has a large number of aged or handicapped people who would need help in leaving the building. If some streets are torn up for construction, the computer can give the fire trucks the best route to the blaze (turning around a hook-and-ladder truck on a narrow street is not something one does in a hurry).

In many cities, fire departments are plagued by false alarms, usually from the same call boxes. In some districts, the firemen don't even leave the station until a police officer checks that a fire exists, or a second alarm is turned in. This practice may cause a delay in getting to a real fire, and here the computer might help by indicating the number of false alarms from the box in question over the last ten years. If the false alarm rate is low, there is a good chance that this is a bona fide fire.

Police departments can and have used computers to check lists of criminals, the type of crimes that they commit, and their present location. Criminals, like the rest of us, are creatures of habit; a safecracker seldom takes up strong-arm robbery. If a safe is "blown," and the evidence indicates that an expert did the job, the computer can call up the names of all the possible suspects and eliminate those who are deceased, have gone straight, or are already "doing time."

Computer Graphics

This is an area that is changing and expanding so rapidly that it is almost impossible to write something that won't be out of date before it gets to the printer. The best we can do is discuss come applications and suggest that you read the literature for current details. (*Machine Design* is great for mechanical engineering types, *Spectrum* for electrical engineers; aeronautical engineers might look at the journals of the American Institute of Aeronautics and Astronautics while civil engineers have the journals of the American Society for Civil Engineers.)

One important application is generating three-dimensional views and solid models of two-dimensional drawings. This is especially important when designing complex three-dimensional systems like refineries, industrial plants, ships or multi-story warehouses where belts, pipes, conveyors, and sewage and electrical lines all pass through the same area. In the past the *only* way to avoid having the contractor come in with the sad news that "two pipes cannot occupy the same space at the same time" was to actually build a three-dimensional model in wood and plastic. This was time-consuming and expensive. A more modern system involves giving the dimensions to the computer and having it "draw," on the cathode ray screen, a model of all the elements. You can even have things in color, with water green and sewage red, while other programs allow the system to optimize the piping layout to save money. The point is that you don't end up with a sewage pipe in the middle of an office or an electric line snaked inside a heating duct.

Another application of computers involves three-dimensional stress analysis. Again, the computer generates a perspective image of the object, lays out the grid for solution by finite analysis techniques, and then solves the problem. If the structure is too weak, the computer will predict the possible point of failure and might even be programmed to determine where and how the structure should be improved. This does not suggest that engineers will be out of jobs, because someone has to conceive the idea in the first place. The people who will be out of work are the detail draftsmen who did routine layouts and converted one view of a structure into another.

Computer animation is another area of interest to anyone who has to explain things to someone else. Don't think that limits it to teachers and employee-training programs. Every engineer has to "teach" the corporate managers that "his" ideas are the ones that the stockholders' money should be spent to support. Managers have a tight time schedule, a repugnance for technical details (that obscure the real facts), and you can't force them to listen with threats that "it will be on the next exam." Animation is great, but in the past it has been costly. Now you can draw a series of animated stick figures showing how passengers will get out of an airplane in an emergency. The computer can put in all the intermediate figures so that the motion is not "jerky." The computer can show the effects on passengers of canes, tight skirts, excessive weight, and so on. Best of all, the passengers aren't just "stick figures"; they have real faces, legs, and arms that can get "caught" on seat projections that are in the wrong place.

For some details on how computer graphics can be used for auto crash simulation, you might want to look at the article by F. S. Lavoie, "Computerized Cartoons," *Machine Design, 43* (9), 1971, p. 71. Other applications to general engineering problems are discussed by L. Teschler, "New Realism for Computer Graphics," *Machine Design, 50* (4), 1978, pp. 93-99. A few examples of what can be done will be given here. Some of these examples were taken from the Teschler article referred to above. Figures 5-1, 5-2 and 5-3 show a series of views of a spacecraft support frame. Figure 5-1 shows an overall view. If the engineer wants to see more detail in a specific area, he commands the computer to "bring it closer." The results are shown in Figures 5-2 and 5-3. The engineer can get any detail and orientation he likes.

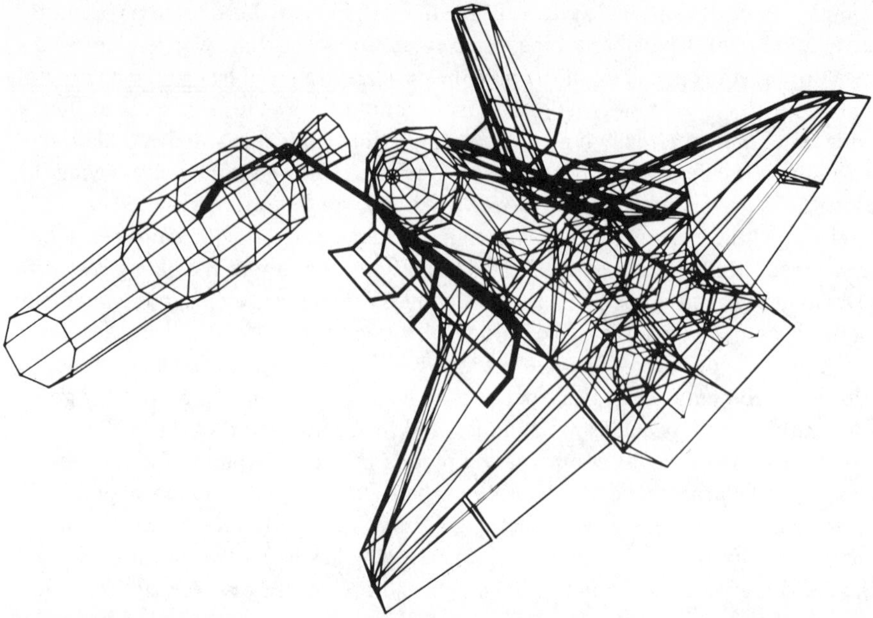

Figure 5-1. Overall view of spacecraft support system. (Photo courtesy of Evans and Sutherland Computer Corporation, Salt Lake City, UT 84108.)

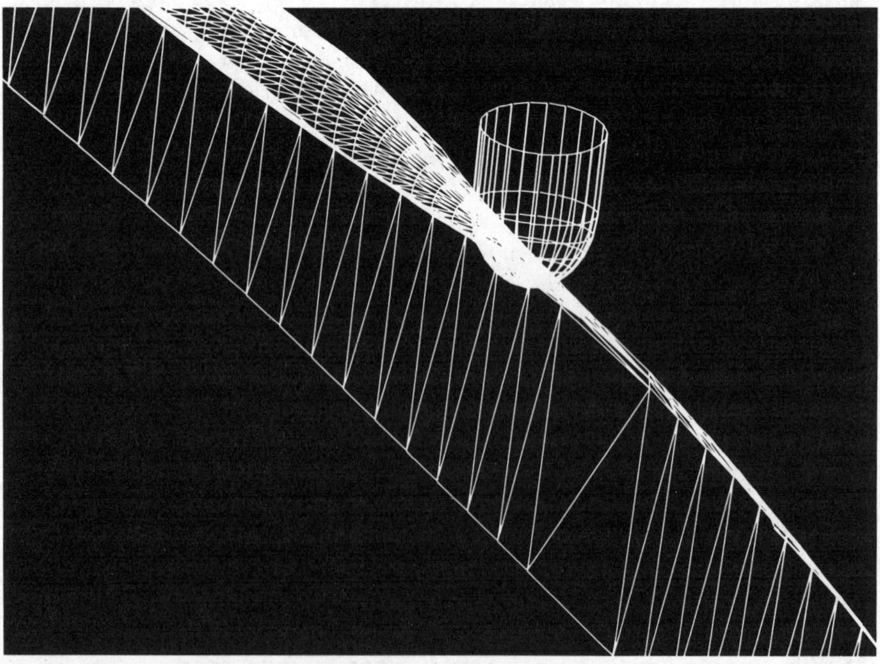

Figure 5-2. Side view close-up of spacecraft solar panel support. (Photo courtesy of Evans and Sutherland Computer Corporation, Salt Lake City, UT 84108.)

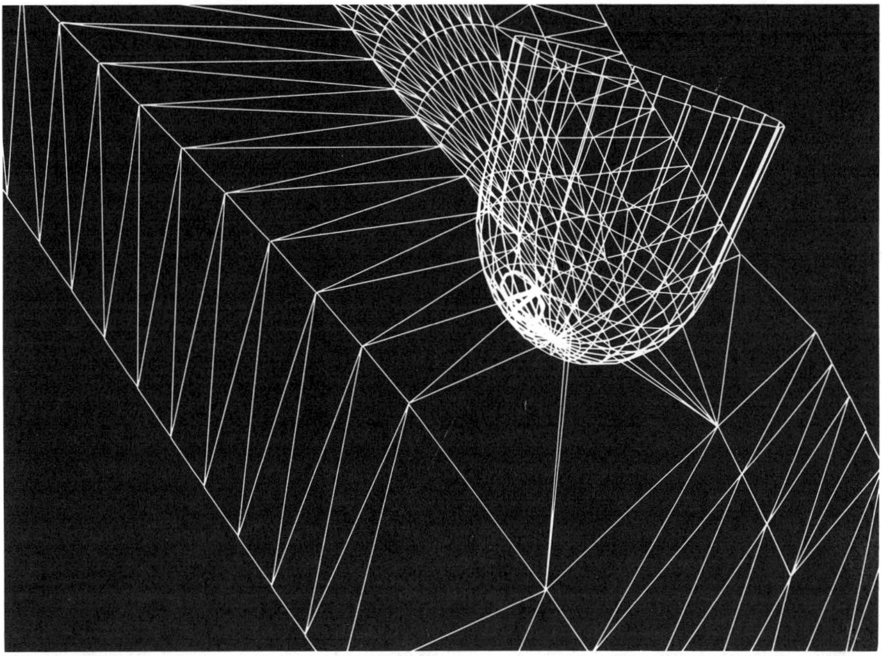

Figure 5-3. Overhead view close-up of spacecraft solar panel support. (Photo courtesy of Evans and Sutherland Computer Corporation, Salt Lake City, UT 84108.)

Another example of computer applications for general engineering problems is shown in Figure 5-4. The product is a forging that would be extremely difficult to render on two-dimensional drawing paper. The point to appreciate is that this product does not really exist except in the computer. What Figure 5-4 shows is only a photograph of a cathode ray tube screen. The engineer can call for the part to be rotated, stretched, have larger holes drilled and so on – all with the computer. He can run stress tests to see if the part will stand up to the expected loads. He can see what the effects of wear or a change in materials (i.e., aluminum versus steel) will be. Needless to say, "The practice of design engineering will be revolutionized by the computer."

One last example of automotive applications includes a mention of "fitting the automobile to the customer" or, as it is called, human engineering. A typical question arises when an armrest is to be located in a new car design. The armrest can be moved forward or backward, rotated at an angle, or made wider if that best suits the potential customer. At one time this was a matter of guesswork and hope-for-the-best, now it is simple to store the dimensions of the 95th percentile of the population (95% of the population will fall into the group) and locate their arms with respect to the proposed armrest. The computer shows the optimum position, angle, and width – the job is done. A computer has helped fit the device to a human

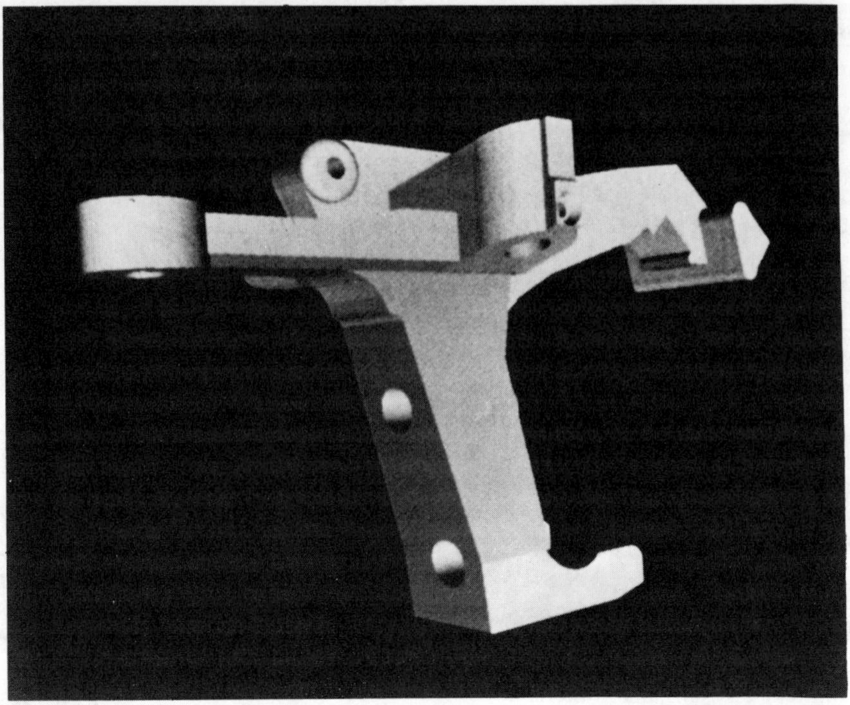

Figure 5-4. Proposed forging with hole locations. (Photo courtesy of Mathematical Applications Group, Inc., Elmsford, NY 10523.)

being. For more details you may want to look at the article by C. Wise, "Chrysler Puts Cyberman in the Driver's Seat," *Machine Design, 50* (8), 1978, p. 46.

Another interesting area for computer display involves numerically controlled machines that will cut, mill, weld, turn, and fasten parts together. The problem here is that all the actuating arms have limited movement, and the computer that drives them has no common sense at all. If you tell the machine to smooth one surface of a block, you had better indicate which side of the block the tool should come in from. If you don't, the machine might try smoothing the surface from inside the block, with disastrous results. Here again, you can program the computer to display the block and then have the tool(s) come in and do their thing. If one operation requires that the tool holder exceed its limits or work from the inside of a solid surface, the computer will stop and tell you to "fix" things. It is much easier to change a cathode ray tube display than fix a broken tool.

In this connection there has been a problem with industrial controllers in that they required the user to "know" some sort of computer language in order to "write" the program. As you might guess, that barrier to the use of computers has not lasted long, and there are now industrial control systems that can be programmed in "English." The translation of English into computer language is done by a built-in assembler.

Typical commands that might be given to an industrial washing machine would require the user to type in commands like "presoak 5 minutes, wash 5 minutes, rinse hot, rinse cold, spin dry." This may not seem like much to someone used to a home washing machine but, to an industrial user who might wash 50 different items in 30 different ways, it could be important. You can even see the day when the materials, say, "circuit boards" (they get washed too, you know), are loaded and the machine is commanded to "wash sequence number 3," or "wash Orlon" and away it would go.

Similar techniques might be applied to machines doing other repetitive operations; the point is that the controller understands English. For more details see "Product Controller is Programmed in English," *Machine Design 50* (13), 1978, p. 8, or contact Rockwell International, 3430 E. Mira Loma Avenue, Anaheim, CA 92806.

Finding the Right Path by Computer

In a sense, this is a subset of the piping problem discussed earlier, but it is so interesting and there are so many applications that it is worth discussing further. In the layout of printed circuit boards for electrical applications, there is a need to route wires from one component to the other. There is also the requirement that the components to be interconnected be near one another. At the same time, the heavier items, such as transformers, should be at the edge of the board so that they will be supported against vibration. Shielding of stray electrical signals and separation of components that might be affected by leakage is another thing that must be weighed to get the optimum design.

All of these factors could in principle be taken into account by a human designer, but in practice the job is just too tedious and nonrewarding. However, the computer is perfect at that sort of thing. It can go through every possible combination and permutation to discover the best shielding method or wire routing system. For details of one such program you might want to consult "Designing Circuit Boards the Easy Way," *Machine Design 50* (5), 1978, pp. 110-115.

Computer Controlled Check-In at Airline Terminals

Nothing is more annoying than arriving at the airline ticket counter with a valid ticket and having to wait 15 minutes while some mixed-up traveler has his itinerary rewritten. The same wait might occur at the gate where boarding passes are issued.

All this annoyance could be avoided by printing airline tickets that are computer readable. On arrival at the check-in counter one would insert the ticket. The computer would verify the reservation and come up with "Do you have any baggage to check?"* The customer would press the "yes" or "no" button to answer. If the

*For a recent reference in this area, you might want to look at the two-part article by A. L. Robinson, "More People are Talking to Computers as Speech Recognition Enters the Real World," *Science 203* (4381), 1979, p. 634.

answer was "yes," the computer would say, "How many pieces?" and, "Please put them on the scales." The computer would print the baggage checks, give the passenger the receipts, and allow a human employee to put the tags on the luggage.

The next step might involve the computer asking, "Would you like to choose your seat?" and displaying a diagram of the aircraft with available seats in green and filled seats in red. The passenger pushes the appropriate button, and the boarding pass is issued. The computer takes the seat in question off the "open list" and marks it "taken."

This type of operation would speed up the check-in process and free human beings to help one another, rewrite tickets, etc.

Computers That Respond to Voice Commands

Let us begin by saying that we will not consider a real "talking computer" like the machine HAL in *2001, A Space Odyssey;* we are operating at a much lower level. Think about a mechanic who has a small garage with perhaps one assistant in the office. The phone rings and his hands are covered with heavy grease. Instead of having to wash up or get the phone full of grease, he yells "32" at a loudspeaker-microphone on the wall. The computer recognizes that "32" means "put the phone on the loudspeaker." The computer does this and the mechanic can tell the customer what the situation is. If the customer wants to talk to the office, the command "51" will switch the call to the office phone. If the conversation is complete, the cry of "75" will hang up the phone.

One point here is that the objective is not a computer that will recognize a particular human voice, though that could be done; it is simply a machine that can recognize a spoken number and provide the appropriate response. In theory the machine should be trained by the person it will have to listen to. This is easy enough in a one-man garage, but may be a problem when many voices must be recognized. Once again, we can expect rapid development of this sort of system when the potential begins to be appreciated. For the latest news in this area, you might want to look at *Popular Science 213* (2), August 1978, p. 62. The article is called "Voice Recognition Systems," and the interesting thing is that gear of this type is *already* available for "home computers."

Advanced Verbal Communication and Response Systems

Here again the numerous applications would fill a book, and we can discuss only a few. A potential user will think of many more. One example might be an automated checklist for airline pilots. Another might be a turn-on procedure for some complex system like a paper mill. Human beings tend to forget, and in some cases one missed step can mean disaster.

In the case of an aircraft checklist the pilot might say "computer, institute takeoff check." The computer answers, "affirmative, begin takeoff checklist." The pilot replies "checklist begin" just to be sure the computer knows what to do. The use of a special sequence of words is insurance against an accidental start-up or aircraft use by unauthorized personnel.

The computer might say "control system check." The pilot responds by performing the indicated operations; the computer notes that the proper action has been taken and indicates "control system OK." This sequence allows the computer to provide the next statement "radio system check," and so on. The point is that the computer doesn't forget; it ensures that the proper action has been taken and informs the pilot while making a record of what was done. This precludes arguments about who did what.

At this point someone will ask, "Why have the human pilot at all?" The human is there to handle the unpredictable and unexpected from a blown tire to a stuck nose wheel. The routine tasks are left to the idiot computer. Human beings are available to solve problems, not memorize long lists. For the latest details on low-cost computers that both "talk and listen," you might want to look at the article by L. Teschler, "Computers for the Home," *Machine Design 50* (17), July 1978, p. 68.

The Talking Computer as a Helpful Device

Another and somewhat more advanced application of the computer would allow the public to interact with the computer whenever information, assistance, etc., are needed.* One example might be a green pillar box that people of all ages would know "you can talk to." If a child were lost, he or she could go to the pillar box and say "I am lost" or simply begin to cry. The computer comes "on," activates its infrared sensor, and measures the child's height. If the speaker is under 3.5 feet (1.07 m), the computer decides the subject is a child (or a dwarf) and turns on the "child tape." A kindly voice (everyone's grandfather), begins to talk to the child, finds out what his name is, where he lives, and so on. At the same time the appropriate authorities can be notified to look for the parent and pick up the child. As more people travel, the pillar box could be a traveler's aid at every airport and train terminal. When spoken to in French, it would turn on the French tape and be prepared to reply or summon a French-speaking person, physician, dentist, lawyer, policeman, etc.

A similar system could provide weather information, give directions to the nearest public toilet, and so forth. Here again the computer is an ideal device for remembering things that humans forget. It is great at "standing guard" 24 hours a day (at low cost) and then summoning a human being when judgment is needed. We don't need a French speaker at every airport and railroad station 24 hours a day. What we need is a system (the computer) that can recognize when a French speaker is needed and know how to contact one.

*If you think this is a long way in the future, look at Texas Instruments "Speak and Spell" device. It is programmed to speak more than 200 words in a human voice with inflection and fidelity, price $47.95.

The Computer as a "Person to Talk to"

There is evidence that much of what psychiatrists, psychologists, and counselors actually "do" is listen to people talk out their problems. The difficulty is that the time of these trained workers is expensive, and they are not available to the general public or the poor.

This obvious need has hatched out a large number of "self-trained" or diploma mill counselors, many of whom are more in need of therapy than their customers. The point is that computers can, and have, provided much the same conversational therapy that professionals provide, and at far lower cost. This is not to suggest that a computer will replace the trained professional, but rather that the computer can be programmed to "talk" to the patient and encourage the patient to discuss his or her problems. As the details develop, the computer may discuss similar cases where patients with the same type of problem have recovered after appropriate therapy. Another situation might have the computer recommending one or more group therapy situations as appropriate to the patient's needs.

We must warn you that the whole area of psychotherapy is subject to a hash of vague definitions and that fundamental differences of opinion exist between equally reputable practitioners. There are real questions about how human beings might relate to computers for therapy, and the best we can suggest is that you consult the letter and references given by J. Weizenbaum, *Science 198* (4315), October 28, 1977.

Eyes for the Computer

The simplest way to input data into a computer is the familiar typewriter or punched card; however, we recognize that it would be very convenient if the computer could see for itself. Some applications, i.e., counting blood cells or analyzing astronomical photographs to provide a readout of light intensity for 10,000 stars, are obvious. However, these are by no means the only things that can be done. One such application involves the inspection of minute printed circuits to see which are good and which are bad. In the past, human operators did this (with considerable eyestrain) and recorded the location of each circuit for "use" or the "trash pile." This process can be done much faster by a computer, which immediately records the location of each circuit and designs the appropriate connections (assuming that it is not one for the trash pile). The computer can be "set" to look for ragged edges, off-color coatings, or whatever problem is most serious.

Another application, in automobile crash testing, involves the computer's "reading" the motion pictures taken of the event and plotting the speed, movement, etc., of all the parts of the dummy (to simulate the human response) and the adjacent parts of the automobile. In some cases, there are more than two hundred objects to be kept track of; but if some piece of plastic is going to break off and hit the driver, the manufacturer wants to know so he can change the design. To do this sort of film reading with the human eye would be "awful," but for the computer it is routine.

Other applications where the computer "looks" at the number of autos at various major intersections and adjusts the traffic lights for optimum flow are now under test in some European cities. The advantages in having optical data available to the computer are so large that there will be many other applications of this type in the near future.

Using the Computer to Detect Electrical Power Problems and Take Proper Action

As our society becomes more dependent on electrical power, we tend to forget that, like all human enterprises, the power company is subject to a variety of problems: shortage of fuel; effects of rain, snow, and ice; vandalism; and federal regulations. All of these things (especially snow and ice) tend to interfere with the constant voltage and frequency that we have become accustomed to. This is particularly troublesome where computers are used, since they suffer from even small variations in voltage and frequency. There are all sorts of devices that we can buy to prevent or reduce this problem. The question is, "What to use and when?"

Let us consider the simplest situation – a sudden surge in line voltage and frequency as someone turns off a large piece of machinery. This can be taken care of by a constant voltage transformer permanently installed in the power line, as long as the surge doesn't last more than about two cycles (1/30 second). For details contact the Sola Electric Division, General Signal Corp., Elk Grove Village, IN 60007. If the surge lasts longer than 1/30 second, the computer will have to be disconnected from the power line and run from batteries, a gasoline or diesel generator, or a flywheel system. Design and purchase of these systems is out of our range. The point here is that the computer has to detect that

1. Some trouble has occurred.
2. The problem will last too long to count on the constant voltage transformer.
3. Make decision on appropriate action.
4. If power is limited, make decisions on what data can be "lost" and what must be "saved."

This is important. In many cases the user can live without the computer for five minutes or so until the system, generator, etc., can pick up the load. In this case all that is necessary is enough "immediate" power to hold the computer in the resting or storage mode so that important data are not lost. If the computer is operating an airport control system, there is *no possibility* of turning it off for a moment. In that case the immediate power (usually batteries or a flywheel) must be able to pick up the whole load in 1/30 second or less. If worse comes to worst (as it usually does), the computer will have to start turning off lights in the building to "save" itself. (For more information on how a system of this type was adapted to a pulp and paper plant, you might want to consult the article by A. Kaya, "Industrial Energy

Control, the Computer Takes Charge," *Spectrum 15* (7), July 1978, p. 48. The complexity of control will surprise you.)

All of this suggests that the computer must have the ability to "sense" the line voltage and frequency on a continuous basis. If a problem occurs the computer will have to

1. Provide an alarm.
2. Measure the duration and seriousness of the surge (a total loss of power is the worst case).
3. Start getting ready to turn on the auxillary generator.
4. Make plans to bring elevators to the nearest floor and open the doors.
5. Provide power to stop electric transit vehicles at stations where passengers can alight.
6. Decide which electrical appliances (lights, fans, etc.) can be shut off and which must be kept on.

Meeting all of these requirements is not easy; it requires a good deal of programming and testing. Nevertheless, as cities and airports become more complex there will be no escape from the problem. If the computer must protect itself, it might as well protect us too.

The Computer as an Industrial Safety Device

Here again there is a host of applications, and we can discuss only a few to whet your appetite. Industry has a basic need for security against fire, theft, vandalism, etc. There is a need to monitor machines whose failure may be catastrophic. Energy conservation is necessary, if the company is to operate these days. All of these things are applications for the computer.

SECURITY APPLICATIONS

Some of these are obvious. The plant will install fire and intrusion sensors for the computer to scan sequentially, so that every sensor is on line for perhaps 10 milliseconds once a minute. This allows the computer to scan 100 sensors continuously at the rate of one every ten milliseconds. If the computer detects some kind of signal (the seismic sensor might pick up a distant truck), it might compare the signal level or frequency spectrum with a previously programmed pattern. If the amplitude and/or frequency is that of a distant truck, all is well; if it is a human foot in an area where people are not supposed to be, the alarm will sound.

A very similar system can be used with TV scanners in areas where no one is supposed to be; the alarm will sound if any movement is detected. The alarm must be set at a level high enough to ignore the occasional rat or cat but that presents no problem. There are already neighborhoods (of the wealthy kind) where every house is wired to a fire and burglar alarm system that rings at the local police station. In these days of high crime levels such a system provides a real sense of security.

MONITORING MACHINERY

One example that is already under study involves fitting a large airplane with sensors to detect the formation of cracks or defects in major structural members. In essence, this system listens to the creaking noises that the aircraft makes in response to varying stresses and strains. (Humans can't hear these noises directly – the frequency is too high.) Once again, the computer scans all the sensors on a sequential basis looking for sudden changes in sound level, changes in frequency, anything different. When something is detected, the pilot can be warned.

The computer has all sorts of advantages in this area. For one thing, it can "average" the normal noises over some period of time and simply ignore any signal within that "average band." If a signal that is larger than average is observed ("larger" here might mean 10% or more), the computer gives the pilot an alarm signal. Similar systems might compare the signals from the sensors in a given area, i.e., left landing gear; if several sensors indicate an excessive signal during a landing, it is a "sign" that the gear is due for close inspection.

Another good spot for sensors is in aircraft gas turbines where the highest temperatures are reached during takeoff. Each engine has limits on temperature and time, but in some cases the pilot may not be able to watch everything at once. The computer will watch the sensors to monitor the time–temperature curve, and when it reaches a pre-chosen value the computer may signal "time to check the engine."

In this connection, we might note that in many cases perfectly good aircraft engines are removed for inspection, because the allotted hours of use have been expended. This policy is a waste of money to the company, and it is not a guarantee that a "sick" engine will be pulled before it fails in flight. What is needed is a sensor and a computer to tell the pilot, "Because of excessive wear your time on No. 3 engine will be out in 10 more hours." The problem here is that, while the computer part of the job is relatively "easy," the sensors for the sick engine do not yet exist. Here we have to have faith that the sensors under development will do the job.

Another and more current (because the sensors are already available) application involves monitoring the oil in the engine and transmission for sudden increases in acid level, water content, or trace metals. Trace metals (i.e., iron, copper, lead) may be a sign of excessive engine wear or incipient gear failure. Once again, a warning that trouble is imminent can be used to "fix" a problem before it gets serious.

For more details you might want to write to one of the addresses below. Ask for information about the Spectrometric Oil Analysis Program (code name SOAP).

Commander, Naval Air Systems Command
(Attn: AIR-4031A)
Department of the Navy
Washington, DC 20460

Commanding Officer, Naval Air Rework Facility
(Attn: Code 0340)
Naval Air Station
Pensacola, FL 32508

The Computer as an Aid to Better Use of Transport Vehicles

It has been known for years that the "better" drivers did less damage to loads, got better fuel mileage, and had less vehicle maintenance than the "hot rod" types. In the past it was difficult to find and reward the good drivers, but recently a mileage recorder and shock load sensor system has become available. This unit provides a tape showing gas mileage from point to point and the number of "panic stops" or "jack-rabbit starts" per trip.

These data together with maintenance information on an annual basis is fed into a computer. The better drivers are easily identified and rewarded. Driver response to the plan has been excellent – everyone likes to be recognized for doing a good job. For more details, contact Aware Systems, 4110 N. 70th St., Scottsdale, AZ 85251

Computers to Control Drilling

One interesting application of computers involves control of a drilling process in a production line setting. Engineers have known for years that the torque required to turn a drill increases as the hole gets deeper; this is due to the increased friction against the sides of the hole and the need to raise the chips a greater distance before they break off. The problem here, especially with a small drill, is that excessive torque will break the drill. (Having a steel drill broken off inside a hole in an aluminum block is not much fun.)

Another problem in drilling involves the rate of feed, which is a function of the downward force exerted on the drill. If this force is too high the drill will bend, friction against the sides of the hole will increase, and breakage is likely to occur.

For best operation, the computer system senses the torque required to turn the drill and the horsepower required by the drilling system. The rate of feed is then adjusted for optimum drilling as a function of the size of the drill, the depth of the hole, and the material involved. The net increase in production is significant. For more details you might want to look at "Optimizing Drill Torque on the Production Line," *IEEE Spectrum 15* (3), 1978, p. 22.

Cement Plant Problems

This is another example of how the computer might be used in an industrial setting. Consider a cement plant that burns coal; the coal is stored as large lumps to reduce dusting and oxidation losses, but it is actually burned as a ground powder. So we have the problem of grinding enough coal to keep the kiln hot, while at the same time not grinding it faster than it is consumed.

The rate of coal usage is a function primarily of the flow of material from the quarry to the grinding train. The rate at which the ore (mostly limestone) is mined depends upon the number of men working, the drilling/blasting schedules, and the number of equipment breakdowns. The problem is made somewhat more complex because the moisture content of the ore varies with the weather (rain), and the ore itself is by no means pure limestone. There is a time lag of some 45 minutes between the loading of the ore into a truck and the arrival of the material at the grinder; this lag allows for analysis of the ore for impurities and water content. Impurities may require that the ore be "doctored" by adding other chemicals. The computer can make sure that these materials are available in the proper quantity for addition at the optimum time.

The ore is transported to the grinder by a mixture of trucks and conveyors. If there is a breakdown, the computer will have to order the system to draw ore from the "ready reserve" while the problem is being repaired. If the reserve begins to run out before the ore begins to flow again, the coal grinder and transport system will have to slow down or stop to allow the kiln to cool to standby temperature.

A similar situation exists in the grinding train that crushes the limestone before it gets to the kiln. If a breakdown occurs, the computer calls upon the ready reserve (usually stored in the *surge building*). The objective is to keep the plant running – that is the way to make money for the stockholders.

Another consideration in the grinding process is the variation in ore hardness, with the "soft" material grinding rapidly and vice versa. The computer must have information on the grinding rate so it can control the rate (via the feeding system) at which ore is fed to the grinder. If the rate of grinding exceeds the capacity of the kiln, the overflow is sent to the ready reserve for use in case of a breakdown.

All of this must be coordinated with the cement bagging and transportation system and the predictions of how much cement will be needed by the customers. Certainly human judgment is needed to set up the software that tells the computer what to do; once the system is set up, however, the computer can monitor and control a large number of factors that no human could possibly remember. For more details you might want to look at the book edited by T. J. Harrison, *Minicomputers in Industrial Control,* Instrument Society of America, 400 Stanwix Street, Pittsburg, PA 15222 (1978).

APPLICATIONS OF COMPUTERS TO LABORATORY INSTRUMENTATION

Once again we can mention only a few of the many applications that are available. To keep up to date in this area, you should read the publications of the Instrument Society of America and the *Review of Scientific Instruments.*

One item of interest involves the use of a microprocessor chip to average a set of weighings automatically on a balance sensitive to noise or vibration. In the past, it was possible to "damp" a balance to remove noise and vibration effects, but this made the unit very slow in approaching the final value. The newer system is fast in response and allows an accurate determination of the true average weight.

Another balance feature is a printout; this allows you to review all the data to see what sort of data the average was based on. Given 10 readings of which the greatest was 110 grams and the smallest, say, 90 grams, we can have some confidence in an average of 100 grams. If the highest value was 150 grams and the lowest 60 grams, an average of 100 grams might not mean much.

Another application that is not without its pitfalls is curve smoothing. Erratic data are not improved by smoothing, and sometimes an anomaly can be a sign of some significant phenomenon. This is another case of "The computer can help you with the data, but it can't help you think."

Computer programs that subtract background signals are also often useful. Even in the dark a photodetector will yield some signal due to random cosmic rays and noise. This background can be removed by a computer programmed to yield only the "light on" signal. The thing to watch for is a poorly designed program. If the background signal is of value 1 and the light on signal is 10, the difference is 9 for a signal-to-background ratio of $9/1 = 9$, which is very good.

However, suppose the background goes up with the signal to the point where the difference is 100,000 - 99,991 for a difference of 9. Now the signal-to-background ratio is 9/99,991; obviously this information is meaningless! The program must be written to avoid this sort of error.

Computers can be used for timing experiment cycles, turning equipment on and off when a process is finished, and even monitoring the product. Usually the biggest problem lies in the sensors and the conversion of data into a form that the computer will accept. As time goes on, more and more sensors will have data outputs suitable for computer input.

One application that saved us time and trouble arose when an experiment generated six feet of chart paper with four traces. Trying to plot this on one 8½" × 11" sheet was tough, because the interesting events occurred at random intervals. After we found it possible to convert the analog traces into digital form, the computer plotted the data in a format suitable for reproduction on an 8½" × 11" sheet. One interesting thing was the ability of the computer to "try" different scales for plotting the data and showing the results on a cathode ray tube. If we didn't like one format we could just "wipe it" and try another; when we got what we wanted we simply asked for printed copy.

Using the Computer to Monitor Animal Movements

Back in the Dark Ages (when graduate assistants were cheap) it was common practice to set up a rat maze problem and have a graduate student count the time the rat took to run the maze under various conditions. Other experiments involved counting how often an animal pressed a lever in response to a stimulus. In one case, a rat received an electric shock if it went to a given area of the cage. The next step involved placing the rat's food in *that* area and observing how long it took for the rat either to forget about the shock or get up his courage to take another crack at getting the food.

The point of all this is that these observations took a good deal of time, most of which involved simple observing and recording. These days, graduate students have other (we hope, more profitable) things to do, so that mere observation and recording are the task of the computer.

In the case of the rat that must go to a particular area for food, we would set up a light and a photocell (see Chapter 7 for details) that would generate a signal whenever the rat entered a particular area. If the light were thought to be a disturbing factor, we could go to an infrared source and detector; the rat would never know it was there. The data on how often the rat entered the test area could be recorded and correlated with age, sex, shock level, or whatever parameter the experimenter thinks might be important. In many cases, significant correlations have been missed because it took too much time to plot them; with the computer the time is minimal. The computer can easily apply statistical tests to demonstrate that correlations are accurate within known confidence limits, or significant at the *X* level.

MEDICAL APPLICATIONS OF THE COMPUTER*

As the United States moves closer to socialized medicine, we must make plans to handle the vast increase in patient load associated with federally funded health care. Computers will play a major part in this process, and it is of some interest to examine a few selected applications.

At this point we must recognize that computer-assisted health centers are very expensive, and a large patient flow is needed to justify the equipment costs. Installations of this type are usually called multi-phasic health testing centers (MHTC) and are generally associated with medical centers or large hospitals.

One example involves the computer as a system for taking patient histories. We should warn you that something of this sort has been tried before with rather poor results. (See Engineering Concepts Curricular Project, *The Man-Made World.* New York: McGraw-Hill, 1971, p. 19.) The difficulty was primarily a lack of proper communication between the program writer and the patient. One part of the program asked, "Have you had abdominal pain within the last year?" Even a physician might have some question about how to answer that one. A much better example might be something like, "In the past year, have you had bad pain, pressure, or a tight feeling in the chest that forced you to stop walking?" The point of all this discussion is that questions must be carefully phrased to be useful.

The best patient history system might involve the patient and a nurse, aide, or assistant to explain the process, show the patient how to use the computer, and be available if there are any problems. Once the patient has started and is doing well, he or she can be left alone.

As we think about our computer checklist, we must recognize that many patients will conceal the fact that they have had problems with or diseases of the reproductive organs. A question like, "Have you had V.D.?" is likely to get the wrong answer. Here is where the computer might give place to a human who could get the data in a more delicate way, or, in suspicious cases, flag the patient for a blood test.

*Interested readers should be aware that there is a special organization for people interested in the medical applications of computers. For more details, contact the Society for Computer Medicine, 1901 N. Fort Meyer Drive, Suite 602, Arlington, VA 22209.

Once again we emphasize that the questions must be phrased in the language that the patient understands and presented in a way that does not threaten or embarrass. The best way to generate the questions might be to work with someone who has taken many medical histories and then try the system out with volunteers from a variety of social strata. The best computer system is worthless if the data are no good.

Having obtained a patient history, the computer can go into its branching modes. If the patient indicates a history of a problem with respiratory disease, the computer might schedule a more comprehensive series of respiratory tests and X-rays. In any case, the patient would be flagged for the attention of the respiratory clinic. If the patient has no specific problems, the computer proceeds to split patients by age and sex after the usual vision, hearing, weight, height, blood, and urine tests have been done.

One example of the split by sex might involve a Pap smear for all women and an X-ray mammography test for women over 50 years of age or women who have a family history of cancer. Women outside of this group might be scheduled for ultrasonic or thermal scans to detect breast cancer, if the facility has such equipment.

Another example of the sex split might involve the typical electrocardiogram (ECG). There is good evidence that the typical resting ECG does not stress the heart and is a poor indicator of incipient cardiac disease. A stress test where the patient runs on a treadmill or pedals a bicycle until the heart rate increases significantly is a better test. Males, who are more prone to heart disease, might be scheduled for this type of test, especially after age 35 or 40. (In case you are worried, the hazard of a stress test is quite low; in any case, every test is supervised by trained nurses and attendants with a physician available in case of problems.)

These are only examples of what might become an extremely complex process, and here we must add a note of warning again. If the computer isn't carefully programmed, it will put the patient through every test in the book at fantastic expense. It will then present the physician with an enormous mass of almost useless data which he is very likely to throw away without reading. Programming must involve what might be called stop points where the data are presented to a physician with some suggestions about possible diagnosis, or simply the question, "What do you want to do next?"

The Computer as an Aid in Patient Evaluation

We wrote about the computer as a device to flag unusual items. One application in cardiology involves the detection of life-threatening electrocardiographic events. The procedure involves taking an ECG of the patient. This produces about 10 meters of graph paper with from 5 to 14 traces, all of which have to be scanned by the physician.

This "scanning" is a great waste of the doctor's time because most of the traces will be perfectly normal for much of their length. (If there are abnormal traces every inch of the way, the patient may be past medical assistance.) What is needed is a computer that can be programmed to scan the traces and pick out, or flag, possible problems that the physician should examine more closely. Needless to say, this is a difficult problem for the programmer and to date (1980) systems of this type have not been 100% successful. The encouraging part is that the potential is there, and as the computer develops, things will get better. (One example is the chess-playing computer. The first versions played at a child's level, but now a chess computer can give anyone but a master a pretty good game.)

The Computer as a Library Tool in Medical Diagnosis and Treatment

Here again the field is so broad that we can't possibly cover all of it. If you are really interested, you might want to look at the book edited by M. F. Collen, *Hospital Computer Systems,* New York: Wiley, 1974. One application of interest would allow the physician to draw on the ability and memory of many people outside his own domain. The objective is *not* to tell him "how" to treat the patient, but to provide information that may help with diagnosis and treatment.

For example, suppose a patient came in with such symptoms as high fever, muscular pain, and swellings in the armpit area. There could be all sorts of diagnoses, but the idea of "plague" might not occur unless the physician had seen it some time before. Then the symptoms were coded into the computer, all the possible causes would appear. Once the decision was made, the physican could put that into the computer and get back the latest information on prognosis, treatment, etc. (No one can keep up with the rapidly growing medical literature, FDA approval on drugs, etc. Why not let the computer do it?)

At this point your reaction might be "that is for the future," but you're wrong again. At the Arizona Health Sciences Center (Tucson) a program called MYCIN is being developed "on the computer." This system will allow a physician to contact the computer by telephone, describe the patient, the tests run to date, and other pertinent diagnostic information. The computer comes back with appropriate requests for more information, i.e., is the patient an alcoholic, what drug sensitivities exist, can other tests be done, and so on. When these data are in, the computer will suggest a diagnosis and an appropriate treatment, with a request that the outcome be reported to improve the computer's knowledge bank.

Here again the computer is not a replacement for the physician. It is an assistant for his memory and judgment. In these days of rapid medical advances, no source of help can be neglected.

For the latest information on this topic you might want to look at the paper by H. M. Schoolman and L. M. Bernstein, "Computer Use in Diagnosis, Prognosis, and Therapy," *Science 200* (4344), May 26, 1978, pp. 926-931.

The Computer for Hospital Instrument and Apparatus Inventory Control and Exchange

Instrument costs are now a major factor in all industries, but this is especially apparent in hospitals where equipment is very costly, failure may be life-threatening to a patient, and obsolescence is often a matter of three to five years. All of these factors suggest that apparatus be used with the utmost efficiency and that hospitals in the same area share equipment whenever possible.

One example of computer application occurs during preparation for a heart bypass operation that requires the use of an intra-aortic balloon pump. Once the procedure is scheduled, the pump technician would code it into the computer and call for a printout of the hospital's available blood pumps. Some will be in use, some in repair, and others available. If there is nothing available the technician might call for a printout of the equipment at other nearby hospitals. If hospital B has a unit, the technician can call and arrange to borrow it.

One advantage of having all these data in the computer is that in emergency situations the computer can be asked to contact all the local hospitals and locate the needed equipment in a hurry. When a patient is really sick, the staff has more to do than play telephone clerk.

Another advantage of the computer systems is that all the data on maintenance, condition, date of purchase, repair record, vendor service, etc., are available. There is no question about when the apparatus is due for a checkup or what kind of help the vendor provided the last time a problem occurred. This is important when new equipment is to be ordered. The fact that all the data on costs are available makes it easier to justify the billing process. Hospitals are being asked to prove that a particular procedure should cost *X* dollars. Having all the cost and maintenance data on hand makes life much simpler.

The Computer as Part of an Emergency Medical Network

Two of the problems in emergency medical care are: first, getting the patient to a primary medical care unit; and second, transporting the seriously injured to a hospital that has the necessary surgical, medical, or burn facilities. The present systems have grown out of the MASH (Military Air Surgical Hospital) system developed during the Korean War.

The application of this facility might involve a policeman, nurse, or other personnel observing an accident, injury, or heart attack. A quick call brings out the primary (emergency) medical services that in a city might come by ambulance and in rural areas via helicopter.

Let us consider an accident in which several people are burned. As the chopper lifts out the injured, the crewman tells his computer what the problem is. The computer surveys its file of hospitals having burn units, the number of beds available, and the distances involved. The decision is made to go to hospital A, and at the same time the computer puts in a beeper call to the burn specialists so that they will be on hand when the chopper arrives.

The ability to decide which hospital to go to is most important when a number of people are injured in the same accident. There is a tendency to take all the patients to the nearest hospital, especially if a number of ambulances or helicopters are used. This may result in their having to wait for emergency care. With the computer network, the patients would be distributed in the most efficient way.

Another application of the computer would involve the medics in the helicopter who would enter the patient data, condition, and injuries, in the computer as the vehicle traveled to the hospital. The computer might well search its memory and come up with some possible diagnosis. The appropriate hospital personnel, blood plasma, etc., can be on hand when needed.

The computer could keep up with which physicians, technicians, and nurses are at each facility so that equipment and personnel are used with maximum efficiency. For information on how this *has* been done you might want to look at "Regionalization of Trauma Patient Care: The Illinois Experience," *Surgery Annual* 7, 1975, pp. 25–52.

Hospital Security and Accounting Problems

Hospitals have all sorts of doors and exits so that patients can be evacuated in an emergency. They also have a lot of expensive equipment and narcotic drugs on hand, both of which may be targets for theft.

It should be apparent that we are trying to do two things by using the computer for drug dispersement: first, keep drugs and equipment out of the wrong hands, and second, record every drug usage. Obviously this system will not protect a nurse when someone has a gun at her head, but it will prevent much of the unauthorized use of drugs. The second factor in the computer system is very important, since it allows the hospital to keep track of who got what medication and when. In many hospital systems some 20% of the medications cannot be charged to a specific patient. With the computer system, every medication must have the name of the patient involved, who ordered the drug, and when it was given. This will help in the equitable distribution of hospital costs. It also reduces the "I'll take some of that home" syndrome among the hospital staff.

The Computer as a Means of Preventing Adverse Drug Reactions

Another application of the computer would allow it to keep track of the latest data on drugs, applications, and adverse effects. A physician might prescribe a blood pressure control drug for a patient with a bleeding ulcer. The computer would have an up-to-date file on all adverse drug reactions, and, if there were information suggesting that this drug should not be used on a bleeding ulcer patient, the physician and pharmacist would be alerted. Here again we use the memory and correlation functions of the computer to do something that no human being could or would do – keep up with all the drug literature and correlate it to the patients and prescriptions in the hospital.

Once again, this computer application is in active use at one facility. For more details, see "Receptivity of Physicians in a Teaching Hospital to a Computerized Drug Interaction Monitoring and Reporting System," *Medical Care 15* (1), 1977, pp. 68–78.

OTHER COMPUTER APPLICATIONS

The Computer as an Aid to Agriculture

One thing farmers need if they are to be successful is information, about present and future crop prices, trends in commodities, new fertilizers, insecticides, and the like. In the past, this information has been available only from farm magazines or USDA bulletins that arrived too late or when there was no time to read them.

This system has been updated by radio transmissions available through a "Green Thumb Box" via the USDA. With a relatively simple system, the farmer or rancher will be able to call a local network center and get reports on weather predictions, commodity trends, market prices, and agricultural advice. If he has a hard-copy printer (available from Radio Shack and other sources), he can get copy to read or store in order to read at his leisure.

It is easy to see that this system will expand to provide other services, including sales of used machinery, exchange of farm products, and warnings of potential disasters, i.e., hail or windstorms. By using such information, the farmer will be able to make the best decision about what, when, and where to plant. For more details see M. Stenzler, "Computers Will Offer Agricultural Advice to Farmers," *Electronic Engineering Times,* June 12, 1978, p. 12. Another good source of information would be the local USDA extension office.

Computers for Professional Sports

A computer stores and processes information, and, as such, is ideally suited to something as complex and financially risky as professional sports. The Dallas Cowboys use a computer to do a number of things. They tabulate data for every player and every possible opponent on ball handling ability, speed, special characteristics, ability to run certain types of plays, possible blind spots, history of injuries, etc. This allows the coaches to use the Cowboy players to best advantage and to play to the weak side of the opponents.

Other applications include monitoring possible draft choices and deciding whom to pick up for what position. The computer can be helpful at contract time when a player asks, "Why don't I rate more money?" If you can show him how he did with respect to other players at similar salaries, he can appreciate why he got the offer he did. The system helps the coach evaluate players on a more equitable basis since all the data from several seasons are available rather than whatever the coach happens to have seen and remembered.

The computer can be coupled to a radar scanner and TV system that allows the speed and path of the ball to be accurately monitored. This can be correlated with a player's arm motions so that mistakes can be corrected before they become bad habits. A player with a history of a particular type of injury can be flagged by the computer for special preventative treatments or careful monitoring to catch any problem before it becomes serious. Once again the computer shows its ability as a memory and storage system.

Computers for People Matching

The idea here is not one of those "make a date with your ideal" gimmicks, which have generally fallen on their faces, but instead, a more realistic system that allows an employer to find the exact candidate he might be looking for. One application, in Hollywood, is called Quick Cast, where prospective actors make TV tapes that are available to possible employers. If someone needs a crippled Indian who can speak Hopi he need only code this into the computer. The tapes can be viewed rapidly without wasting the actor's time in actually appearing to "read." The computer can provide information on the actor's present location, work experience, picture and TV credits, contractual status with other companies, and the like. The objective, which is working very well, is the matching of talent with someone who needs it. What better purpose could a computer serve?

You can easily see further developments in this area. An engineer is told that his project will end in 60 days. He asks for and receives a computer printout that provides information about other openings, locations, cost of living, provision for moving expenses, housing, and so on. Once again the computer has served human beings. The opposite side of this coin is the ability of an employer to find just the person he needs even though that individual may not be actually looking for a job. Having your resumé in the computer will be like being in *American Men in Science;* it tells potential employers about you. It certainly is an improvement over the familiar "old boy network."

The Computer as a Lie Detector

This is an area that many people would like to avoid, but we engineers have a reputation for getting into things, so we shall not hesitate. Certainly the desire for getting at the truth is ancient in human society, especially when something like a trial is scheduled. There have been trials by combat, appeals to the gods, divination, oracles, witches, and of course torture in many forms – all attempts to get at the truth. Most of these techniques have been abandoned in relatively civilized nations (primarily because they were not effective rather than for any humanitarian reasons). The more up-to-date countries use things like lie detectors and voice analyzers.

All of these modern devices depend on the idea that a person telling a lie is aware that lying is wrong or dangerous (in terms of a perjury conviction). This knowledge produces a "stress" that the lie detector or voice analyzer can detect. Effects of stress might be perspiration, fast breathing, or changes in the pattern of frequencies of the human voice. These techniques are generally successful when used by trained operators, although the HAGOTH Voice Analyser (HAGOTH Corp., Issaquah, WA 98027) claims that no operator training is needed; red and green lights indicate whether the speaker is lying or telling the truth.

Without examining the moral aspects* of such devices (HAGOTH suggests that its unit can be used to monitor telephone conversations), we must accept the idea that they may very well detect when a person is under stress. If this is the case, one can conceive of a lie detection system in the courtroom to provide the judge and jury with an indication of the stress state of a witness. If the unit were connected to a computer that could record the various versions of the story, from different witnesses, it might become possible to separate the truth from a variety of lies – this is what justice is all about and as such must be regarded as an improvement.

One difficulty with this system is that it is useless if the person under test really believes what he is saying. A mentally ill patient who is sure he is Napoleon will suffer no stress when he tells people about his victories at Marengo and Jena. Another problem is that the courtroom environment is itself a source of stress, and the stress may increase as a witness describes some traumatic event. This is where the human element must be used to differentiate between stress related to the memory of being mugged and the stress of a person who lies when "claiming" to have been attacked.

This application of computers and electronic machines in this area is so new that it is difficult to predict the results. Once again, we suggest that when it comes to memory and associations there is nothing to beat a computer. Drawing conclusions from the data may be something else unless the computer is carefully programmed.

The Computer in the Courtroom

Here the computer can serve to assist the memory of court officials. One example might be a criminal who has been convicted of assault for the second time. His lawyer claims that the man has reformed, but, when the judge consults the computer, he gets the criminal's entire legal history and the fact that of some 5000

*It is the author's firm belief that moral questions have no place in a book devoted to engineering, on the basis that the book is designed to teach the reader what apparatus is available, what will be coming on the market, and how it works. The uses to which it is put *must* be decided by the individual and the society in which he lives.

The fact is that morals change with time. In 1930 it was a felony for a physician to suggest that a Connecticut resident obtain a birth control device. In 1978 the same state is making birth control information and devices available to teenagers. In view of this change over a period of some 40 odd years, the suggestion of lie detectors in the courtroom does not seem too far out.

people convicted of this crime as a second offense, only 10% actually reformed and "went straight." On that basis the judge decides that what the public needs is the removal of this danger and says, "15 years."

At this point some people may claim that every person is different and it is somehow unfair to use statistics of this sort. The facts are that people behave in ways that can be predicted statistically. If this were not true, it would be impossible to have life insurance, ready-made clothes, or traffic lights that provide a yellow signal that allows "law abiding"citizens to stop for a red light. In law as in medicine, there are no 100% answers; the judge must act on the basis of society's experience about what criminals do under certain circumstances.

The computer in court might involve the use of lie detectors on everyone who enters the witness box. The data would go to a computer that could check one story against another and display a reading of the probability that the witness is telling the truth. The jury could weigh that evidence as they listened to the witness. Here again, we may have complaints that the lie detector test is not perfect, mistakes may be made, etc. The response to that is, "Nothing made by human hands is perfect, criminal justice least of all." The law exists to protect the law abiding citizen; anything that helps it do that job is to be encouraged. In this process we must defend the innocent, but we must recognize that, just as even the best doctors will lose some patients, so in court some innocents will be convicted. The computer is used to keep the mistakes to a minimum.

Another important application of the computer in court might be called the *library function,* which could assist the judge and lawyers to keep abreast of the relevant state and federal decisions that might apply to the case under consideration. Rather than relying on some overworked clerk's memory, or an annual register of court decisions (that may be a year behind) it would be possible to ask the computer to find out what a judge or the Supreme Court did last week.

A similar function would aid juries in accident cases or awards for damages. Once the liability has been demonstrated, the jury could see what award was given for that particular injury in 50 previous cases. That background data could be used as a guide to the present action. Once again it is the computer's job to store and present the data that no human being could remember. The computer doesn't make the decision, it provides all the facts to the person(s) who makes the decision.

SUMMING IT ALL UP

At this point, we have to bring this discussion of computer applications to a halt, not for lack of applications but due to the limits of space. By now you should be convinced (we hope) that the computer is a useful tool in our struggle to survive in an ever more complex society. Since there are no signs of a return to the simple life, we must expect that computers will play a larger role as time goes on. The computer may well be the most important invention of the 20th century. For an expert's view of this, you might want to look at the article by L. M. Branscomb, "Information the Ultimate Frontier," *Science 203* (4376), 1979, p. 143.

GOOD BOOKS ON "BASIC" AND LABORATORY APPLICATIONS OF COMPUTERS

1. Lien, D. A., *BASIC Computer Language,* Cat. No. 62-2016 from Radio Shack Corporation, Fort Worth, TX 76102, 1977.
2. Abrecht, R. L., Finkel, L., Brown, J. R. *BASIC 2nd Edition Self-Teaching Guide,* New York: Wiley, 1978.
3. Finkel, J., *Computer-Aided Experimentation,* New York: Wiley, 1975.
4. Bennett, W. R., Jr., *Scientific and Engineering Problem-Solving with the Computer,* Englewood Cliffs, NJ: Prentice-Hall, 1976.

6

Op-Amp Problems and How to Fix Them

Up to this point, we have more or less treated op-amps as we *wish* they were. Of course, many useful results can be obtained by assuming that an op-amp is an ideal amplifier. However, you should be aware of the limitations of op-amps, if only to know your chances of success for a given project. It's also desirable to be able to estimate and, if possible, to minimize errors induced by the op-amps. The ideal op-amp has been characterized as having infinite input impedance and gain as well as zero output impedance. An infinite frequency response would also be nice. We have, however, already mentioned output roll-off at higher frequencies, and you might as well realize that nothing is ever infinite or perfect.

There is room for discussion about whether a book of this type even needs a chapter like this. Many workers go for years without ever worrying about the non-ideal character of the op-amps they use. However, most op-amp books do have a chapter like this, so we bow to convention. If you want to postpone reading this chapter until you have trouble, feel free to go on to Chapter 7.

OP-AMP PARAMETERS

We will begin this section by defining certain op-amp parameters. (Don't be frightened by the complexity; usually only one parameter will have a predominant effect on a particular circuit. If you can identify this parameter and minimize the resulting error, you're in great shape.) Some of these definitions were covered before in Chapter 2, but in this case repetition is important. This time around you will *know* what we are talking about.

One thing we will introduce is the use of *models.* We will model various effects in terms of things that we can understand and analyze. It is a convenient way of doing things, but nothing more than that. The use of modeling is like the use of a computer; the results are no better than the person who uses it. As they say in the MIT Computer Lab, *GIGO* (garbage in, garbage out). Sometimes there is even some gain, so what you get out is worse than what you put in.

OFFSET VOLTAGE AND INPUT IMPEDANCE

On a real op-amp, a voltage will exist between the input terminals even without the effects of finite input impedance and gain. Think of a battery and resistor in series: the battery is the *offset voltage* and the resistor the *input impedance* R_{in}, which was

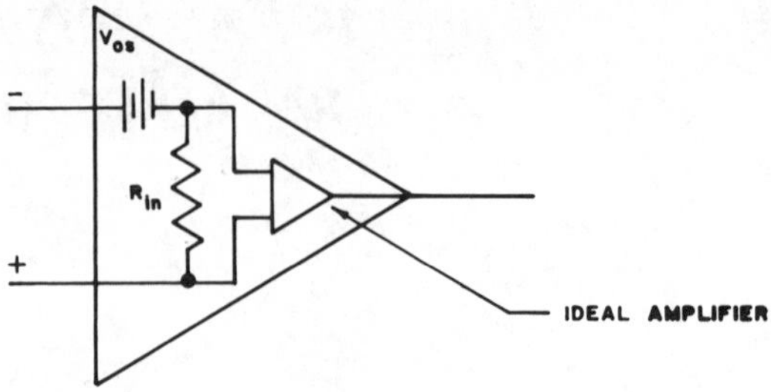

Figure 6-1. Offset voltage and impedance model.

called R_{DM} when we defined the op-amp input impedance back in Figure 2-16. Figure 6-1 is a model showing the offset voltage V_{OS} and the input impedance R_{in}. For single-input op-amps, the noninverting (+) or (2) input of the model may be considered grounded.* The change in V_{OS} with temperature is usually specified in volts per degree centigrade. If you are working at low signal levels, you need an op-amp with a low temperature coefficient for V_{OS}.

BIAS CURRENT

Each input will draw some current from the external circuitry. This will be true regardless of input voltages. These input currents in general will *not* be equal, and they will vary with temperature.

OFFSET CURRENT

Bias currents at the two inputs will only be approximately equal. The difference is the *offset current.* Bias and offset currents can be modeled by two current generators with current values of I_{BA} and I_{BB} (Figure 6-2). The difference between I_{BA} and I_{BB} is I_{OS}: the differential offset current or just the offset current. (***Beware:*** some manufacturers refer to our "bias currents" as "offset currents," and our "offset current" as the "differential offset current" or "difference current." Life is like that!)

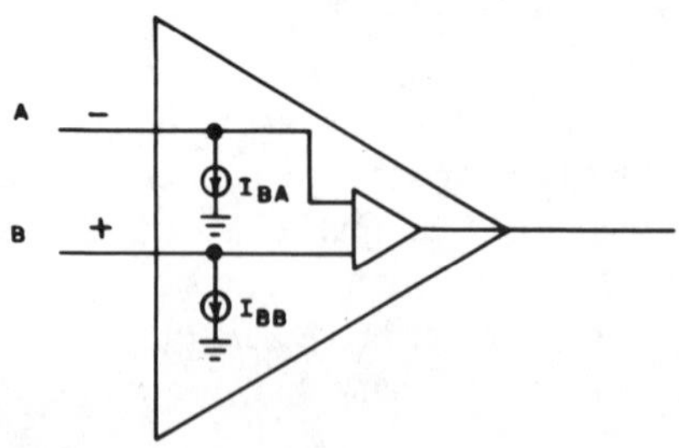

Figure 6-2. Bias current model.

*You should note that certain types of op-amps *always* have input (2) connected internally to ground. Chopper-stabilized op-amps are typical examples of this. The advantages of this type of op-amp were discussed earlier; among other things, their change in V_{OS} with temperature is almost zero.

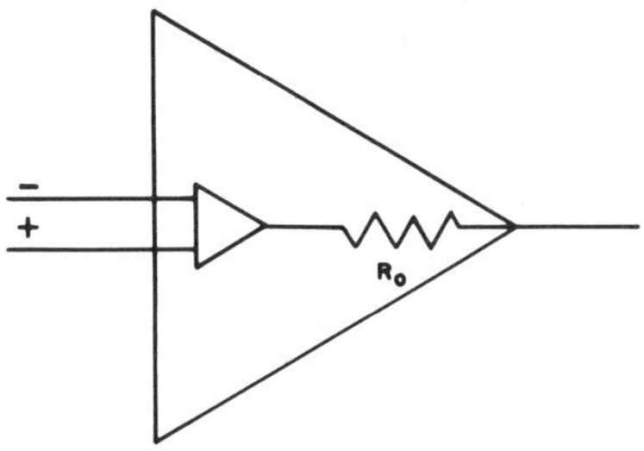

Figure 6-3. Output impedance model.

GAIN

Of course, the *gain* of a real op-amp is not really infinite; it will usually be some large number between 10^3 and 10^5. Furthermore, it will decrease from its DC value as the frequency increases. For this reason, no model will be given here. Additional explanation will be given on pages 293 to 295.

OUTPUT IMPEDANCE

The *output impedance* is modeled by a resistor in series with the output of the ideal amplifier (Figure 6-3). The manufacturer's data sheet will give maximum values for the output impedance.

COMMON MODE IMPEDANCE

This parameter is modeled by a resistance (R_{CM}) or an impedance (Z_{CM})* from either input (-) or (+) to ground, as shown in Figure 6-4. You must remember that for bipolar transistor op-amps, Z_{CM} is higher than Z_{DM}. However, field-effect transistor input op-amps have equal and very high common mode and differential input impedances (see Chapter 7).

DIFFERENTIAL INPUT IMPEDANCE

This impedance is modeled by a single resistor between the two inputs, as shown in Figure 6-4. Obviously, if the (+) input is grounded, Z_{CM} is approximately equal to Z_{DM}, but *only* in that case.

"RATED" VERSUS "EFFECTIVE" OUTPUT IMPEDANCE

While we are on this question of op-amp impedance, we must introduce another concept that we left out of Chapter 2 (on purpose) because it seemed likely to confuse.

The manufacturer gives input and output impedances for the op-amp, and you should note that these are the "worst case" values. In some instances we can do

*The use of R_{CM} implies that the impedance is a pure resistance whereas Z_{CM} suggests that there may be capacitative or inductive characteristics as well. Actually, EEs are sloppy and often use both terms for the same thing. You just have to get used to it.

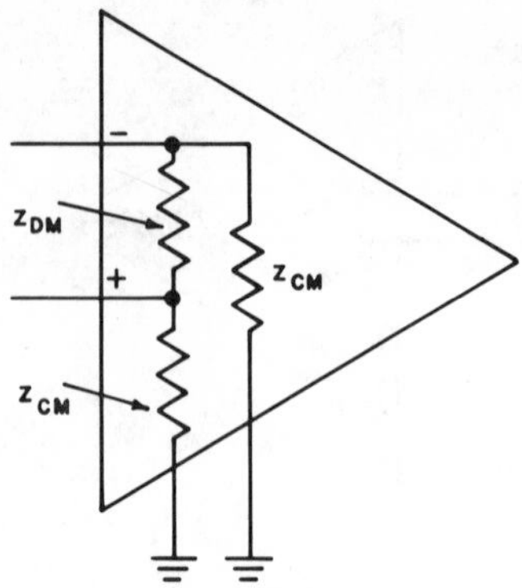

Figure 6-4. Common mode and differential mode impedance model.

much better and this is why we kept saying "don't worry, think of the op-amp as an ideal device." As an example, consider the voltage follower of Figure 2-15. The rated input impedance is R_{CM} or Z_{CM}. If the output voltage is V_0, we note that a small change is reflected back to the input (Figure 2-15) as

$$\Delta V_a = \frac{\Delta V_0}{A}$$

where A is the open loop gain.

This change in ΔV_a changes the input current *to the op-amp* by a factor

$$\Delta I = \frac{\Delta V_a}{Z_{CM}} = \frac{\Delta V_0}{A Z_{CM}}$$

and since we are in the voltage follower configuration where $V_0 \approx V_2$ we can write

$$\Delta I = \frac{\Delta V_2}{A Z_{CM}}$$

Now if we define the *closed loop impedance* Z_{CL} as

$$Z_{CL} = \frac{\Delta V_2}{\Delta I} = A Z_{CM}$$

the "effective" closed loop impedance is higher than the "rated" impedance by the value A where A is the open loop gain. This is a significant factor (like 10^5) and is a good reason for saying that real op-amps can approach the "ideal." We did pick a special case, namely the voltage follower, but the analysis would be the same for any noninverting amplifier circuit.

The *"rated" output impedance* of the op-amp is given as Z_0. For an op-amp in an inverting circuit, the *"effective" output impedance* is

$$Z_{out} = Z_0 \Big/ \left[1 + \left(\frac{R_1}{R_1 + R_2}\right) A\right]^*$$

Here we have used the subscript 2 for the feedback resistor instead of the usual subscript 0. This was done to avoid confusion with the use of Z_0 for the "rated" output impedance.

Looking at the above formula, we see that as A goes to infinity the "effective" output impedance is zero. The op-amp behaves like an "ideal" amplifier.

SLEW RATE

This parameter is usually specified in volts per microsecond; it is the *maximum rate of change of output voltage* that the op-amp can produce. This is illustrated in Figure 6-5.

Slew rate is a further limitation on the amplifier's performance at higher frequencies and with large output voltage swings. It is important to note that the *open loop* frequency response is determined by a *small signal* measurement. This means that the output voltage is not allowed to exceed a value that would cause the slew

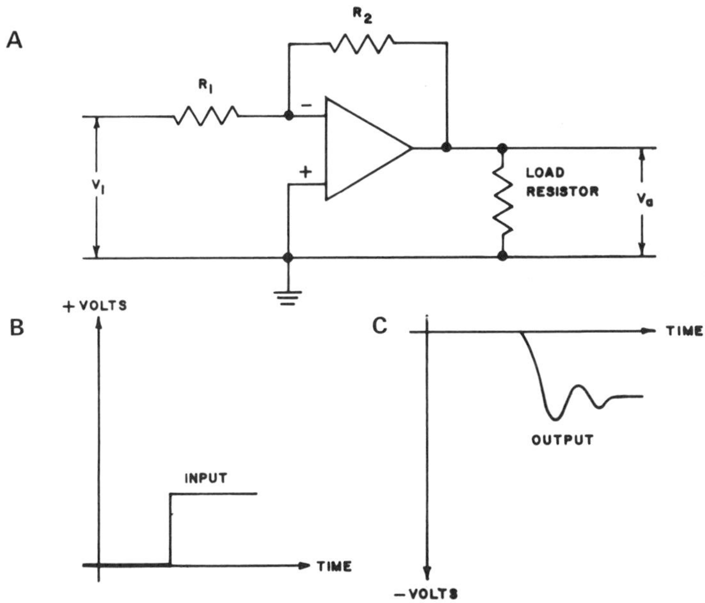

Figure 6-5. Slew rate parameters. *A.* Test circuit. *B.* Input signal. *C.* Output signal.

*The derivation of this formula is given in the book edited by J. Graeme et al. listed in the Selected Reading List.

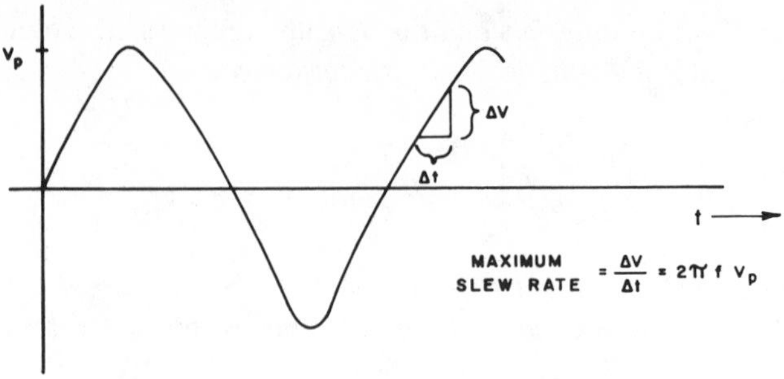

Figure 6-6. Slew rate dependence on amplitude.

rate limit to be encountered. To understand this, it is necessary to realize that the maximum slope or rate of change of a sinusoidal waveform is a product of the frequency and the peak amplitude value: $\Delta /V\Delta\, t = 2\pi f\, V_P$ (see Figure 6-6). Thus, if you were measuring the open-loop frequency response and allowed $2\pi f\, V_P$ to exceed the maximum slew rate, the waveform would be distorted and the *true* small-signal frequency response would not be obtained.

The maximum slew rate is determined by nonlinearities in the internal circuits of the op-amp. These usually result from current-limiting or saturation of the internal circuits. As a result, the maximum slew rate is usually measured for conditions that require full rated voltage and current.

COMMON MODE REJECTION RATIO

This parameter is usually written as *CMR* or *CMRR*. It was defined earlier in Chapter 2 (Figure 2-17).

You can observe the CMR capability of an op-amp by connecting both inputs together and driving them with a signal generator while observing the output signal. One word of warning to note: op-amps have a *common mode voltage limit.* Check the data sheet of any op-amp you are considering for a circuit such as a follower or noninverting amplifier. The common mode voltage limit may be as small as ±1 to 3 volts for some op-amps.

NOISE

Some data sheets specify a *noise voltage* and *noise current* that are referred to the input and measured for some specified bandwidth. This means that there are some unavoidable sources of AC voltage and current in the op-amp. This is modeled by the circuit of Figure 6-7.

These sources represent unavoidable phenomena within the op-amp, such as thermal noise in resistors and random diffusion and generation of current carriers in semiconductor devices. This type of noise, when viewed on an oscilloscope, would have a hash-like appearance. This represents energy at all frequencies and is often called *white noise* in analogy with white light, which contains all frequencies. Op-amp noise data is only measured over a specific bandwidth and, for this reason, a

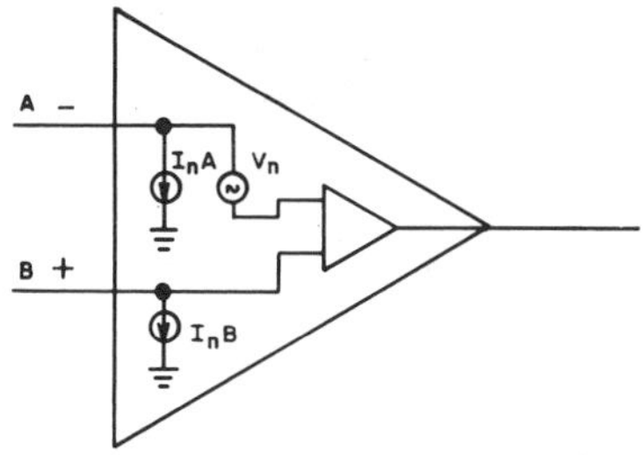

Figure 6-7. Noise voltage and current model.

bandwidth must be given with noise specifications (all frequencies outside the bandwidth were rejected by a filter when the measurement was made, thus reducing the RMS value of the noise by some amount).

OPEN-LOOP GAIN FREQUENCY RESPONSE

This characteristic represents the *roll-off of the output signal* as the *input signal is increased in frequency* when the op-amp is set up with *no external feedback path.* Many op-amps will start rolling off at 6 dB per octave at some low frequency, perhaps 10 to 100 Hz, and continue at that rate until the unity gain point ($V_{out} = V_{in}$) is reached. The frequency at the unity gain point is called the *unity gain bandwidth* and is often denoted by f_T. Op-amps having this continuous 6 dB roll-off are said to be *internally compensated* (this will be discussed on pp. 294–295). Such a response will ensure stability. In order to achieve high gain at high frequencies, some op-amps do not have this feature.

It is important to remember that the open-loop gain frequency response is a *small signal* characteristic. Therefore, the output signal must be kept small enough to ensure that the slew rate limit is not reached.

There is also a *phase shift* associated with amplitude roll-off. This phase shift is over and above the regular 180-degree shift built into the op-amp. If this additional phase shift reaches another 180 degrees, the total amplifier phase shift will be 360 degrees. This is positive feedback, and the circuit becomes an oscillator. For this reason, phase compensation is necessary (see pp. 293–295).

BASIC OP-AMP TYPES

Obtaining a reasonable op-amp for a particular application requires knowing something about the different types of these devices. The three fundamental types are referred to as *discrete, hybrid,* and *monolithic.* A *discrete* amplifier is built from individual transistors, resistors, and capacitors; hence its name. A *monolithic* amplifier is a completely integrated circuit. This means that the fabrication techniques used for making transistors are used for fabricating – on the same semiconductor chip – all transistors, resistors, and capacitors necessary for the complete amplifier. *Hybrid* amplifiers contain an integrated circuit as well as discrete components.

If cost is no object, discrete op-amps offer the best performance. Usually, the monolithic units are cheapest and have better temperature drift performance than all but the best discrete types. This is an important point; the data sheet for a particular op-amp will give temperature coefficients for offset voltage, bias current, and offset current. These temperature drift coefficients will impose a lower limit on the level of DC signal that can be amplified. For example, if you want to amplify 1 mV to 1 volt by a × 1000 inverter circuit and the op-amp has an offset voltage drift of 15 μV per degree Centigrade, the output error due to a 1°C temperature change is 0.015 volt (1.5%). Things might even be worse if a temperature gradient exists in the op-amp. The manufacturer's specifications refer to uniform temperature changes.

The input terminals of most op-amps are connected directly to two transistor base terminals. The emitter-to-base voltage (see Chapter 7) of transistors has a temperature coefficient of about 2.5 mV per degree Centigrade. This means that if the temperature of one input transistor were to change by 1°C with respect to the other, the offset voltages of the amplifier would change by 2.5 mV. In the previous example, this would produce an output error of 2.5 volts – an error of 250%! This is where monolithic units have it over discrete devices. In the monolithic op-amp, both input transistors are on the *same* tiny piece of semiconductor material, and there is much less chance for temperature gradients to occur.

In addition to the three types of construction, there are some other distinctions you should be aware of. *Field effect transistors* (see Chapter 7) are sometimes used for input transistors. Field effect transistors (FETs) have very high input impedances, and an op-amp featuring FET inputs will have very high common and differential input impedances, leading to very low bias currents. Although the bias current is quite low, it *doubles* every 10°C, which could make for a bad drift problem in a very critical circuit. As of 1980, you can avoid these problems by purchasing bipolar-FET op-amps. They have the high input impedance of FETs and the good thermal stability of bipolars. Such is progress!

Other op-amps are *chopper-stabilized.* This means that the input voltage is turned on and off (i.e., chopped) by an internal oscillator to form an AC voltage (see Figure 3-99). The AC voltage is amplified and then reconverted into a DC voltage at the output. This technique avoids the drift problems of DC amplification. Chopper-stabilized amplifiers have very low offset voltages for both initial and drift values. Internally, they are considerably more complex and, as a result, more expensive than other types.

ERROR CALCULATIONS – OR HOW BAD IT REALLY IS

In this section we will try to indicate some of the more critical parameters for some particular circuit types, provide formulas for calculating possible "worst case" errors, and suggest circuits to compensate or "null out" the imperfections in your op-amps. This is by no means an exhaustive treatment of the subject, but it should represent the more important considerations.

Consider the inverting amplifier shown in Figure 6-8, which has offset voltage

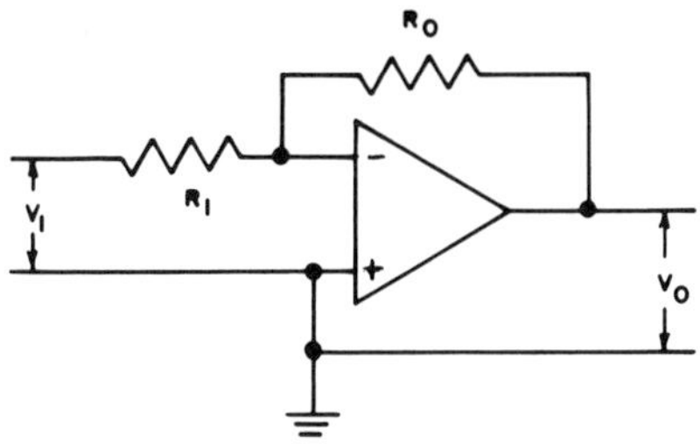

Figure 6-8. Typical inverting amplifier circuit.

V_{OS}, bias current I_B, and finite gain A. These voltage, current, and gain parameters introduce errors of the values listed below. The error due to V_{OS} is

$$\Delta V_0 = V_{OS}\left(\frac{R_0}{R_1} + 1\right)$$

The error due to I_B is

$$\Delta V_0 = I_B R_0$$

And the finite gain error is

$$\frac{V_0}{V_1} = -\frac{R_0}{R_1}\left[\frac{1}{1 + (1/A)(1 + R_0/R_1)}\right]$$

If we modify the circuit as shown in Figure 6-9B, the error due to I_B will be reduced to approximately $I_{OS} R_0$ (recall that $I_{OS} = I_{BA} - I_{BB}$; see Figure 6-2). The offset current I_{OS} will probably be several times less than I_{BA} or I_{BB}. To obtain some idea of the numerical values involved, let's choose $R_0 = 10\ \text{k}\Omega$, $R_1 = 1\ \text{k}\Omega$, $V_{OS} = 10$ mV, $A = 10^4$, and $I_{OS} = 0.05\ \mu\text{A}$. For the V_{OS} error, we have

$$\Delta V_0 = 10\text{ mV}\left(\frac{10}{1} + 1\right) = 110\text{ mV} = 0.11\text{ volt}$$

The effect of I_{OS} on V_0 is

$$\Delta V_0 = 0.05\ \mu\text{A} \times 10\ \text{k}\Omega = 0.5\text{ mV}$$

Allowing for finite open-loop gain ($A = 10^4$) the gain error is

$$\frac{V_0}{V_1} = -\frac{10}{1}\left[\frac{1}{1 + (1/10^4)[1 + (10/1)}\right] = -9.09$$

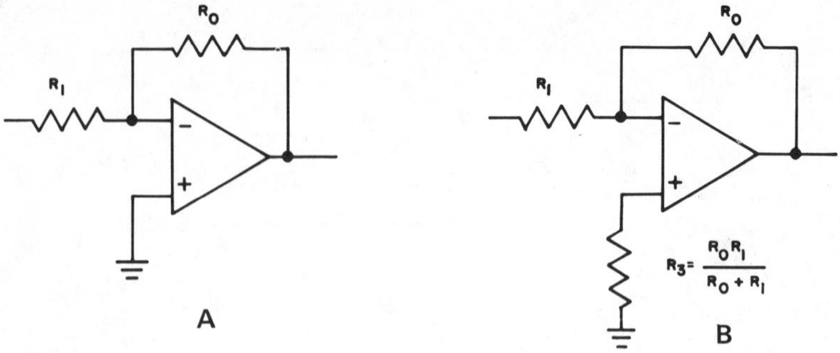

Figure 6-9. Bias current compensation. *A*. Original circuit. *B*. Bias current compensated circuit.

If we assume that there is a 1 volt input, the output error due to finite gain is 0.011 volt. The ideal output, V_0, should be 10 volts. Either V_{OS}, I_{OS}, and A could cause errors of 1.1%, 0.005%, or 0.11%, respectively, in the output voltage. In the "worst case," the errors would be additive, giving a total error of about 1.215%.

The effect of our R_3 resistor in Figure 6-9B is to reduce the error due to I_B ($V_0 = I_B R_0$) to a lower value, $V_0 = I_{OS} R_0$, where I_{OS} is usually less than 0.3 I_B. We have not changed the offset voltage or finite gain errors at all; these require other circuits, as shown in Figure 6-10.

Many op-amps – for example, the Burr-Brown units – will have terminals especially for the purpose of offset adjustment. In such a case, the data sheet will suggest particular circuits to use. The exact resistance values to use in the circuits of Figure 6-10, however, are best demonstrated by example. Suppose we have an op-amp with maximum V_{OS}, I_B, and I_{OS} values of 5 mV, 1 μA, and 0.3 μA, respectively. In addition, assume that R_1 and R_0 have already been specified as 5 kΩ and 20 kΩ.

With the circuit of Figure 6-10A, we could make $R_d = R_1 R_0/(R_1 + R_0) = 4\,k\Omega$ to give the approximate bias current compensation. With the maximum bias current of 1 μA, we would have 4 mV dropped across the resistor R_d. Thus the voltage across resistor R_c should be at least 2 (4 mV + V_{OS}) = 18 mV. Since we have made $R_d = R_1 R_0/(R_1 + R_0)$, we should make $\frac{1}{2}R_c + R_b \ll R_d$. Also, R_b should be several times smaller than R_c in order to make the voltage across the variable resistor R_c fairly independent of the setting of R_c. To satisfy these conditions, we could use R_b = 10 ohms and R_c = 100 ohms. All that remains is to choose R_a such that about 18 or more millivolts appear across R_c. For a ±15 volt power supply, we have about 20 mV across R_c under the following conditions:

$$R_a = \frac{15 \text{ volts} - 20 \text{ mV}}{20 \text{ mV } [(1/R_c) + (1/2\,R_b)]} \approx 12\,k\Omega$$

The nearest standard 5% resistor size happens to be 12 kΩ.

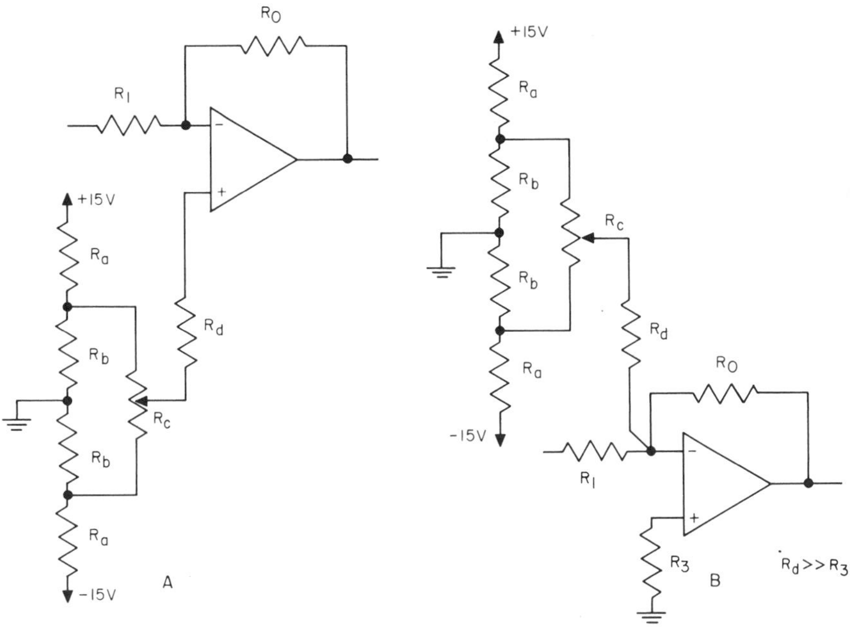

Figure 6-10. Additional compensation circuits. *A*. Offset voltage adjustment circuit. *B*. Offset current adjustment.

For the circuit of Figure 6-10B, R_3 should equal the parallel combination of R_1 and R_0. The resistance R_d must be considerably larger than R_3 if the adjustment circuit is not to affect the amplifier. In setting up this circuit, we could make R_d = 400 kΩ, R_c = 10 kΩ, R_a = 0, and $R_b = \infty$. This would give a maximum current of 15 volts/400 kΩ = 37.5 μA through R_d, which is many times in excess of the maximum difference current of 0.3 μA. We might find this circuit too sensitive for convenient adjustment, so increasing R_d to 4 MΩ would make things more reasonable.

It has been mentioned that the high frequency roll-off op-amps, which is determined by the open loop frequency response, implies that the open loop phase shift is changing from its low-frequency value of 180 degrees. It can be shown mathematically that when the feedback path is closed and certain phase shift conditions are satisfied, an oscillator results. Fortunately, there exist simple criteria for determining when such a condition is possible. Figure 6-11 shows the open and closed loop responses of an inverter circuit (if Figure 6-11 confuses you, go back and read the material on Figure 2-7 again). The closed loop response remains determined by the feedback circuit until the open loop gain becomes less than the closed loop gain; then the response follows the open loop curve.

Our main *stability criterion* is as follows: if the difference between the open loop slope and the closed loop slope is 12 dB per octave at their intersection, the circuit is at the edge of a region of instability. This means that if the difference in-

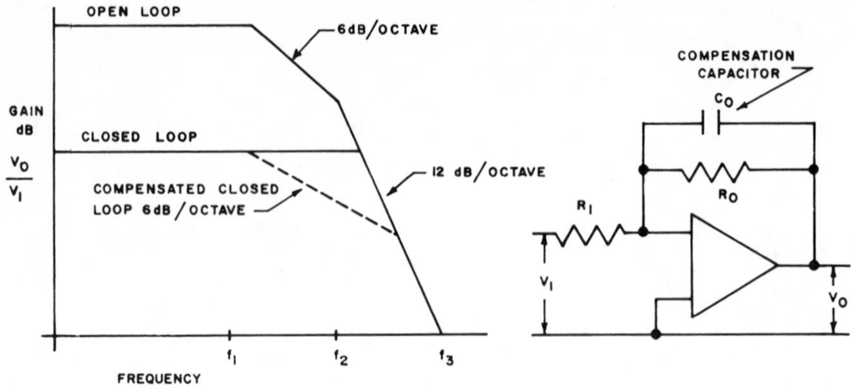

Figure 6-11. Frequency compensation. *A*. Bode plot; open loop and closed loop responses. *B*. Frequency-compensated inverting amplifier.

creases, oscillations are certain to result. To ensure stability, the closed loop response should be *compensated* to make the difference in slopes 6 dB or less. The open loop response shown in Figure 6-11 is 12 dB per octave at the intersection; according to the stability criterion, we have a dangerous situation. If we shunt R_0 with a capacitor C_0, a compensated response will be produced. With C_0 added, the open and closed loop responses differ by only 6 dB per octave. This will ensure a stable system.

Before selecting a value for C_0, check the data sheet for the particular op-amp; the manufacturer may suggest a compensation circuit. Many op-amps have special terminals for this function. When this is the case, the data sheet may contain recommended circuits. When this is *not* the case and the op-amp is *not* stated to be internally compensated to ensure that it rolls off at a constant 6 dB per octave, you're on your own! You're not lost in this case, but you must choose f_1 to be less than f_2. The data sheet might not provide the open loop curve, so f_2 will be uncertain. The data sheet, however, will provide the unity gain frequency, f_3. You then choose f_1 to be as much less than f_3 as possible, consistent with the highest frequency at which you expect to operate the op-amp. Then C_0 is given the following formula:

$$C_0 = \frac{1}{2\pi f_1 R_0}$$

where f_1 has now been chosen as the highest desired operating frequency.

This would be fine for pure sine or cosine wave inputs, but any other waveform, no matter how complex, is actually composed of a fundamental frequency and higher harmonics. These higher frequencies will be attenuated by the compensated roll-off and will cause the waveform to be distorted. If this is your problem, or if you wish to operate at frequencies close to the open loop curve, decrease the size of C_0 and check for any tendency for the circuit to oscillate with an oscilloscope. By varying C_0 downward, you can move f_1 closer to f_2. If f_1 exceeds the unknown f_2, you may be alerted by oscillations. While doing this, you should excite the system with your anticipated high frequency input signal.

It might be found that C_0 is completely unnecessary. In Figure 6-11B, if R_0/R_1 had been larger, interception would have occurred on the 6 dB per octave portion of the open loop curve and no compensation would be required. However, the situation could also be worse: in the example of Figure 6-11, the maximum slope of the open loop curve is 12 dB per octave, but it could be as much as 18 dB per octave or more. In this case, making f_1 less than f_2 with a simple single capacitor circuit will *not* be sufficient for all values of R_0/R_1. If our 6 dB per octave curve intersected with an 18 dB per octave portion of the open loop response, there would still be a dangerous 12 dB per octave difference. Then, either we could make the compensation circuit more complicated, or we could make the compensated response come to unity gain at some frequency less than f_3. The latter method is simpler, although it does restrict high frequency operation needlessly. To use the latter method, look up f_3 from the data sheet and calculate f_1 by the following formula (C_0 is calculated as before):

$$f_1 = \frac{f_3}{R_0/R_1}$$

This is really all we need to say about compensation because the problems usually aren't too serious. If you must learn more about it, we suggest that you look at some of the books in the Selected Reading List. In the next few pages, we will show you a few other circuits to make op-amps look better than they really are. Have fun!

BOOTSTRAPPING CIRCUITS

A useful technique for fixing things is applicable when a very high input impedance op-amp is needed but a field effect transistor (FET) op-amp is too expensive. This technique is called *bootstrapping,* and it involves taking some of the output current and returning it to the input to restore any current taken from the signal source. This provides an output without drawing any current from the source, hence the term "bootstrapping."

Many circuits can utilize this technique, but we will give only the example of an inverting amplifier circuit shown in Figure 6-12. To understand how this circuit works, recall that the current taken from the source is $I_1 = V_1/R_1$ and $V_{01} = -(R_0/R_1)\, V_1$. The output voltage of op-amp 2 is

$$V_{02} = -\left(\frac{R_4}{R_3}\right) V_{01} = \left(\frac{R_4}{R_3}\right)\left(\frac{R_0}{R_1}\right) V_1$$

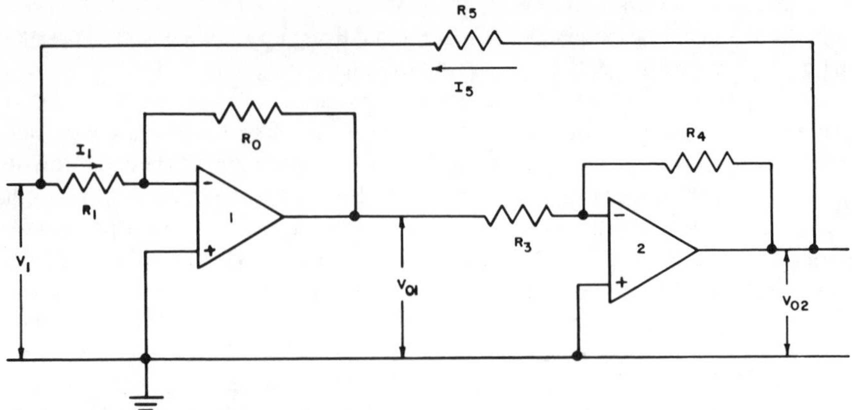

Figure 6-12. Inverting amplifier circuit with bootstrapping.

and we choose R_5 such that

$$I_5 = \frac{V_{02}}{R_5 + R_1} = \frac{V_1}{R_1} = I_1$$

This means that any current taken from the source is replaced by current from op-amp 2. As a result, the effective input impedance of op-amp 1 can be as high as 10^9 ohms, which is almost as good (by a factor of 10^2) as the best FET op-amps.

This can be a fun circuit to play games with. If R_5 is set at a value *below* that needed for the proper value of I_1, the circuit has negative impedance. You might get oscillation, depending on what the source impedance looked like. This point is worth belaboring a bit because of the strange things that happen if you feed back too much current to the source. To help you appreciate the details, we have revised Figure 6-12 as Figure 6-13; note the changes. (For more details, see E. Katell, "Positive Feedback Provides Infinite Input Impedance," *Electronics*, November 18, 1960, p. 102.)

The analysis of the circuit of Figure 6-13 is simple once we recognize that now I_0 does *not* equal I_1; in fact,

$$I_0 = I_1 - I_2$$

The (-) input of op-amp 1 is now at ground potential, so it follows that

$$I_1 = V_1/x R_1$$

and

$$I_2 = (V_{02} - V_1)/y R_1$$

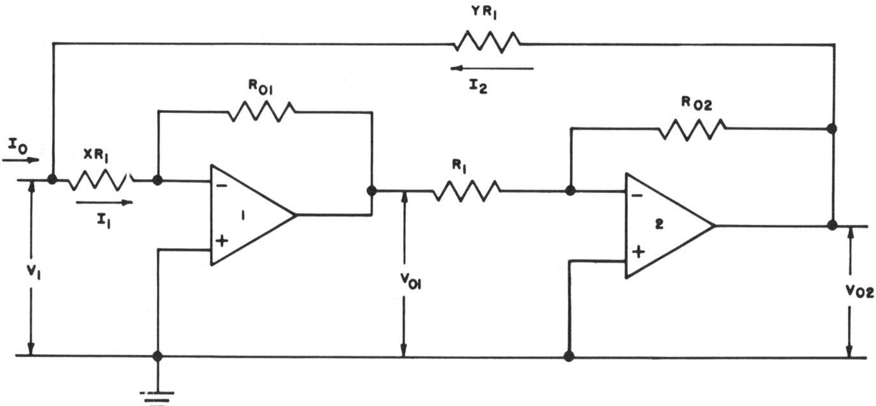

Figure 6-13. Revised bootstrap circuit (see Figure 6-12).

Since $V_{02} = -V_{01}\,R_{02}/R_1$ and $V_{01} = -V_1\,R_{01}/x\,R_1$,

$$V_{02} = V_1\,(R_{01}/x\,R_1)\,(R_{02}/R_1)$$

Substitution in the equation for I_0 yields

$$I_0 = \frac{V_1}{x\,R_1} - \frac{V_1}{y\,R_1}\left(\frac{R_{01}\,R_{02}}{x\,R_1^{\,2}} - 1\right)$$

and for infinite input impedance, $I_0 = 0$ is the criterion. If we set $R_{01} = R_{02} = R_1$, the equation reduces to

$$I_0 = \frac{V_1}{R_1}\left(\frac{1}{x} - \frac{1}{xy} + \frac{1}{y}\right)$$

or

$$I_0 = \frac{V_1}{R_1}\,(y - 1 + x)/xy = 0$$

Then $x + y = 1$ satisfies the $I_0 = 0$ requirement!

This may be correct math, but it might not be correct op-ampery. (Don't gasp at that, "One must be the master of one's words," said the White Queen to Alice.) Certain choices of x and y might require more gain than the op-amps can provide. Conversely, poor choices of x and y might induce saturation when $V_1 = 0$.

In Figure 6-14 we show a circuit that prevents saturation by putting a pot, R_s, across $x\,R_1$. The pots R_3 and $x\,R_1$ and the change of R_1 on op-amp 2 to $R_1/3$ are

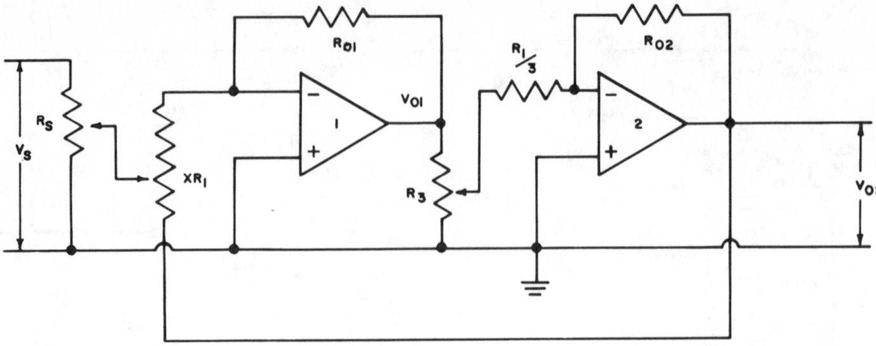

Figure 6-14. Another revised bootstrap circuit with potentiometers (see Figures 6-12 and 6-13).

for gain adjustment in order to obtain $I_0 = 0$. This is a good circuit to set up and play with. If the loop gain is too large, the circuit acts like a negative impedance.

Regarding the circuit of Figure 6-14, you might wonder how we know when the potentiometers are set correctly. This is not easy to answer, but there are a couple of tricks you can use:

1. The total circuit (op-amps 1 and 2) must have unity gain. Put in a known voltage V_1 and adjust it to get $V_{02} = V_1$. Then you can put on the unknown voltage, V_s.
2. Put in a known voltage, say V_1, through a high resistance, e.g., 10 MΩ, and adjust the gains until $V_{02} = V_1$. At that point, you *know* the voltage drop across the 10 MΩ resistance must be less than, say 1 mV, so the current taken from V_1 must be $10^{-4}\mu$A or less.

Techniques of this type can give you effective input impedances of 10^9 ohms, and sometimes *you need it!*

PUTTING TWO CHEAP OP-AMPS TOGETHER SO THEY LOOK LIKE ONE GOOD ONE

You may upon occasion have lots of low cost op-amps – the kind that drift quite a bit – and have an experiment that requires a *low drift unit.* Of course, you could buy a low drift unit, but you can also put two identical cheapies together as shown in Figure 6-15. This is a clever circuit and deserves a little discussion. We put op-amps 1 and 2 together so that the combination looks like one big op-amp. The gain of this super op-amp is the usual $V_0/V_1 = -R_0/R_1$. The external resistor R_0 gives us our normal stability.

Now let's assume that V_1 is unhooked and the system begins to drift. Suppose, too, that the (-) input on op-amp 1 goes to +1 mV; the output then goes to a -1 mV value. This in turn drives the (-) input on op-amp 2 with a -1 mV voltage. However,

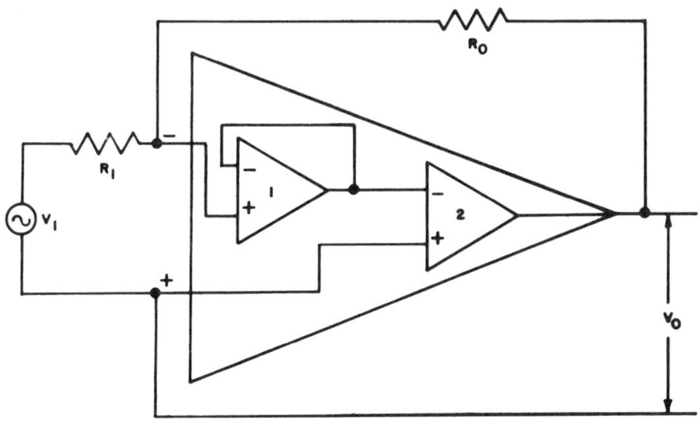

Figure 6-15. Low drift inverting amplifier using two op-amps.

we can assume that both op-amps drift at the same rate, so the (-) input of op-amp 2 has been drifting +1 mV. The application of -1 mV from op-amp 1 gives a net drift voltage of zero at op-amp 2. That's what we wanted: a zero drift op-amp.

This idea can be applied in several different ways, and if you have a dual inline package like the Motorola 1437P, it is especially useful. With the Motorola unit, the two op-amps are packaged together on a single chip and are almost certain to drift at the same rate. You could set up a special circuit to measure drift and check all your op-amps to find two that are alike, but the dual inline package route is much easier. Give both units the same compensation values, then hook them up in series. (For further details, see L. Choice, "Series Connected Op-Amps Null Offset Voltage," *Electronics,* March 27, 1972, p. 92.)

POSTSCRIPT

Once again, that is all (*really* all) we care to say about op-amp problems, because they are really few and far between. Go ahead and hook them up. Most of the troubles you have read about come under the heading of nit-picking.

7 Discrete Devices (If You Must Use Them)

BIPOLAR TRANSISTORS

The most common type of transistor is called a *bipolar transistor.* It has three terminals called the *emitter* (e), the *collector* (c), and the *base* (b). In most (but not all) applications, a small current (I_b) flows into the base and induces a large current (I_c) from the collector to the emitter. The ratio I_c/I_b we call h_{fe} or β (beta); this is the DC gain. Are you wondering why there are two symbols for the same thing? This is only the beginning of the confusion among small signal gain, large signal gain, hybrid parameters, and God knows what else. It makes life hard for anyone who wants to learn about transistors. To be an EE you have to suffer, but don't give up hope! Hoenig will straighten it all out for you. Forward, with vigor!

Transistors are made of semiconductor materials that are either *electron conductors (N type)* or *hole conductors (P type).* A typical transistor is shown schematically in Figure 7-1. In Figure 7-2, we show a model of the transistor as two diodes back to back. We can think of each diode as a point where two semiconductors of different characters meet. This point is called a *junction.* In normal operation, the collector-to-base bias is what we have called *reverse bias.* This means that the collector-to-base current flow will be small (essentially leakage current) unless we inject charge carriers or change the reverse bias to something like forward bias. (This doesn't mean the collector-to-base junction is forward biased; it is an approximation to hint at what actually does occur.)

The emitter-to-base junction is forward biased so that the current I_{be} is proportional to some function of the applied voltage V_{be}, such as

$$I_{be} \approx e^{kV_{be}}$$

Actually, k is a function of temperature and the kind of semiconductor material involved, but these are details that you need not fret about.

Figure 7-3 shows a typical diode voltage-current (V-I) curve. Note that in reverse bias, the current is small until breakdown occurs. In forward bias, there is a "turn-on" voltage that must be reached before the diode begins conducting; when that voltage is exceeded, the current goes up very rapidly with voltage.

There are two types of transistors called NPN and PNP. The circuit symbols for these are illustrated in Figure 7-4. Note that the arrow identifies the emitter, and *always* points toward the N-type material.

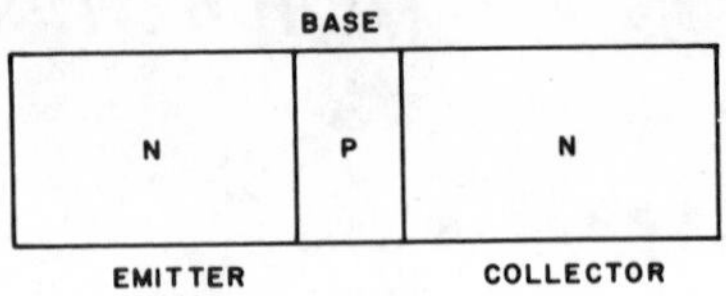

Figure 7-1. Schematic transistor (NPN type).

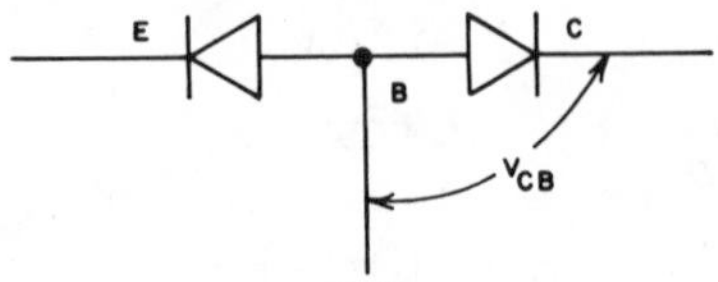

Figure 7-2. Two-diode model of an NPN transistor.

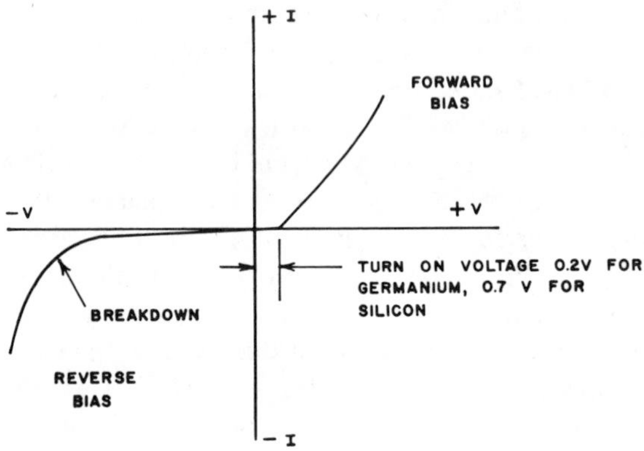

Figure 7-3. Diode *V-I* curve.

Common Emitter Transistor Circuit

A typical NPN circuit is shown in Figure 7-5. A PNP transistor circuit would be exactly the same except that the batteries would be reversed.

The operation of the circuit of Figure 7-5 is qualitatively as follows: The collector-base junction is always reverse biased (note the polarity of V_{cc}). Therefore, I_c is small and so is $V_c = R_c I_c$. The base is forward biased by V_{bb} and R_b. This forward bias sets the steady-state level of I_b. The current I_b tends to remove the collector-base reverse bias and thereby increase I_c. The ratio I_c/I_b is called h_{fe}* (the *current gain* of the transistor). Normally, I_c is set at some fixed value called the *DC*

*This ratio is sometimes called β (Greek letter beta), but we up-to-date types use h_{fe} most of the time.

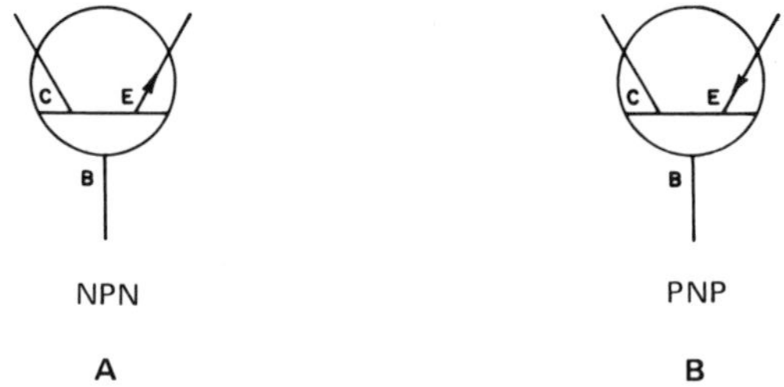

Figure 7-4. Transistor circuit symbols. *A*. NPN type. *B*. PNP type.

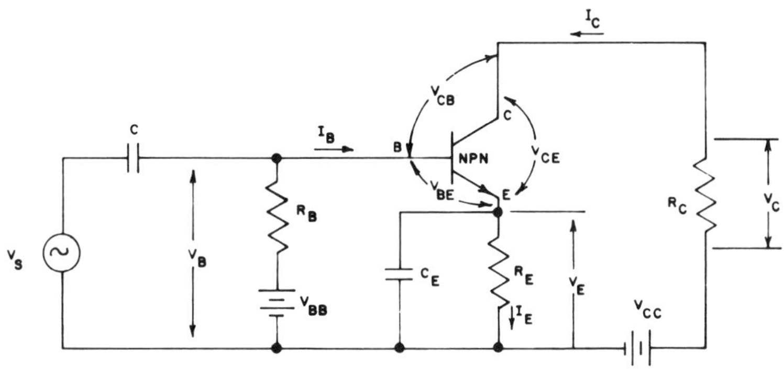

Figure 7-5. Common emitter transistor circuit (NPN type).

level. Now, if V_s (an AC signal) increases the forward bias, I_b goes up. This makes I_c go up, and we have amplified the signal because $h_{fe} \gg 1$. From all this mention of currents in relation to bipolar transistors, you may have gotten the idea that a transistor is a current amplification device. This is correct, at least for bipolars, which is what we are into at the moment. A small current into the base allows a large current to flow from the collector to the emitter.

To understand the quantitative details of this wonder, we have to do a little circuit analysis. Don't worry; the complex formulas will reduce to simple, usable results. We will make use of Kirchoff's law (recall Chapter 1), but you should realize that it is a convention for circuit analysis and nothing more. As a quick review, we apply Kirchoff's law to the circuit of Figure 7-6 and find that $+V - I_1R1 - I_1R_2 = 0$ and $I_1 = V/(R_1 + R_2)$. Remember that we take all currents from (+) to (-). If we go across a battery from (-) to (+), we will call the voltage (+), and if we go across a resistor from (+) to (-), we will call it (-). Then the sum of all the voltages around the loop must equal zero.

Kirchoff's voltage rule when applied around the loop containing the *c* and *e* terminals in Figure 7-5 yields equation (7-1):

$$V_{cc} - I_cR_c - V_{ce} - (I_c + I_b)R_e = 0 \qquad (7\text{-}1)$$

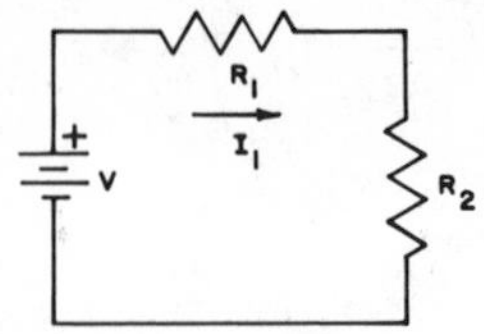

Figure 7-6. Kirchoff's law again.

where $I_c + I_b = I_e$. Using $I_c = h_{fe} I_b$ in equation (7-1) and rearranging the terms gives

$$V_{ce} = V_{cc} - I_c \left[R_c + R_e \left(\frac{h_{fe} + 1}{h_{fe}} \right) \right] \tag{7-2}$$

or

$$I_c = \frac{V_{cc} - V_{ce}}{R_c + \left(\frac{h_{fe} + 1}{h_{fe}} \right) R_e} \tag{7-3}$$

For almost all transistors, $h_{fe} \gg 1$; therefore $(h_{fe} + 1)/h_{fe} \approx 1$ is a reasonable approximation for most transistors. Equation (7-3) may be graphed as shown in Figure 7-7.

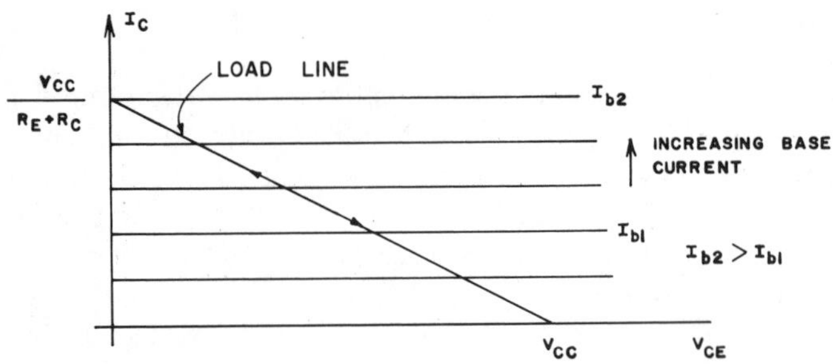

Figure 7-7. Ideal transistor characteristics.

The horizontal lines in Figure 7-7 represent lines of constant base current, I_b. Of course, this relies on the assumption that h_{fe} is constant for all conditions. Nature, however, is not so considerate as to provide such an ideal device; more realistic characteristics of a typical transistor are shown in Figure 7-8.

Transistor Characteristic Curves, Voltage Gain, and h_{fe}

The graph in Figure 7-8 is called the *characteristic curve* for the common emitter circuit of Figure 7-5. If the input is taken as V_b and the output as the voltage at the collector with respect to the common point, an increase in V_b results in an increase

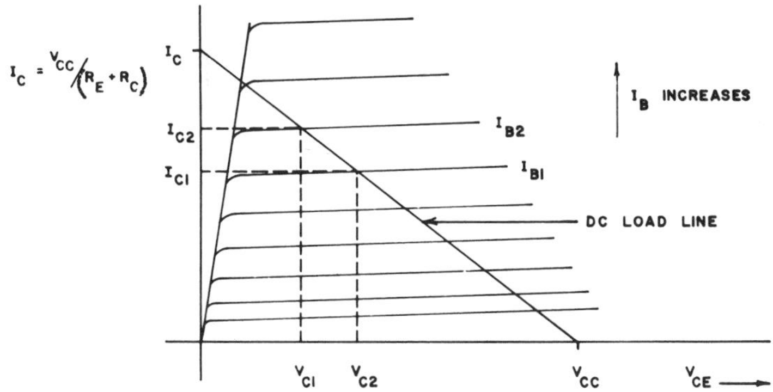

Figure 7-8. Real transistor characteristic curves.

in I_b. This in turn results in an increase in I_c because $I_c = h_{fe} I_b$. The effect of I_c on V_{ce} is given by the equation

$$V_{ce} = V_{cc} - I_c (R_c + R_e) \tag{7-4}$$

Since V_{cc}, R_c, and R_e are constants, V_{ce} will change in proportion to I_c. Figure 7-8 provides a quick means of determining the change in I_c caused by a given change in I_b. Corresponding values of I_c may be read from the vertical scale, and the new V_{ce} voltage can be calculated from equation (7-4).

The plot of equation (7-4) is called the *direct current load line* of the circuit. The DC voltage gain,

$$\frac{V_{ce(2)} - V_{ce(1)}}{V_{bb(2)} - V_{bb(1)}}$$

is calculated as follows.* Applying Kirchoff's voltage rule around the loop containing the base and emitter, we get equation (7-5) and the following:

$$V_{bb} = I_b R_b + V_{be} + (I_c + I_b) R_e \tag{7-5}$$

$$V_{bb} = \left(\frac{I_c}{h_{fe}}\right) R_b + \left(I_c + \frac{I_c}{h_{fe}}\right) R_e + V_{be}$$

$$V_{bb} = \left[(R_b/h_{fe}) + \left(\frac{h_{fe} + 1}{h_{fe}}\right) R_e\right] I_c + V_{be}$$

*By the time you have seen all these equations, you might be ready to give up transistors forever. Fear not! On page 309 we will reduce all this mess to seven simple steps for transistor design.

$$V_{bb} \approx [(R_b/h_{fe}) + R_e]\, I_c + V_{be}$$

$$I_c = \frac{V_{bb} - V_{be}}{(R_b/h_{fe}) + R_e}$$

Therefore, by equation (7-4),

$$V_{ce(2)} - V_{ce(1)} = -(R_c + R_e)\, I_{c(2)} + (R_c + R_e)\, I_{c(1)}$$

or

$$V_{ce(2)} - V_{ce(1)} = -(R_c + R_e)\left[\frac{V_{bb(2)} - V_{be(2)}}{(R_b/h_{fe}) + R_e}\right] + (R_c + R_e)\left[\frac{V_{bb(1)} - V_{be(1)}}{(R_b/h_{fe}) + R_e}\right]$$

We will discuss V_{be} below; for now, we will assume that its change is negligible. By assuming this, we can simplify the above monstrosity to

$$V_{ce(2)} - V_{ce(1)} = -\left[\frac{R_c + R_e}{(R_b/h_{fe}) + R_e}\right](V_{bb(2)} - V_{bb(1)})$$

Now we can write the *voltage gain*, A_V, as

$$A_V = -\left[\frac{R_c + R_e}{(R_b/h_{fe}) + R_e}\right] \tag{7-6}$$

The important thing to note in equation (7-6) is that if R_e is *small*, A_V depends very strongly on h_{fe}. However, if we choose a *large* enough value for R_e, then h_{fe} has very little effect on A_V. This is how we stabilize circuits against changes in h_{fe} (more about that later).

You should recall that the base-to-emitter junction is forward biased, so V_{be} is the voltage across a forward-biased diode. All transistors are made from either germanium or silicon. Roughly speaking, V_{be} will be 0.7 volt for silicon and 0.2 volt for germanium. This information is necessary if you are to understand the operation of many transistor circuits.

Things are not quite as simple as they might seem at this point. The transistor gain, h_{fe}, is *highly temperature dependent*, as well as varying considerably for different values of I_c. In addition, transistors manufactured by an identical process will exhibit wide variations in h_{fe}. Also, there is a *leakage current*, I_{cbo}, that will flow from the collector to base even when there is *no* current in the emitter. This leakage current will increase by several orders of magnitude at high operating temperatures. We must note that I_{cbo} is the sum of two components: I_S and I_L.* I_L increases with

*For our physicist readers, we will mention that I_S is due to thermal excitation of charge carriers in the junction itself. The voltage-controlled current I_L exists because of the increase in width of the depletion region as the applied voltage goes up. So there!

applied voltage, V_{cb}, whereas I_S increases with temperature. The effect of I_S usually dominates at high temperatures. When the total circuit resistance between the base and emitter is large enough, then I_S, which flows on the surface of the transistor material, will add to I_b and be amplified by the transistor. However, the presence of R_e helps mitigate this leakage amplification, which is one of the reasons why it is used. When R_e is not zero, a significant increase in the emitter current $(I_b + I_c)$ due to I_S tends to reverse bias the emitter-base junction and thereby reduce the forward bias on the base; this limits the increase in I_c. In effect, R_e stabilizes the circuit against thermal runaway and variations in h_{fe} or I_S.

Because of these problems with transistors, a simple, single transistor amplifier is *never* used to amplify direct current (DC). It is used in switching applications, but in the present discussion we're assuming that AC analog operation is desired.

Biasing for Stabilization

Our next objective is to set up a bias system that mitigates the temperature effects just discussed. Just in passing, we might note that the problem of biasing causes more misery to new EEs than anything except, perhaps, the opposite sex. To make transistor biasing easier for you, we hereby give you our handy method for biasing transistors. In the design of our bias system, we have two goals: (1) we want to stabilize the circuit, and (2) we want to replace the two-battery system shown in Figure 7-5 with one battery.

We attack the second goal first by noting that a two-battery bias circuit can be converted to a single-battery system using the circuit of Figure 7-9A. Given the voltages of the circuit in Figure 7-9A, we obtain the equivalent one-battery circuit shown in Figure 7-9B. To find R_1 and R_2 in terms of R_b and V_{bb}, we write two equations:

$$V_{bb} = V_b - I_b R_b = V_{cc}\left(\frac{R_2}{R_1 + R_2}\right) - I_b R_b$$

$$R_b = \frac{R_1 R_2}{R_1 + R_2}$$

In a later section we shall demonstrate that

$$I_b R_b \ll V_{cc}\left(\frac{R_2}{R_1 + R_2}\right)$$

so that we can approximate V_{bb} as

$$V_{bb} = V_{cc}\left(\frac{R_2}{R_1 + R_2}\right)$$

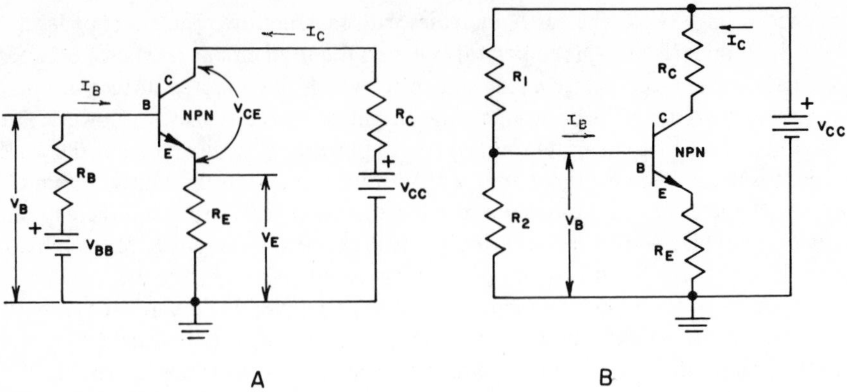

Figure 7-9. Equivalent biasing circuits. *A*. Two-battery bias circuit. *B*. Single-battery circuit.

If we rewrite the second equation as

$$\frac{R_b}{R_1} = \frac{R_2}{R_1 + R_2}$$

and substitute this in the first equation, we obtain

$$V_{bb} = \left(\frac{R_b}{R_1}\right) V_{cc}$$

or

$$R_1 = R_b \left(\frac{V_{cc}}{V_{bb}}\right)$$

Once R_1 is known, we can solve the second equation for R_2:

$$R_2 = \frac{R_b R_1}{R_1 - R_b}$$

Having solved the battery problem (that was easy), the bias stabilization problem comes next. Look at Figure 7-10 and note that we can adjust I_b (and thus I_c) by changing the ratio R_1/R_2. It is particularly important to note that moving *up* on the variable resistor turns the transistor on, and vice versa! Also note that here we used a PNP transistor just to mix you up; the only change is that V_{cc} is turned around. (This rule applies to NPN transistors, too – just try it! You only need to reverse the V_{cc} battery.) To continue with our bias stabilization design, we must

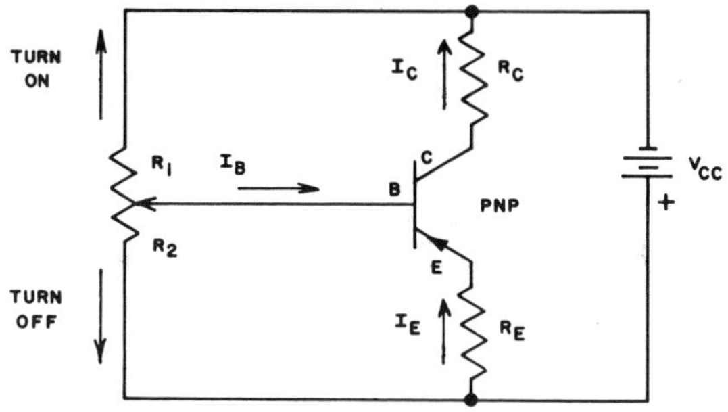

Figure 7-10. Transistor bias control circuit.

stabilize the circuit against changes in I_c or h_{fe} with temperature. This is done by putting a resistor R_e between the emitter and ground, as shown in Figure 7-10. As temperature goes up, so does transistor leakage, but now I_c does not go up nearly as much because any increase in I_c increases the voltage drop across R_e. This tends to reverse bias the emitter-to-base junction, which in turn reduces I_c, so a large change in the transistor temperature does not change I_c appreciably.

What about the effect on the base signals that we want to amplify? If they don't increase I_c, then we have no gain. Therefore, we don't want R_e to affect AC signals. We will short circuit R_e with a capacitor C_e so that for DC signals, $R_e = R_c$, whereas for AC signals, $R_e = 0$. The capacitor C_e is not shown in Figure 7-10, but don't worry about it; you'll see it in Figure 7-11.

Seven Steps to Transistor Design

How do we pick values for the various resistors and capacitors in Figure 7-11? Well, at this point we introduce the famous Hoenig "Seven Steps to Transistor Design." It is an approximate method, but it relatively easy and it works, so stay with us.

The Design Procedure:

Step 1. *The transistor type must be known.* This in turn allows us to use the Allied Catalog to find I_c, I_b, and V_{ce}. As an example, let's choose a Motorola 2N3903 NPN transistor. The catalog gives the data for the parameters defined below:

V_{ceo} = 40 volts (maximum collector-emitter voltage with base open)
f_T (min) = 250 MHz (frequency at which current gain falls to unity)
P_D = 310 mW (maximum power dissipation)
h_{fe} = 50/150 at I_c = 10 mA (current gain at an I_c value in the center of the characteristics diagram; when written as 50/150, these are minimum/maximum values.)

For design purposes, we need I_c, I_b, and V_{ce}. For this transistor, I_c = 10 mA is a

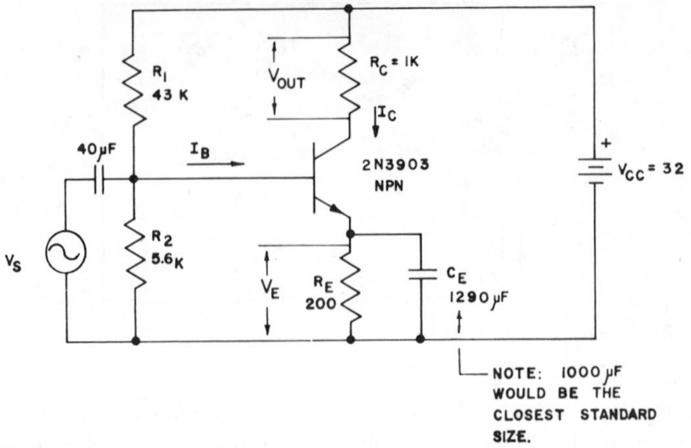

Figure 7-11. Transistor amplifier circuit.

good value. What value of I_b should we use? Well, $I_c = h_{fe} I_b$, so, being conservative, let's make $I_b = 10$ mA/50 = 0.2 mA (note that we have used the minimum value of h_{fe}). Again, we emphasize this is conservative design.* Now let's continue with the sequence of steps below while referring back to Figure 7-9A as necessary.

Step 2. *Pick I_c, I_b, and V_{ce}.* In this case, we have chosen $I_c = 10$ mA and $I_b = 0.2$ mA. Since we know $V_{ceo} = 40$ volts, $V_{ce} = 30$ seems like a safe choice. But the thing to watch is the power limitation: $P_D = 310$ mW. The product, $V_{ce} I_c = 30(10) = 300$ mW is a little too close, so we drop to $V_{ce} = 20$ volts. The "worst case" power product is now 200 mW.

Step 3. *We need V_e, the bias on the emitter.* A choice of about one-tenth V_{ce} is a good one, so in this case, $V_e = 2$ volts is a good value. (By the way, always make a list showing each value for I_c, V_e, and so on as you get them; it makes it easy to find them later in the analysis.) Since $I_e = I_b + I_c$ and we know that $I_b = I_c/h_{fe}$, we can assume that $I_e \approx I_c$. Then, since $V_e = 2$ volts $= R_e I_c$, we calculate $R_e = 2$ volts/10 mA = 200 ohms.

Step 4. *The next thing we need is R_b.* We want R_b to be as large as possible so that incoming signals will not be lost by R_b short-circuiting them. Now we must pick one of the rules of thumb for estimating R_b, and, as you might guess, we pick the most conservative one: $R_b = 15R_e$, which in this case yields $R_b = 3000$ ohms.

Remember that the values you choose are not critical here. Once you understand a few simple rules, you can make almost any transistor work with parts from the junk box. All transistor designs have ±20% accuracy at best, so don't let the dif-

*Arizona is a very conservative state. In fact, some of us actually believe in the economic theories of Edmund Burke. So "conservative" is the way we go!

ference between 3000 and 5000 ohms throw you. We ended up using $R_b = 5000$ ohms in our example circuit because it improved the operation, but 3000 ohms worked quite nicely, too. Again we repeat, don't waste time fretting! Pick approximate values; then breadboard the circuit and test it. Since two transistors with the *same* number and from the *same* factory can differ by ±30% or more, it is a waste of time to be too accurate in these estimates.

Step 5. *Now you can pick V_{cc} or R_c, but NOT both.* In most cases, R_c is chosen to be as large as possible, depending upon what sort of battery (V_{cc}) is available. Given R_c, then from equation (7-1) we obtain

$$V_{cc} = V_e + V_{ce} + I_c R_c \tag{7-7}$$

(Note that $V_e = I_e R_e = (I_c + I_b) R_e = I_c R_e$ because we have used the approximation, $I_e \approx I_c$.)

Using equation (7-7), if you pick V_{cc}, then

$$R_c = \frac{V_{cc} - V_e - V_{ce}}{I_c}$$

On the other hand, choosing R_c as, say, 1 kΩ, we have

$$V_{cc} = 2 + 20 + 0.01(10^3) = 32 \text{ volts}$$

for our example case. If this value seems too high, you should realize that a much smaller value of V_{ce} could have worked, too. As long as the collector voltage is about 2.0 volts above the emitter voltage, all is well. Here, $V_e = 2$ volts, so if $V_{ce} =$ 7 volts, things would still be okay; V_{cc} would drop to 15 volts. Usually, we try to operate with $V_{ce} \approx 0.5\, V_{ceo}$, but we don't *have* to.

Step 6. *To find V_{bb}, we must first find V_b (the voltage on the base) from $V_b =$* $V_e + V_{be}$. Note that V_{be} is the voltage of a forward-biased silicon pn junction, which is about 0.7 volt. We figured V_e to be 2 volts in Step 3, so in our present example, $V_b = 2.7$ volts. We can then find V_{bb} from

$$V_{bb} = V_b + I_B R_B$$

In this case, $V_b = 2.7$ volts, $I_b = 0.2$ mA, and $R_b = 5000$ ohms, so $V_{bb} = 3.7$ volts.

Step 7. *To find R_1 and R_2 in our "real" circuit, we use*

$$R_1 = R_b \left(\frac{V_{cc}}{V_{bb}} \right)$$

and

$$R_2 = \frac{R_b R_1}{R_1 - R_b}$$

In our example, sticking in numbers yields $R_1 = 43{,}200$ ohms and $R_2 = 5650$ ohms. The thing to remember is the ratio R_1/R_2 controls what fraction of V_{cc} that is applied to the base.

Now our simple design is finished, and a few last comments will be all that is needed:

1. R_b is *not* a real resistor, because there is no R_b in Figure 7-9B. *It is just a step in the analysis.*
2. When confronted with an unlabeled transistor, assume that $h_{fe} = 40$.
3. To choose C_e, we note that the voltage drop across C_e should be very low for *any* frequency (f_0) that we wish to amplify. One criterion for this is

   $$I_c X_c = 0.1\ V_e$$

   where $X_c = 1/2\pi f_0 C_e$This is *almost* good enough, and many electronic types use the relation

   $$C_e = I_c/0.1\ V_e\ 2\pi f_0$$

   However, one of us thinks that isn't enough, and he suggests using the relation

   $$C_e = \frac{I_c}{2\pi f_0 k}$$

   where $k = 3.5 \times 10^{-2}$ for silicon and 2.5×10^{-2} for germanium. For our case, $I_c = 10$ mA, so with $f_0 = 50$ Hz and a silicon transistor, C_e comes out to be 910 μF. Since 1000 μF is the nearest commercial size, we will use that value.

Transistor Amplifier Circuit

Let's return to our transistor circuit for another quick look to see what happens when we put in a signal at the input of the circuit in Figure 7-9B. First, let's redraw the circuit of Figure 7-9B as Figure 7-11 with the resistor values obtained above and a capacitor, C_e, chosen for $f_0 = 50$ Hz. Note that we are using a transistor of the NPN type again. More important, note that we have set up the DC bias on this circuit, but any signal V_s that is to be amplified *must* be AC. The resistor R_e effectively prevents amplification of DC signals, and we have inserted a 40 μF capacitor between the signal source, V_s, and the transistor. This will stop any DC signal from V_s that might otherwise mess up your amplifier. How did we get 40 μF? Well, one rule would be to divide C_e by h_{fe}, which in this case yields 26 μF. However, we used a 40 μF capacitor because that is what we found in the junk box – this was a real circuit!

Now the question is, what is our AC gain? You will recall that our DC voltage gain equation is (7-6):

$$A_V = -\frac{R_c + R_e}{(R_b/h_{fe}) + R_e} \tag{7-6}$$

Equation (7-6) can't be applied to AC signals because $R_e = 0$ for AC (the capacitor C_e passes AC signals to ground).

To find our AC gain, we must use h_{ie}, the AC emitter-to-base impedance. Typical values of h_{ie} vary widely, but a choice of 500 is a good guess. The effective input impedance, which we will assume to be a pure resistance, is h_{ie} and R_b in parallel:

$$R_{in} = \frac{R_b\, h_{ie}}{R_b + h_{ie}} \approx 450 \text{ ohms} \tag{7-8}$$

The AC output impedance is R_c, and, since we define *AC voltage gain* as V_{out}/V_{in}, we can write

$$A_{V(AC)} = \frac{R_c\, h_{fe}}{h_{ie}} = 100 \tag{7-9}$$

Now we must limit the swing of V_s to a range of, say, ±0.07 volt (this is a small-signal circuit design, and the swing in V_s must be held to something less than 0.1 V_{be}). The current taken from V_s by the input resistance R_{in} is

$$I_s = \frac{0.07}{R_{in}} \approx \frac{0.07}{450} \approx 150\ \mu\text{A}$$

The output voltage swing is $A_V\ V_s$, i.e., $V_{out} = \pm 7.0$ volts. The output current swing across R_c is 7.0/1000 = ±7.0 mA, so the current gain is

$$\frac{\Delta I_c}{\Delta I_b} \approx \frac{7.0 \text{ mA}}{150\ \mu\text{A}} \approx 47$$

which is just about equal to the minimum value of h_{fe} for this transistor.

You should note that we couldn't possibly have any larger input value for V_s, because if we did, we would swing I_c close to zero (recall that we set the DC value of I_c at 10 mA).

The Three Types of Transistor Circuits

That's all there is to transistor circuit design. Now we go on to greater things (other neat circuits), but first a few words as to what circuit to use when. A transistor can be hooked up in one of three ways: *common emitter, common base,* or *common collector.* Each type of circuit has its own specific properties, as listed below. You just pick the one that fits your application (remember, all these descriptions are comparative).*

1. *Common emitter:* High current gain, low input impedance, and adjustable output impedance.
2. *Common base:* No current gain, good for voltage gain at high frequencies, low input impedance, and high output impedance.
3. *Common collector* (more often called an *emitter follower*): High input impedance, low output impedance, and voltage gain less than unity. For typical applications and circuits, see Selected Reading List.

How to Find Whether Your Transistor is NPN or PNP

Suppose you have a transistor with *no* numbers and you want to know if it is a PNP or NPN type. First, look at the bottom and identify the emitter, collector, and base from Figure 7-12. There are some manufacturers who make epoxy transistors with the leads *e-c-b* instead of *e-b-c; watch for this.* Also, high power transistors come in different packages. A good reference for packages is the *G.E. Transistor Manual.*

Another way of finding out whether an unidentified transistor is NPN or PNP is just to take two of the three leads and make sure one of your two leads is the base. Call one lead *c* and the base lead *b*. Then take an ohmmeter and put the (+) terminal to the *c* lead and the (-) terminal to the *b* lead. Measure the resistance, R_a. Now reverse the leads and measure the resistance again; call it R_b. If R_a is *less* than R_b, the transistor is PNP; if R_a is greater than R_b, it is NPN. If R_a equals R_b or if R_a and R_b are "infinite" (i.e., over 10 MΩ), throw it away!

When doing this test, remember that for some ohmmeters, the terminal marked "plus" (or the one *not* marked "common") might be negative on the ohms scale, whereas it is positive for voltage measurements. (We don't know why, but instrument makers are malicious.) To find out which output on the ohmmeter (VOM) is (+), take a diode, put the leads across it, read the resistance, then reverse the leads, and do it again. The ohmmeter lead that was on the (+) side of the diode when you read the *lower* resistance is the (+) one.

Once you know whether your transistor is PNP or NPN, you only need to assume that $h_{fe} = 40$ (conservative again), $I_c = 5$ mA, and $V_{ceo} = 20$ volts. Most of the time, these figures should be okay. If they are too low, you can make V_{ceo} and I_c bigger. If they are too high, you have lost a worthless transistor.

*For more details on design of transistor circuits, consult Motorola's *Semiconductor Power Circuits Handbook* (see the Appendix).

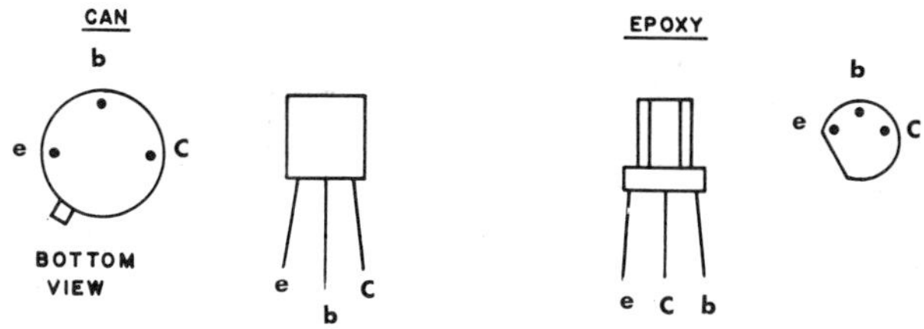

Figure 7-12. Different types of transistor packages.

Bipolar Transistors versus Other Types of Electronic Devices

Assuming that you might want to build more transistor circuits, we recommend the following for your library or workbench:

Semiconductor Power Circuits Handbook ($2.50)

and

Semiconductor Data Library ($9.00)
Motorola Semiconductor Products Inc.
5005 E. McDowell Ave.
Phoenix, Ariz. 85008

Other works on transistors can be found in the Appendix for those of you who really get into such circuits.

The remainder of this chapter will be devoted to describing some other types of electronic devices that you might want to use – or *have* to use. You should appreciate by now that it isn't as easy with transistors as it was with op-amps; this is why op-amps are taking over rapidly in many areas. Of course, there will always be areas of high current, high voltage, or very high frequencies where discrete devices are better than op-amps. Because of that, we will discuss certain other types of electronic gadgets and their applications.

FIELD EFFECT TRANSISTORS

A *field effect transistor (FET)* is another solid state device that can be used to amplify analog signals or to turn currents on or off. This device has completely different characteristics from the usual transistor, which is called a *bipolar transistor* to differentiate it from FET devices. The bipolar transistor is said to be a *current-controlled* current source, since $I_c = h_{fe} I_b$. The FET is a *voltage-controlled* current source over most of its useful operating region.

The FET has three terminals called the *drain, source,* and *gate.* In circuit operation, these terminals roughly correspond to the collector, emitter, and base of bipolar transistors. The principal operating difference between the two types of

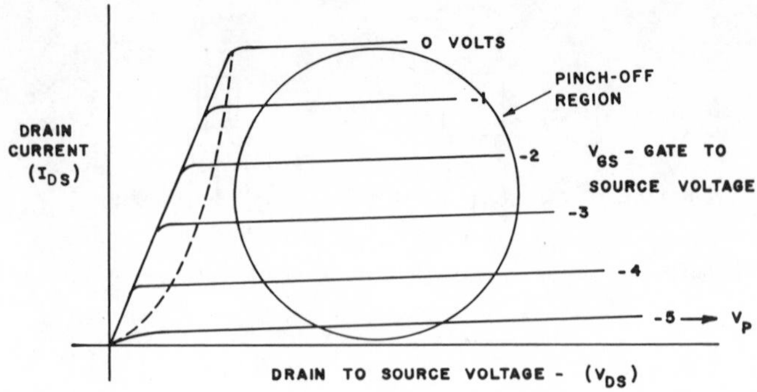

Figure 7-13. FET characteristic curves.

transistors is that with the FET, the gate draws very little current during operation, yet the voltage applied between the gate and the source controls the current flowing from the drain to the source. Typical FET characteristics are shown in Figure 7-13.

There are two types of *junction FETs,* which are referred to as *N-channel* and *P-channel* FETs. This distinction corresponds to that between NPN and PNP bipolar transistors. Schematic symbols and the usual direction of current flow for these two types of FETs are shown in Figure 7-14. There is one thing to watch here: the circuit symbol for an FET is very similar to that for a unijunction.* *Try to be careful in drawing the symbols and watch out for other people's carelessness in reading their circuit diagrams.*

It is most important to remember that an *N-channel FET has a P-type gate, and vice versa.* This means that with an N-channel FET, the gate-to-channel diode is forward biased when the gate is positive. The opposite is the case for a P-channel device with an N-type gate.

The arrowhead in the circuit symbols for FETs points in the direction of the gate-to-channel diode. Since the gate is one terminal of a diode, current would flow in the gate if a voltage of the right polarity were applied; however, for analog operation the gate is *always* operated reverse-biased. As a result, there is only a small reverse leakage current flowing in the gate. For this reason, the FET has a very high input impedance, which may be as high as several thousand megohms for some junction FETs. This is distinctly different from the bipolar transistor, which has only several hundred to several thousand ohms of input impedance.

There is at least one other type of FET device that we should mention so that you will know they exist. Of course, all FETs have certain unique characteristics, but some of them are better than others for certain applications. The *MOS-FET*† is a metal-oxide semiconductor FET with an insulating layer (the oxide) beneath the

*A *unijunction* is a three-terminal device that is used as a high frequency oscillator. Its characteristics are therefore somewhat similar to the neon bulb oscillator shown in Figure 3-120. For details, see the *T. I. (Texas Instruments) Transistor and Diode Data Book.* Unijunctions are also ideal oscillators for FM telemetry (see p. 215).

†Sometimes called an *IGFET* by the IGnorant. (That's a joke, son!)

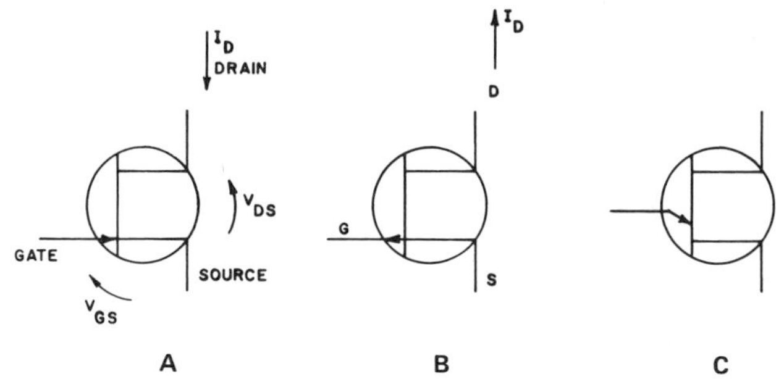

Figure 7-14. FET circuit symbols and unijunction symbol for comparison. *A*. N-channel FET. *B*. P-channel FET. *C*. Unijunction.

gate. This means that the gate impedance is even higher than that of the junction FET discussed above. MOS-FET devices do not have much current-handling capacity, but they are great for devices that cannot take significant current from the source. We won't discuss MOS-FETs in further detail because they are still relatively rare and are more difficult to use than junction FETs. For the rest of this book, it is junction FETs all the way (MOS-FETs will be saved for our next book).

It is interesting to note that FET devices behave much more like vacuum tubes than do bipolar transistors. This is of no great consequence here, because we haven't said anything about vacuum tubes. However, for those of you who learned about vacuum tubes in the "good old days," FET devices will be like coming home.

FET Circuit Design

In Figure 7-15, we show two typical FET circuits that you might build while learning about these devices. Figure 7-15*A* shows a two-battery bias circuit, whereas Figure 7-15*B* shows a single-battery bias circuit. This is very much like what we did when learning about the bipolar transistor (recall Figure 7-9). Notice that we have used an N-channel FET. This device has a P-type gate, and it is important to recognize that the gate is *always reverse biased* with respect to the source. If the gate were forward biased, the gate current would be quite large; conversely, when the gate is reverse biased, the gate current is very small – this is the major advantage of the FET.

The normal operating region for the FET is the area in which the I_{ds} versus V_{ds} curves are relatively flat (see Figure 7-13). Note that at *some* V_{gs}, the value of I_{ds} becomes zero. This is called *pinch-off,* and the operating area shown in Figure 7-13 is sometimes called the *pinch-off region,* even though $I_{ds} = 0$ for *only one* of the possible values of V_{gs}.

Among the parameters we need for design is I_{dss}, the *drain current with the gate connected directly to the source.* There is a useful rule that relates I_{ds} to V_{gs}:

$$I_{ds} = I_{dss}\left[1 - \frac{V_{gs}}{V_p}\right]^2$$

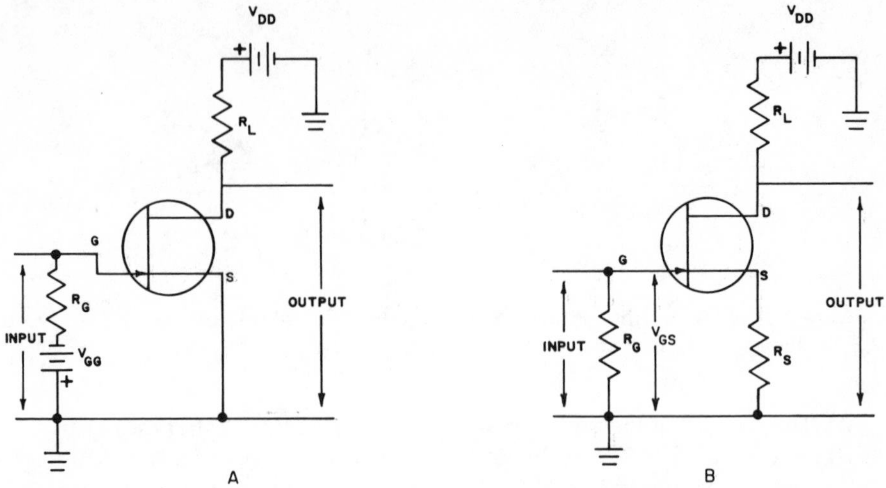

Figure 7-15. FET biasing circuits. *A.* Two-battery circuit. *B.* Single-battery circuit.

The term V_p is the *pinch-off voltage,* or the gate voltage at which I_{ds} is almost zero for all values of V_{ds}. Given the above parameters, we can design an FET switch by simply swinging V_{gs} through the proper range to change I_{ds} from its maximum value (I_{dss}) to almost zero.

This is *not* enough theory if we want to build a linear amplifier, because in that case we must bias the FET in the center of its operating range. Not only that, but we must plan for our old friends, temperature change and device variations. The temperature problem is shown in Figure 7-16, where I_{ds} is plotted versus V_{gs} for two different temperatures. If we pick the *design point* (quiescent point) as shown in Figure 7-16, all is well – the temperature variation cannot hurt us.

The variation in characteristics from one device to the next is shown in Figure 7-17. Here we assume that the manufacturer has given maximum and minimum values of I_{dss} and V_p for some N-channel FET. You should recall that I_{dss} and V_p appear in our first (and most important) FET equation:

$$I_{ds} = I_{dss}\left[1 - \frac{V_{gs}}{V_p}\right]^2$$

If we are given the maximum and minimum values of I_{ds} that we would like to hold in a circuit, we can pick a *bias line* that keeps us in the proper range. A bias line of this type is shown in Figure 7-17, for which the allowed values of I_{ds} are I_{ds1} and I_{ds2}.

This completes our discussion of the theory; now for the practice. Just to make sure that you know what is happening, there is a current from the source to the drain (I_{ds}) due to the voltage V_{ds}. The magnitude of I_{ds} is controlled by the gate voltage referred to the source, V_{gs}. As V_{gs} goes up (to a larger absolute value), I_{ds} goes down, and vice versa.

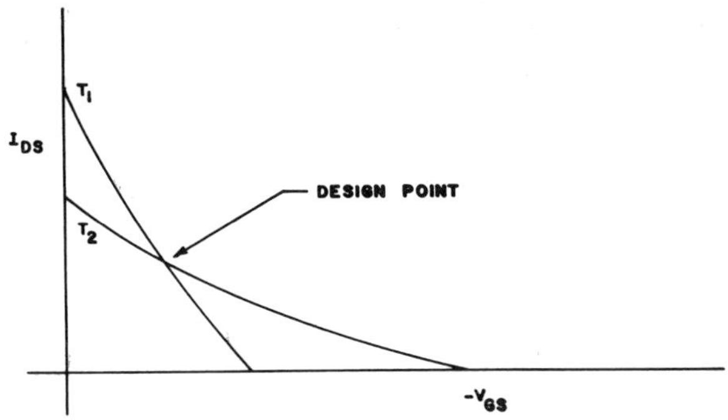

Figure 7-16. Variation of I_{ds} and V_{gs} with temperature.

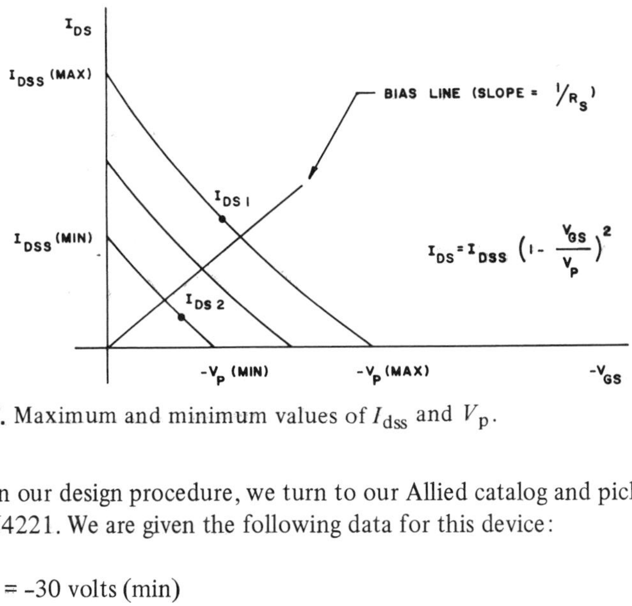

Figure 7-17. Maximum and minimum values of I_{dss} and V_p.

To begin our design procedure, we turn to our Allied catalog and pick a Motorola FET 2N4221. We are given the following data for this device:

1. BV_{gss} = -30 volts (min)
2. I_{dss} = 2.0/6.0 mA (min/max)
3. I_{gss} = -0.1 μA (max)
4. R_{ds}(on)= 400 ohms
5. Y_{fs} = 2000/5000 μ/ohms (min/max)*

These parameters tell you:

1. BV_{gss} = -30 volts is the maximum *gate-to-source breakdown voltage* with the

*Here you might be asking, μ what per ohm? Since Y_{fs} is in terms of $\Delta I/\Delta V$, which will be explained, it must have units of $1/R$. These units are sometimes called *mhos,* or in this case, *μmhos,* where mhos = 1/ohms. The unit that you will usually find in the literature, though, is μ/ohms (μmhos = μ/ohms).

source connected directly to the drain; 30 volts (min) means that the device will withstand at least 30 volts but no more.

2. The values for I_{dss} are simply the maximum and minimum values of I_{dss} in our equation

$$I_{ds} = I_{dss}\left[1 - \frac{V_{gs}}{V_p}\right]^2$$

3. The term I_{gss} is the *gate leakage current* with the drain short-circuited to the source. In effect, it tells you how little current the FET takes from the signal source.
4. The term R_{ds} (*on*) usually refers to the ratio V_{ds}/I_{ds} when $V_{gs} = 0$. In effect, it is the "on" resistance of the device.
5. Y_{fs} is the *forward transfer admittance,* which is just a fancy name for the change in drain current for a given change in gate-to-source voltage: $Y_{fs} = \Delta I_{ds}/\Delta V_{gs}$ when V_{ds} is constant. It is the amplification factor, g_m, to you vacuum-tube types.

For our design, you must realize that any decent EE would have the *Motorola Semiconductor Data Book,* which gives all the data on this device. All you have is Allied's catalog, so this is an *approximate* design. (Isn't everything these days?)

Let's pick a 15 volt battery for our power source. We then decide to work with $V_{ds} = 5.0$ volts (it's a safe value because the maximum BV_{gss} was 30 volts). We will try to drop about one-third of the supply voltage across the device; that's why we picked a 15 volt battery.

Our FET circuit is shown in Figure 7-18. Note that for a "worst case" design, $I_{dss} = 6$ mA. This means that we must pick $R_d + R_s$ such that $I_{dss}(R_d + R_s) =$ 10 volts. In this case, $R_d + R_s = 1.67$ kΩ. However, before we can pick actual values for R_d and R_s, we must note that $V_{gs} = R_s I_{ds}$. Therefore, R_s is critical if we want to hold the FET in the proper part of the operating region. The problem then is to pick V_{gs}. This would be easy if we knew V_p(max) and V_p(min) because we could then pick $V_{gs} = 0.33\ V_p$(min). But we don't get V_p values from the Allied Radio catalog, so we must guess that V_p(min) = 5 volts. This suggests a quiescent gate voltage of $V_{gs} = 1.7$ volts. This in turn allows us to pick R_s: $R_s = V_{gs}/I_{ds} = 1.7$ volts/6 mA = 283 ohms. For this design, the closest standard resistor is 270 ohms, but that's close enough (remember that all circuit design accuracy is ±20%).

Now if I_{ds} should go down because of temperature, V_{gs} will drop, too, because $V_{gs} = I_{ds}R_s$. This makes I_{ds} go back up because

$$I_{ds} = I_{dss}\left[1 - \frac{V_{gs}}{V_p}\right]^2$$

So we have both stability and a "very worst case" design.

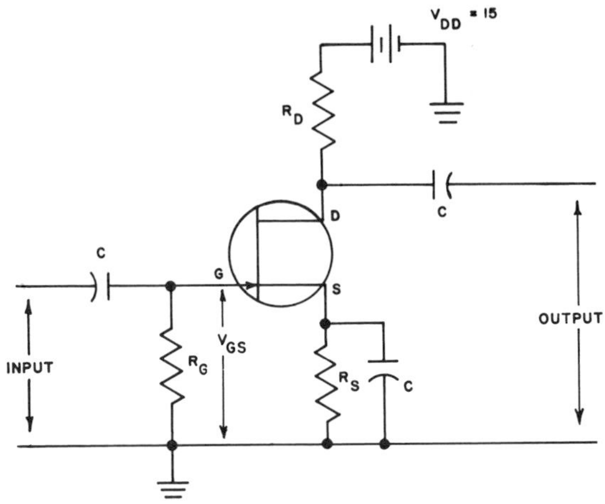

Figure 7-18. An FET circuit.

All we have left to find is R_d. This can be derived from our above computations: we found that $R_d + R_s = 1.67$ kΩ and $R_s = 283$ ohms, so $R_d = 1387$ ohms. Another way to get R_d would be to calculate it as

$$R_d = \frac{V_{dd} - (V_{ds} + V_{gs})}{I_{dss}} = 1.36 \text{ k}\Omega$$

This is just an application of Kirchoff's law around the loop that includes R_d, the D to S part of the FET and V_{gs}. It really amounts to the same thing, but the above equation shows what's happening.

For R_g, the gate resistor, pick almost any value between 5 and 20 MΩ; you can't go wrong.

At this point you might well say, "Okay, your technique worked for *this* FET, but would it work for any other one?" That's a good question, and the answer is really related to our next comment on the circuit discussed above. When you hook up this circuit, you may find that the FET is in the "off" condition. If so, this means that we have set the gate bias voltage too high and must pick a lower value for V_{gs}. This would be due to our using I_{dss}(max) as a design parameter. The FET you have might be closer to I_{dss}(min), so you have to adjust things. The major change should be an increase in R_s and R_d. Use decade resistance boxes, a signal source on the input, and an oscilloscope on the output; change the resistors until you start getting some voltage gain.

We have given you a very conservative design technique so that you can't zap the FET while you're learning. It's easy to become less safe, but harder to go the other way. A similar comment applies to the guess that $V_p = 5$ volts. If you use that value, you won't be in danger with *any* junction FET. The guess of $V_{gs} = 1.7$ volts could also cause all the voltage to be dropped across the resistors. If this happens, decrease the resistor values.

Going back to the circuit of Figure 7-18, if you swing the gate ±0.2 volt, the minimum value for Y_{fs} (2000) indicates a change in drain current of $\Delta I_{ds} = \pm 400$ μA. This may not seem like much, but when you realize that the input current to the gate is about 10^{-4} μA, you can see why FET devices are useful. The voltage swing is ΔI_{ds} times the value of R_d.

Note in Figure 7-18 that we have bypassed the source resistor with a capacitor, C, just as we did for the bipolar transistor circuit. This means that we have built an AC amplifier. Actually, FETs are also good DC amplifying devices, too. For DC amplification, all capacitors are left out; you just have to hook up the FET and try varying R_s and R_d to get the best result.

Biasing FET Circuits

You might be wondering about all that biasing we did on the bipolar transistor device to reduce temperature and device variation effects. Well, if you had the *Motorola Semiconductor Data Book* or the transistor specification sheet for your FET, you would look at the I_{ds} versus V_{gs} versus temperature curves and pick a quiescent I_{ds}, called I_{dq}, where the curves intersect. Next, you would plot the I_{ds}(max) and I_{ds}(min) versus V_{gs} curves and pick your maximum and minimum operating values of I_{dss}, called I_{ds1} and I_{ds2} (see Figure 7-17). Then you would pick R_s so that your bias line – based on $I_{dq}(V_{gs} = I_{dq} R_s)$ – has the proper slope; the *proper slope* means that the bias line hits the I_{dss} (max) curve above I_{ds2} and the I_{dss} (min) curve below I_{ds1}. (That's how EEs do it, or at least how they say they do it.)

If all this hasn't thrown you, there is still another point to be made: you can put other bias voltages on the gate with a resistor string, just the way you did with the bipolar transistor circuit. In this case, the gate voltage is $V_{gs} = V_{bb} - I_{ds} R_s$. The circuit is shown in Figure 7-19. Remember that V_{bb} is a fictitious voltage that is not shown in the *real* circuit of Figure 7-19.

For the circuit of Figure 7-19, let's postulate a hypothetical voltage, V_{bb}, that would exist at the gate *if* we were using a two-battery bias system. Since we are using a single battery,

$$V_{bb} = \left(\frac{R_2}{R_1 + R_2}\right) V_{dd}$$

Our relation for V_{gs} then becomes

$$V_{gs} = \left(\frac{R_2}{R_1 + R_2}\right) V_{dd} - I_{ds} R_s$$

By the proper choice of R_1 and R_2, we can bias the gate to almost any value we

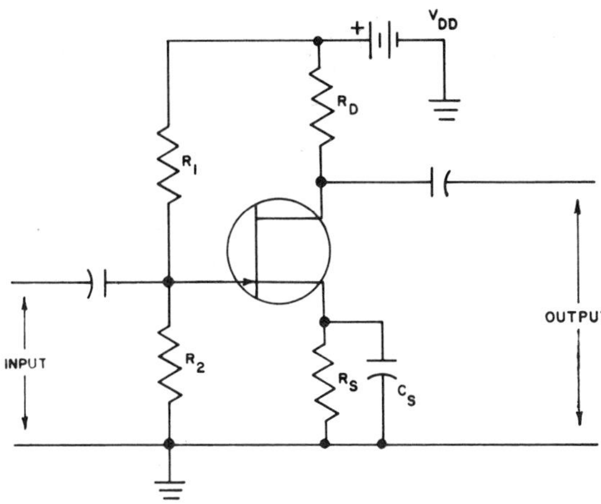

Figure 7-19. FET circuit with fixed bias.

choose without using improper values of I_{ds} or R_s. You should note that with an N-channel, P-gate device,

$$|I_{ds} R_s| > |\left(\frac{R^2}{R_1 + R_2}\right) V_{dd}|$$

and the same rule holds for a P-channel, N-gate FET. You should also note that resistor R_2 in Figure 7-19 is in parallel with the gate impedance. Unless R_2 is greater than 100 MΩ, your input impedance is entirely determined by R_2 in parallel with R_1, and much of the advantage of using an FET has been lost.

Don't let all this buzz-fuzz about complex biasing schemes bother you. Hook up your FET with a source resistor and play with it. Junction FETs are hard to kill (which makes them ideal for beginners) *as long as you don't forward-bias the gate.* MOS-FETs have an even higher input impedance, but they are easily zapped.

You can avoid all this misery by using FET op-amps. On that note, we end this discussion of FETs.

SCR AND TRIAC DEVICES

We used SCR devices previously in Chapter 3. Now we will give that long-awaited-for discussion of SCR and TRIAC devices and some design techniques for them. *SCRs (semiconductor-controlled rectifiers)* and *TRIACs* are solid state devices that can be used to control AC current and, as a result, AC voltage. An SCR can be thought of as a diode that will not conduct forward current until a small momentary current is applied to a third terminal, the gate terminal. A TRIAC is a three-terminal device, like the SCR, with the additional feature of being able to conduct in *either* direction when a gate voltage is applied. SCR and TRIAC devices are particularly

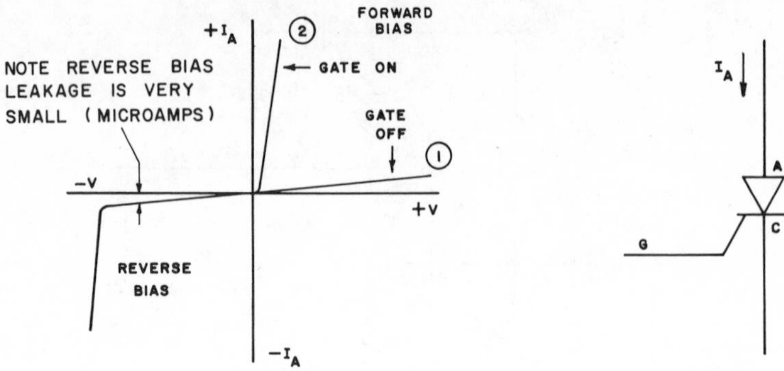

Figure 7-20. SCR characteristic curve and circuit symbol.

suited for high current loads, and devices rated at 500 amps or more are commercially available. Careful heat sinking is required for these devices if correct operation is expected.

Figure 7-20 shows a typical *V-I* curve for an SCR. When no gate signal is present, the *V-I* curve is a straight line (*1*, in Fig. 7-20) through the origin. If a gate signal is applied during the forward-bias phase, the current jumps to the upper curve, *2*, and *stays there after the gate signal is removed.* Current continues to flow until the applied voltage, *V*, is reduced to some low value; then the SCR current drops back to the value determined by the straight line *1* again.

A TRIAC is a device that can "switch on" in either the forward or reverse direction. In a sense there is *no* forward or reverse bias.

SCR Circuit Design

To use these devices, there are only a few simple rules to remember, which we present as *Tricks for Designing with SCR Devices:*

1. Check the peak inverse voltage in your system and V_{rrm} (V_{rrm} is the peak off-state voltage; *don't exceed it* or even work within 20% of it).
2. I_T (rms) is the maximum current during the conducting phase at some ambient temperature, usually 25°C.
3. $I^2 T$ is a measure of the Joulean, or $I^2 R$, heating. As I (use your maximum current) goes up, T (the ambient temperature) *must* go down, or vice versa.
4. A maximum required gate voltage is given for some particular gate current. This is the signal that is *guaranteed* to make the SCR turn on. A smaller signal *might* work, but the listed value *will* work.
5. If you are switching a large voltage from a low impedance source into a low resistance load, you may run into the dI/dt problem: SCR devices have a dI/dt limit. To work this out, you need the manufacturer's SCR manual. We use the General Electric *Replacement Semiconductor Cross-Reference Guide* or

Sylvania ECG Semiconductors (both booklets are free at your neighborhood electronics shop).

6. The gate voltage must *never* go negative with respect to the cathode, because that *zaps* the device. A diode in series with the gate prevents this from occurring. There are SCRs designed for (-) gate signals but they are less common than the (+) type.

A TRIAC is much like a SCR from a design point of view. In the case of a TRIAC, you are given:

1. An "on state" current at a given ambient temperature; this is written as I_T (rms).
2. V_{drom}, which is the maximum "off state" voltage (this is a peak value, *not* rms).
3. V_{gt} and I_{gt}, which are the maximum gate voltage and current, respectively.
4. P_{gm} and $P_{g(ave)}$, which are the maximum values of gate power on a peak (P_{gm}) or an average basis ($P_{g(ave)}$).

For more details, see the *G.E. SCR Manual;* they will provide more details than you will ever need. We suggest that you purchase some surplus SCR devices and start playing with them.

The biggest problem with SCR and TRIAC devices is that you can't shut them off unless you reduce the applied voltage to essentially zero. This problem led to the invention of the *SCS* or *silicon-controlled switch.* The SCS is still a somewhat rare device, and we will let you read about it in the *G.E. Transistor Manual.* They really work quite well, but their "hold-off" voltage is limited.

If you do have to switch higher voltages at low (1 amp or less) currents, the best thing is to go to *reed switches.* Such switches are actuated by moving magnets or a current through a small coil. The best reference on these is the *Machine Design* magazine's issue on electronic controls (Penton Building, Cleveland, OH 44113), which might cost you about $3.50.

Applications of SCRs and TRIACs

A few applications of SCRs and TRIACs are discussed below. To help you appreciate these circuits, we will review the ideas of *how* SCRs and TRIACs work. When forward bias is applied, only negligible current flows *until* a gate signal is applied. When the gate signal arrives the device "turns on" and stays on as long as forward bias is applied to the anode-cathode leads. By controlling the time at which the gate signal is applied, we can control what fraction of a voltage cycle is applied to a load.

This technique is often used for speed control of electric motors (more details on this application will be given in the next few pages). The control circuit of Figure 7-21 is often used with a TRIAC for inductive (electric motor) loads; to understand it, we turn to Figure 7-22. The control circuit changes the rms load voltage by changing the *conduction-angle* or *time of turn-on* of the SCR or TRIAC (see Figure 7-22).

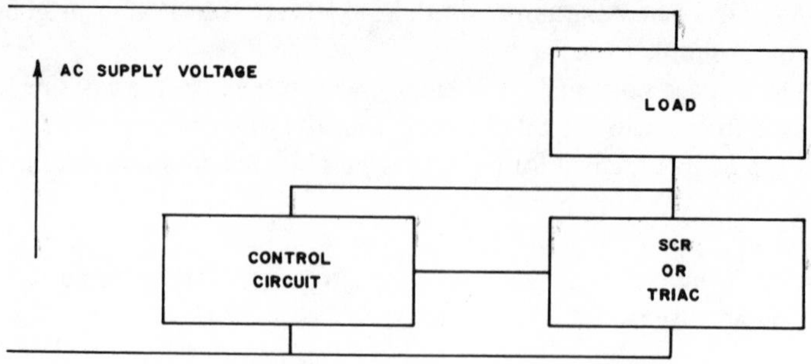

Figure 7-21. SCR or TRIAC control circuit for inductive loads.

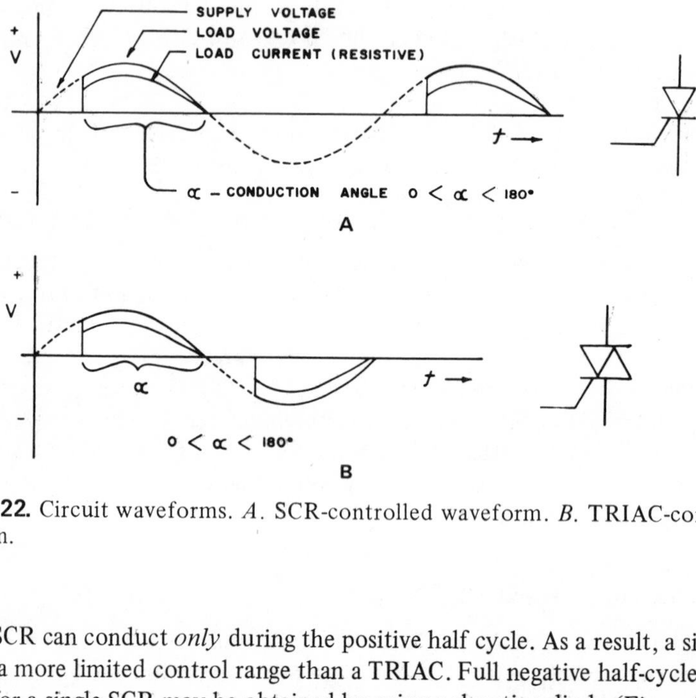

Figure 7-22. Circuit waveforms. *A*. SCR-controlled waveform. *B*. TRIAC-controlled waveform.

The SCR can conduct *only* during the positive half cycle. As a result, a single SCR has a more limited control range than a TRIAC. Full negative half-cycle conduction for a single SCR may be obtained by using a shunting diode (Figure 7-23A). Manual full-wave control could be obtained by switching this diode in or out. To get full-wave control with SCR devices, we have to use two of them in parallel; this circuit is shown in Figure 7-24.

We have noted earlier that the SCR or TRIAC will remain in the conducting state once a certain *latching current* has been reached. Conduction will stop when the current is reduced to zero at the zero crossing of the AC supply voltage. Thus, the gate current only needs to be a short pulse. For this reason, most control circuits take the form of a pulse generator whose firing time may be delayed past the zero crossing of the supply voltage by a variable resistor. However, when some

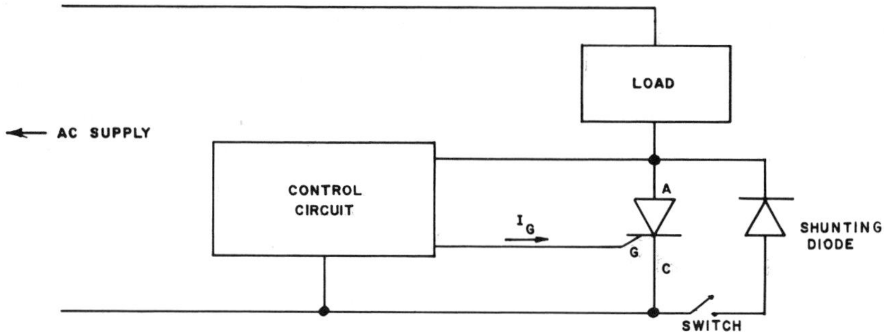

A

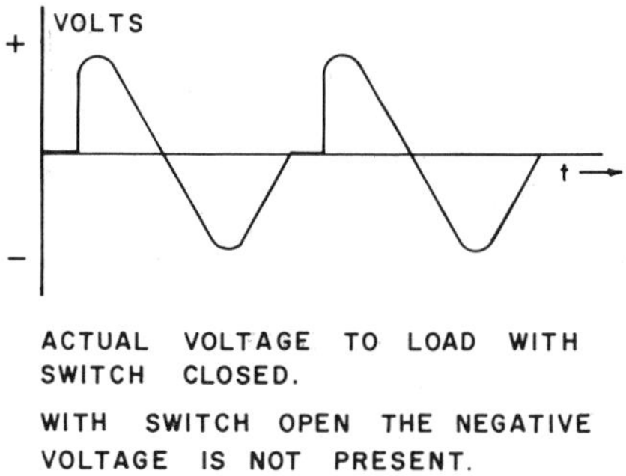

B

Figure 7-23. *A*. Half-wave control with negative half-cycle conduction. *B*. Waveform with control on positive half-cycle and full voltage on negative half-cycle.

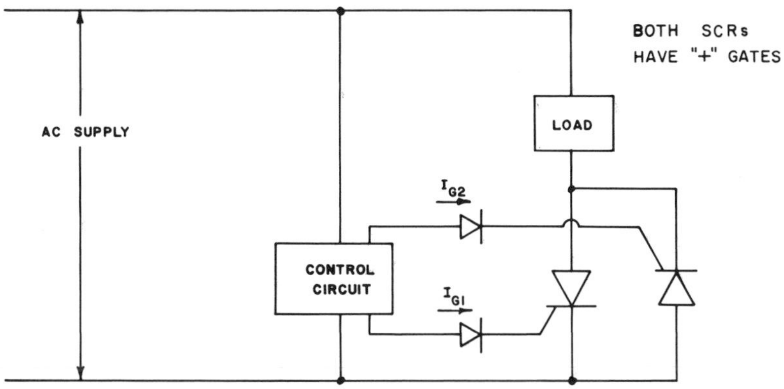

Figure 7-24. Full-wave control with two SCRs.

performance compromise can be made, particularly with regard to repeatability over a temperature range, very simple circuits can be used.

A simple and practical application of a TRIAC is shown in Figure 7-25. Instead of a complicated control circuit, the capacitor charges from the line voltage until the TRIAC gate fires. The voltage across the TRIAC falls to nearly zero, and the capacitor is discharged. The same thing happens on the next half cycle of the supply voltage, except that all voltages and currents are reversed. **Caution**: AC voltages are present throughout the circuit; measure the voltages with a VOM. If an oscilloscope is used, it must be run from an isolation transformer or have a floating input.)

There is one thing to note about the simple motor controller shown in Figure 7-25: There is no feedback to keep the speed of the motor at some chosen value if the load changes. In more advanced motor control circuits, some of the current to the motor is diverted to provide a feedback signal to the gate of the TRIAC. If the motor speed drops, the TRIAC comes on a bit earlier in the cycle to increase the motor speed. If the load decreases and the motor speeds up, the TRIAC comes on somewhat later to reduce the torque, thereby causing the motor to slow down. Advanced circuits of this type may be found in various books, e.g., the RCA *Thyristor and Rectifier Manual* (costs 75¢ from Newark Electronics).

GLOW LAMPS

Glow lamps or neon bulbs consist of two metallic elements spaced a short distance apart in a glass envelope and immersed in neon gas at moderately low pressure. The

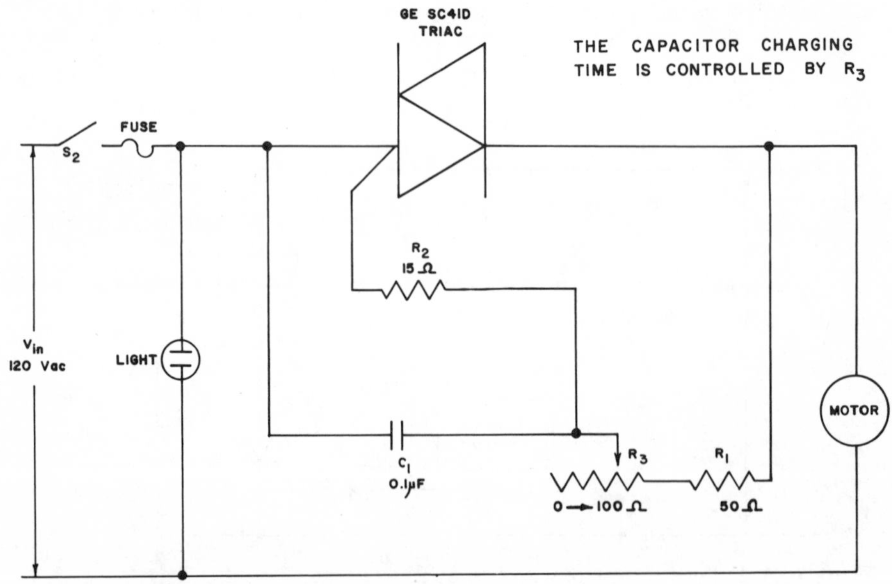

Figure 7-25. TRIAC motor-speed control circuit.

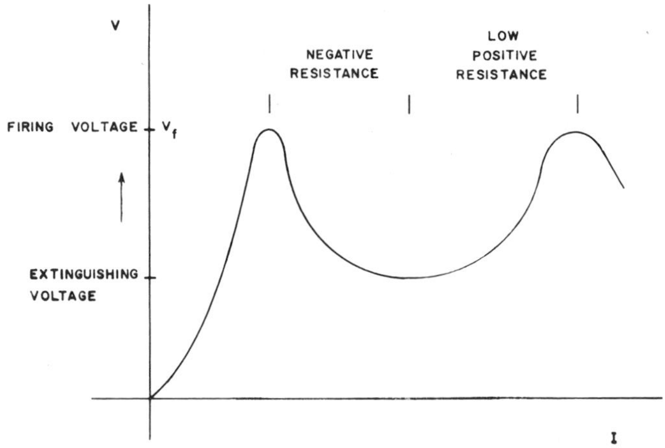

Figure 7-26. Glow lamp characteristic curve.

electrical conduction between these two elements is a complicated process involving electric fields, neutral atoms, excited atoms, positive ions, and electrons. The resulting relation between voltage and current has two useful features: (1) there is an operating region where very small applied voltage changes cause large current changes, and (2) there is also a region of negative resistance where *I* is *inversely* proportional to *V*. These are shown in Figure 7-26. The negative resistance region allows the glow lamp to be used as an oscillator. The low positive resistance region provides for voltage regulator applications, much like the zener diode did.

Glow Lamp Oscillator

A *glow lamp oscillator* may be made by connecting a resistor and capacitor to the lamp (Figure 7-27). The following things happen in the circuit of Figure 7-27 when the switch is closed. The current and voltage of the lamp increase from point *E* to point *B*, passing through *A*. During this time, the capacitor is charged to V_f. At point

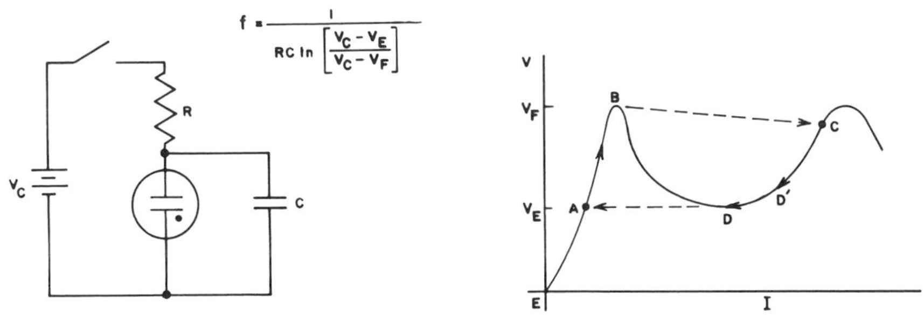

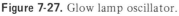

Figure 7-27. Glow lamp oscillator.

B, the tube fires, and the lamp current rapidly increases to point *C*. At this point, the capacitor begins to discharge through the lamp. When the lamp voltage reaches point *D*, the lamp current abruptly drops to point *A* and the lamp goes out. From point *A*, the capacitor again charges to point *B*, thereby restarting the cycle. The voltage and current of the lamp continuously cycle around the path *A–B–C–D*.

It is not hard to see why the circuit is sometimes called a *relaxation oscillator;* the voltage increases to a peak and then relaxes to a lower value. The frequency is determined by the values of *R* and *C*. A complete formula for frequency is given in Figure 2-27; For the circuit to oscillate, *R* must be greater than a certain minimum value. If *R* is too small, the circuit will remain at some point between *C* and *D*. Usually, $V_f = 70$ volts, so a good value for *R* might be 60 kΩ.

Glow Lamp Voltage Regulator

A *voltage regulator* using a glow lamp is even easier to make than an oscillator. If the capacitor is removed and if *R* is adjusted until the circuit is at some point near *D′* in the curve of Figure 7-27, we have a voltage regulator, as is shown in Figure 7-28. To understand how the circuit works, consider that initially $R_L = \infty$ and the current is at point *D′* in Figure 7-27. As R_L drops, current begins to flow through R_L rather than through the glow lamp. This moves the operating point from *D′* to *D* (a decrease in lamp current). However, the voltage across the lamp changes only very slightly, so this circuit acts as a type of voltage regulator. Before Clarence Zener invented his diode, people often used to use glow lamp voltage regulators.

In case you are wondering about how the glow lamp "starts," think about it this way: There are always some free electrons in the argon-neon gas (due to stray cosmic rays and radioactivity); if the voltage between the two electrodes is large enough, the electrons will be accelerated, knock other electrons off the neutral gas molecules, and the lamp "fires." This suggests that neon lamps are light sensitive and, in fact, they can be used to build crude (low-cost) detection systems for visible light and radioactive emissions. To insure that the lamps will fire at some fixed voltage, they can be purchased with a minute amount of radioactive material built into the bulb. Without that, you just have to wait until a stray photon or cosmic ray comes along. This should suggest some experiments to the clever reader.

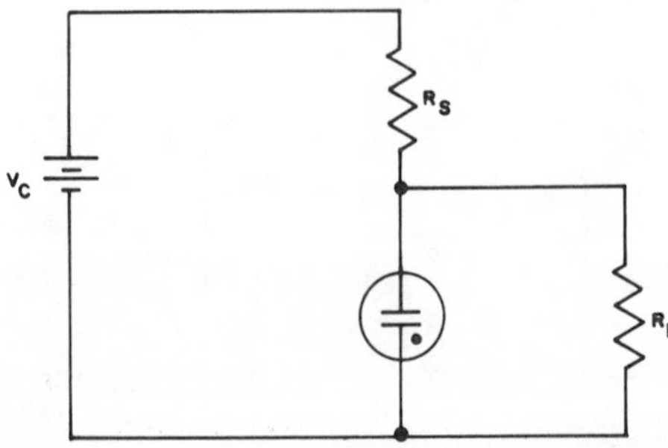

Figure 7-28. Glow lamp voltage regulator.

In this connection, the *G.E. Glow Lamp Manual* has a host of clever circuits – e.g., counters, scalers, level detectors, and so forth – why not look at it? For a copy, write to

G.E. Corporation
Miniature Lamp Dept.
Nela Park
Cleveland, OH 44112

GETTING RID OF TRANSIENTS – OR MEET THE VARISTOR

One of the nasty little things EEs have to learn the hard way is when to expect transients and what to do when you have them. In this book, a *transient* is not a member of the "great unwashed" but rather a sudden voltage pulse that comes along when you don't want it and wipes out your $15 diode.

To see a transient, you only need to wind 100 turns of insulated wire on an iron nail and hook up a circuit consisting of a switch, a 1.5 volt battery, and the coil of wire all in series (Figure 7-29A). Next, hook your trusty oscilloscope across the switch, put it on the DC input, and watch the voltage as you open and close the switch. At first, when the switch is open, you should measure the full 1.5 volts; when it is closed, the voltage will be zero. *Now,* when the switch is opened, you might think that the 1.5 volts would reappear. Well it does, but not right away: first there is a large transient voltage that might be as high as 50 volts if you opened the switch fast enough. That's what blows out $15 diodes.

To get rid of transients, you might try first a 1 μF and then a 10 μF capacitor across the switch. The first thing that EEs do when they run into a transient is to put a capacitor across it.* The problem is that with 115 volt circuits, the necessary capacitors can get pretty big. This brings us to a better device, called the *varistor.*

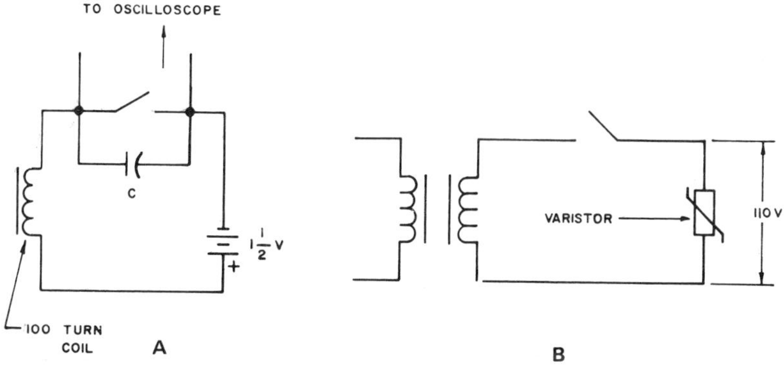

Figure 7-29. *A.* Transient test circuit. *B.* Varistor application circuit.

*In the case shown in Figure 7-29, try 100 μF in parallel with the switch. It will reduce the transient but will not eliminate it.

The *varistor* is sometimes like a zener diode: it passes very little current until the applied voltage reaches some given value, then it acts almost like a short-circuit. The varistor will take more of this short-circuit current than a zener diode will, and for this reason varistors are preferred. How do you learn to use them? Well, you could start by writing to the G.E. Corporation, Semiconductor Products Division, Syracuse, NY, and asking for their *Design Manual on Varistors* or their *Transistor Substitution Book.* These have all sorts of complex design formulas for EEs to play with (they dig that stuff). There are also some useful circuits for scientific types to build.

To save you that trouble, we herewith give our own special lesson in the use of varistors. Let's suppose we have a transformer driven on the primary side by the 110 volt line through a toggle switch. If your house or laboratory power lines are struck by lightning, the voltage on the power line might be 550 volts instead of 115. It won't last long, but then maybe it won't take long to wipe out a $50 transformer as well as the rectifier diodes on the secondary side. To prevent this exciting but costly event, we look for a varistor. The normal line voltage is 110 volts; if the power company has a bad day (stubborn electrons, of course), it might go up to a peak of 125 volts. We therefore select a G.E. 130A10 varistor and put it across the line *ahead* of the switch or transformer. Under normal conditions, it just sits there drawing a very small current. When the lightning surge comes along, the voltage rises above 180 volts and the varistor passes up to 1000 amps of current for 7 milliseconds. If that doesn't stop the lightning surge, you have been the recipient of a direct hit and are dead. The surge test circuit and this application of the varistor are shown in Figure 7-29.

Obviously, there are many other applications that the G.E. people will be glad to tell you about. For now, remember that if you have problems with voltage surges, try a capacitor first (they're always easier to get). If that doesn't work, try a varistor with a voltage rating *above* the peak value that might normally occur. At present, varistors don't come in every voltage you might want – but don't worry, they will!

ELECTRONIC DEVICES: POSTSCRIPT

At this point we are going to have to conclude this discussion of electronic devices, not that we have run out of devices, but because we are running out of pages. If you have stuck with us this far, you should be able to go to the Motorola, G.E., or RCA transistor design manuals and pick out the devices that will be of use to you. The various technical magazines that we give in the Appendix have all sorts of useful circuits and advertisements of new devices; go ahead and make use of them. Don't hesitate to contact manufacturers for free samples of their devices and application notes telling how to use them. Most companies hope that university students will see their gadgets at school and then, when they go to work, they will only design with gadgets from XYZ Company because that is what they are familiar with. Don't laugh! It works quite well, and there is no reason why you should not let this phenomenon work for you. In Chapter 8, we will give some suggestions on how to write begging letters. If you are wealthy, you can ignore them. If not, take our word for the fact that many of the gadgets that *we* use for student projects and research were corporate gifts.

The rest of this chapter will cover optical devices. Optics and the gadgets people use for optical work comprise a special branch of electronics. The first thing you need is a knowledge of the jargon, so away we go.

OPTICAL-ELECTRONIC DEVICES

One of the things most experimenters want to do is to use light-sensitive devices. Either they want to measure light intensity, or they want to use optical devices for counting, sorting, or detection.

Using optical detectors can be either very simple or extremely difficult: it all depends on what you want to do. Part of the problem is that optics is a very old science and there are several systems of units in use. Another difficulty is that all optical devices are packaged in containers that *must* have windows. This means that the properties of the detector are very dependent on the window material. You can save yourself a lot of misery by appreciating that fact.

Figure 7-30 shows the transmission characteristics of various materials and some source spectrums for comparison purposes. For the various light sources, we show only the wavelength of maximum intensity. The tungsten lamp, for example, is

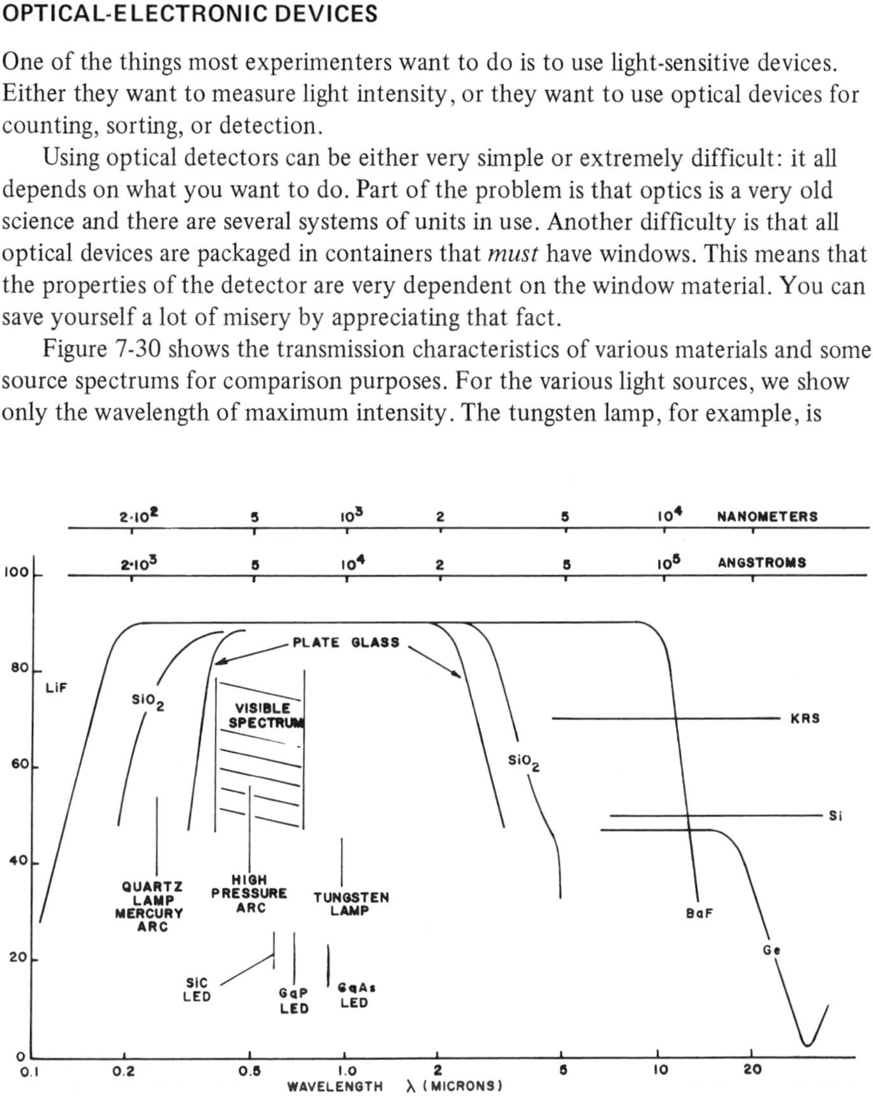

Figure 7-30. Transmission characteristics of optical materials and peak output wave lengths of certain optical sources.

most intense at 1 micron, but there is considerable radiation in the visible region, too. Notice that in the visible region, most materials are quite transparent. In the ultraviolet (UV) region, you must use first quartz (SiO_2) and finally lithium fluoride (LiF) as you go to shorter wavelengths. In the infrared (IR) you can use silicon (Si) or germanium (Ge). All these special materials are expensive; you use them only if you have to. (For more details, spend $2.95 on the *RCA Hobby Circuits Manual,* RCA Commercial Engineering Division, Camden, NJ 08107.

We will next explain the terms used in photometry, and that can get fairly complex. If you want to skip over these definitions and start reading the sections on optical sources or optical detectors, go ahead. If you run into a definition that you don't understand, come back and read about it.

Photometry

This is indeed a topic we would rather avoid because photometry, the measurement of light, is cluttered up with the most awful collection of units and nomenclature. However, we educators never give up, so bear up and stay with us.

There are two systems of units: *radiometric* units and measurements are related to the absolute electromagnetic properties of light; *photometric* units refer to the measurements that are intended to relate to the human eye. A light source that will be detected outside the visible light range might be rated in radiometric units. A visible light source *might,* note *might,* be rated in photometric units. A source for both purposes might be rated either way or both. (The American Standards Association waited too long in this area.)

In radiometric measurements (RM), the standard unit is *watts per meter squared.* In photometric measurements (PM), it is *lumens per meter squared.* The term *lumens* refers to human eye response, so to compare RM units to PM units, we need a curve of lumens per watt as a function of wavelength. This curve just happens to look exactly like an eye sensitivity curve (Figure 7-31). It rises to 700 lumens per watt at 550 nm (recall that $1\ \mu = 10^{-6}$ m, $1\ \text{nm} = 10^{-4}$ m, and $1\ \text{Å} = 10^{-10}$ m) and drops to almost zero at 400 nm in the near UV and at 700 nm in the near IR. (You will notice that at first we are switching back and forth between ångstroms, nanometers, and microns. This is not sloppy editorial work; rather, we think you should get used to the various units that people use. We haven't even mentioned *wave numbers* yet: those are $1/\lambda$ in units of cm^{-1}. For example, 0.5 micron is 20,000 wave numbers in cm^{-1}. We will generally use microns in this text, but don't count on it.)

The photometric rating of optical sources is in terms of *candelas.* A one candela isotropic souce (one radiating equally in all directions) radiates a total flux of 4π lumens (PM system). However, there are no such sources, and radiation intensity is usually given as a function of the angle with respect to the vertical taken on the light source itself. The light intensity of the source in a particular direction, divided by the projected area, is called the *radiant emittance* in the RM system and the *luminous emittance* in the PM system.

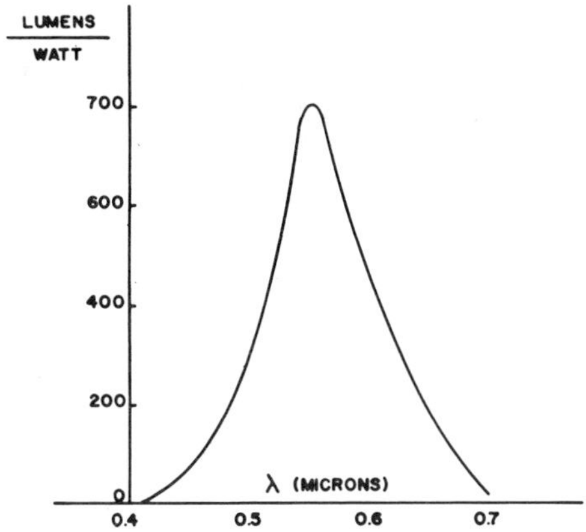

Figure 7-31. Eye sensitivity curve and the lumens per watt relationship.

If the source is being viewed from a distance, the power output per unit solid angle is called *radiant intensity* in the RM system and *luminous intensity* in the PM system. The units are watts or lumens per unit solid angle. Eventually this radiation must fall upon a surface and the incident energy in RM units is watts per meter squared. The PM or visual units are lumens per meter squared except when people use lumens per foot squared (peasant types); it could be worse – e.g., firkins of photons per fortnight per furlong squared.

At this point, you are probably ready to throw up your hands in despair and forget the whole project. Don't give up, we almost did that with this book a few times, but here you are reading it. The best thing to do now is to go on to the subsequent sections on light sources and detectors, pick up a few cheap devices, and go at it. Very few people really worry about irradiance, anyway. If you are one of the purist types who do, sit down and write to the General Electric Company (Lamp Department, 21888 Tungsten Road, Cleveland, OH 44117) and ask for the *Manual on Solid State Lamps.* It has some excellent design procedures if you are inclined that way.

Light Sources

We are all familiar with the incandescent (hot filament) lamp and the fluorescent lamp. There are all sorts of variations and combinations of these, and we will divide them into two basic categories.

THERMAL SOURCES

These devices produce a continuous spectrum from the IR to the UV depending on their temperature and envelope. The hotter you run them, the more the peak of the radiated light moves toward the blue. Quartz iodine-filled lamps come close to normal sunlight, whereas at the other end of the spectrum, the globar lamps radiate in the far IR. Usually cost goes up and lifetime goes down as you go to higher temperatures. Once again, G.E.'s manuals are a good source of information. If you are going to work in this area, you should plan to take *Electro-Optical Systems Design* magazine. (It is free, what can you lose?)

GAS DISCHARGE OR SPECTRAL SOURCES

These devices depend upon an electrical discharge in a gas or a gas mixture. Their output is not continuous over the whole spectrum; rather, it consists of lines or bands depending on what the filling gas is or what the electrodes are made of. If you need just one line for analysis purposes, you can buy these lamps filled with mercury, cesium, helium, and so forth. Edmund Scientific (Barrington, NJ 08007) usually has some in stock. The mercury-filled lamps are good in the UV region if they have a quartz envelope.

In the familiar fluorescent lamp, the inside of the tube is coated with a phosphor that gives off light in the visible region when hit by the UV radiation from the mercury discharge inside the tube. It is a kind of frequency shifter. One thing about all discharge lamps: if you run them on 60 Hz, the output is not constant but a sine wave at 120 Hz. The light output does not quite go to zero as the input voltage goes through zero, but it does drop. If you are going to be working with fluorescent light, you should expect to see AC effects. Your eye tends to integrate over the 120 Hz variation, and sometimes you forget it is there.

ELECTROLUMINESCENCE

New electroluminescent solid-state lamps are coming into use, and several varieties already exist. There is the type that works on AC only and emits a soft green light. This is true electroluminescence, but the light output is low and suitable only for special purposes – they are not yet used as a light source for photometry, but may be in the future. The *light-emitting diodes (LED)* gadgets are true diodes and emit light only when forward-biased. Their output is centered about a particular frequency that depends upon the material used to build the device. At present, red seems to be the most popular color. They work on AC or DC at potentials of a few volts. The light output is surprisingly high, and the cost is quite low. Get the latest details by writing to the Motorola Semiconductor Division (5005 McDowell Rd., Phoenix, AZ 85008) or looking at the G.E. manual on solid state lamps (G.E. Corp., Lamp Dept., 21888 Tungsten Road, Cleveland, OH 44117).

FLASHLAMPS AND PULSED SOURCES

Things of this kind are useful in stopping motion, for example, taking photographs of animals as they enter a trap. The simplest system uses standard photoflash bulbs, but if you want the light to last no more than a few microseconds, you must go to *condenser discharge lamps.* These are gas discharge lamps that are generally filled with xenon and that have three electrodes. A capacitor (condenser) charged to, say, 5 kilovolts is put across the two outer leads. No current flows until the third lead is pulsed to about 50 volts. When the third lead is pulsed, the capacitor dumps its charge into the lamp and an intense flash of light occurs.

There are lots of good circuits for condenser discharge lamps, but one thing is important: don't try to use cheap high voltage capacitors. You should buy only "discharge grade" capacitors; the other kinds have too much inductance and are not made for flashlamp service. Good circuits for devices of this type can be found in *Popular Electronics* (Ziff-Davis Publishing Co., 1 Park Avenue, New York, NY 10016).

WHAT LIGHT SOURCE SHOULD YOU USE?

It is, of course, the same old story: it all depends on what you want to do. In general, if you stick to some consumer-type light source, the cost will be lowest. If you need high intensity in the visible region, the quartz iodine lamps are best; used with a suitable reflector, they are real zappers. J.C. Whitney & Company (P.O. Box 8410, Chicago, IL 60680) sells all sorts of high intensity driving lights for cars. They may not last too long (perhaps 15 hours) but they might be worthwhile for a project. If you run them at low voltages, they will live much longer. Hook them in series while you are getting ready; then switch them to parallel for the picture.

For signaling or indication, the LED devices are just fine. Their power requirements are low, which is nice if you're operating on batteries. For high light output with battery-powered systems, the fluorescent lamps are better. Buy or build an oscillator to turn your battery DC into AC, then boost the AC with a transformer to run your light source. Such gadgets are called *inverters* or *power converters.* (A good circuit is that described by John Colt in "Build a Power Inverter," *Popular Electronics,* May 1969.)

Choosing an Optical Detector

This problem is a little like getting married. Every man wants (or thinks he wants) a rich, beautiful, sexy girl whose hobbies are home-cooking and making her own clothes; and while every woman wants a tall, dark, handsome man who will make lots of money and fix things around the house, while she has a career, if she wants one . Most individuals settle for a good deal less and like it. Well, it's the same with photodetectors; we all want wide spectral response, high quantum efficiency, 100% linearity, fast response time, low noise, zero fatigue, and minimum cost. We will settle for a good deal less and like it.

In order to make the best of the photodevices that are available, we must clarify the meaning of certain terms.

SPECTRAL UNITS AND SPECTRAL RESPONSE

The light entering our device has a wavelength λ and frequency f, where λ is in units of length and f is in reciprocal time ($f = 1/\tau$), cycles per second, or hertz. The product of λf is c, where c is the speed of light (3×10^8 meters per second, or m/s).

Normally, we don't measure light wavelengths in meters but in either ångstroms (1 Å = 10^{-10} meter), microns (1 μ = 10^{-6} meter), or nanometers (1 nm = 10^{-9} meter). The worst thing about it is that all three systems are in use and you just have to get used to it. Also, on older equipment, you might find the wavelength expressed in millimicrons (mμ), which are, of course, equivalent to nanometers.

As an aid to your memory, remember that the middle of the visible spectrum is in the green at about 5000 Å. Our 5000 Å green light may have its wavelength expressed as 500 nm or 0.5 μ, depending upon what you are reading. Once again, we will stick with microns from now on, but we must warn you that this isn't the only notation problem in optics.

Spectral response is just a question of what range of wavelengths your device is sensitive to. If you are working with a wide range of frequencies (e.g., solar spectrometry), you need uniform response over a wide range of wavelengths. If your light source is an LED (light-emitting diode), narrow response is fine provided that it is at the right wavelength. It will be up to you to check on the wavelength-response question.

QUANTUM EFFICIENCY

This is really a question of output divided by input. For a photodetector, we want to know how many electrons we can get per incoming photon. Of course, you want a high quantum efficiency but there may be other limitations (like money). Another type of limitation is linearity.

LINEARITY

This is a point often forgotten until it is too late, and then the weeping and wailing starts. *Linearity* is a simple thing to ask for: if the intensity of the incoming light changes by a factor of, say 10%, the output should also change by 10%. It is easy to define, but you had better read the manufacturer's specs carefully. Sometimes they forget to tell you about linearity, or about the lack of it.

RESPONSE TIME

This can be another tricky one. Suppose a sudden change in light intensity of, say 20%, occurs at the input; then the output should change by exactly 20%. That's linearity. The next question is, how long does it take for the final output value to be achieved? The answer to that is in terms of *response time.* If you are going to use those great chopping techniques that we taught you earlier, fast response time is important. For example, suppose you want to chop the incoming light to reduce noise. You pick a chopping frequency that is about five times the natural noise frequency in your system. Of course, you can't go too high or gain-bandwidth problems will bother you, so you pick, say, 53 Hz. The thing *not* to do is to pick some number that can be divided or multiplied to get 60; you will have enough 60 Hz noise anyway without asking for more.

Having picked 53 Hz as the chopping rate, you can calculate that each pulse will be about 19 milliseconds long. The response time of your detector should be about

1% of that, or 190 microseconds. That way your output electrical signal will have the same shape as the incoming optical signal. (One word of warning: when looking at response time specs, remember that there are two ways to define the term. If you don't recall what "eith time" or "90% time" means, go back and look them up on page 92.

NOISE

This is really a can of worms. There must be dozens of ways to look at this problem, and if you start reading things like the *RCA Electro-Optic Handbook* (RCA Commercial Engineering Div., Camden, NJ 08102), you will see all of them. We hope to provide a simple and improved discussion.

DRIFT. This is a slow change in output, usually over periods of minutes to hours. It is generally due to changes in temperature as the device warms up or as the ambient temperature changes. To get rid of drift, you try to stabilize the temperature, or you chop the light so that drift becomes unimportant, or you do both.

A worse problem is caused by "real noise," the high frequency stuff that simple chopping can't get rid of. We subdivide this into:

DARK CURRENT. This is the output you get even when *no* light is coming into your device. It may be due to the temperature of the gadget itself, which is why most of them are operated below room temperature. It may also be due to leakage, or even to radioactive contamination of the material the device was built of. You have to watch the specs on this *very carefully*. While we are on this question of dark current you should appreciate that the dark current due to thermal effects, cosmic rays, and radioactivity is an absolute limit in the sense that you cannot – repeat, *cannot* – detect any signal below that level. This should explain why astronomers go to such trouble to keep their detectors cool, shield against stray cosmic rays, and buy detectors made of special materials that are "low" in radioactive elements, i.e., uranium-free glass.

SHOT NOISE. This is the jitter in the output due to the electrons banging around inside the device. Shot noise is really like the word "virus," a generic term for something that can't be explained any other way.

NOISE-EQUIVALENT POWER (NEP). We have ignored a few other sources of noise to bring you to the NEP concept, because all the other sources look like shot noise at the output and the differences are only of interest to specialists. (They will sneer at this book anyway: "No one could possibly use optical detectors without an M.S. in Physics and a Ph.D. in Optical Sciences." To Hell with them.)

The *noise-equivalent power* simply means that input which produces an output equal to the rms noise current or voltage. As an example, suppose we have a phototube that has a one electron per photon sensitivity (great), but its output *in the dark* is a current that is equivalent to three photons per second (not so great). It turns out that the minimum number of photons that you can detect per second is not *one* but *three,* because that is the NEP. The NEP is an important concept, and these days most manufacturers rate their devices in that way.

FATIGUE. Fatigue is really a form of noise. It occurs when photodetectors are exposed to intense light for long periods of time. If the output drops when the input intensity does not, you have detector fatigue.

Fatigue is most prominent in photomultipliers. If they are run at high light levels for several hours, their current output goes down and no one knows why. Sometimes they recover if they are left in the dark for a while, but sometimes they don't. The effect is *not* limited to photoelectric devices, so be on your guard.

DETECTIVITY OR D* (D-STAR)

You thought we were done, didn't you? Well, there is still detectivity to be defined. *Detectivity (D*)* is an attempt to put all the concepts of noise, sensitivity, and bandwidth together for the comparison of devices. D* is measured at some particular wavelength, λ, with a light that is chopped at some given frequency, which illuminates a fixed area on the detector. The bandwidth of the amplifying and recording circuits is also given. In theory, a larger value of D* means a better device. Not being optics types, we can't comment on whether this concept is a useful one. However, you will see it in the literature and you should know what it means. After that you are on your own.†

COST

Naturally this is an important question, but the optics market is so fluid and prices are changing so fast that any cost figures we could give you would be meaningless. In general, the photoelectric multiplier devices will be the most expensive because they are complicated to build. Photodiodes, on the other hand, will cost the least because of mass production. As usual, the decision will fall upon the user, which is why you have to know something about the gadgets before you buy them. As the folks at Merrill Lynch, Pierce, Fenner & Smith say, "investigate before you invest."

If you are very green at the photodetector business, the best thing might be to buy an assortment of cheap devices from Edmund Scientific Company (Barrington, NJ 08007). I bought a box of assorted devices 12 years ago, and 12 generations of students have learned on them. The Edmund Company also sells assortments of lenses and light sources, and you will enjoy their catalog. If this "bull moose" approach to the problem of choosing an optical detector turns you off, just continue reading to find out the details about what devices are available. Then you can take on the what-to-buy problem with more (?) confidence.

Photoelectric Devices

The photoelectric cell is the oldest optical-electronic sensor we have. Originally, the term was used *only* for devices in which incident light induced electron emission (photoemission), and the current thus produced was a measure of the light intensity. In recent years, the term *photoelectric device* has been used for all sorts of photo-

†An excellent recent review is that given by R. Keyes and R. H. Kingston in "A Look at Photon Detectors," *Physics Today,* March 1972.

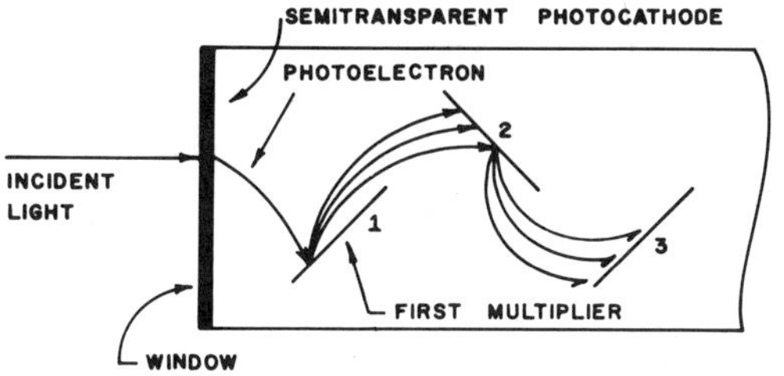

Figure 7-32. Schematic phototube with multipliers.

voltaic and photoconductive systems. (The growth of this type of sloppy language and the decline of precise English is a perfect example of Gresham's law: "Bad money drives out good money every time.")

A schematic photoelectric device with an added multiplier is shown in Figure 7-32. This multiplier increases the gain, the noise, and, of course, the price. A real phototube will have more stages of amplification (up to 10), but the idea is the same. There are also some gas-filled phototubes for sale that provide increased gain; however, the gas-filled types usually have a slower speed of response. The point to remember is that *the output is a current proportionate to the light intensity, wavelength, polarization, and angle of incidence of light on the cathode.* Many people don't know about the effects of angle of incidence and polarization, but these can be useful (see S. Hoenig and A. Cutler, "Polarization Sensitivity of the RCA-6903 Photocathode Tube," *Applied Optics* 5:1091, 1966).

Of course, the window material is important, too. If you want to use phototubes or other optical detectors, you had better rush down to your Allied catalog dealer (or write to RCA, Commercial Engineering Div., Camden, NJ 08102) and order a copy of the *RCA Manual PT-60, Phototubes and Photocells.* It discusses only RCA products (what else?), but it gives a very good review of what you need to know to get the most out of photodetectors. It also has a good discussion of optical detector problems and techniques to solve them.

It is particularly important to reemphasize that the output of almost all photodetectors is *strongly* dependent upon the wavelength of light involved. To demonstrate this, we show in Figure 7-33 a series of spectral response curves for various types of photodetectors. The photomultipliers are usually best in the visible and near UV regions; other devices (to be discussed below) are usually used in the IR. Notice that all curves drop off very rapidly near 0.2 micron (2000 Å). This is *not,* repeat *not,* an intrinsic photocathode phenomenon, but it is due to the fact that no good window materials exist below 0.2 micron. If you are out in space with windowless photomultipliers, you can work at 0.1 micron. Please note that the data in Figure 7-33 were compiled from various sources, and that the "sensitivity" of a device may be measured differently for each type of device. Don't try to read our curves for comparison of devices where they overlap; use them only as a guide to which device to use in what spectral region.

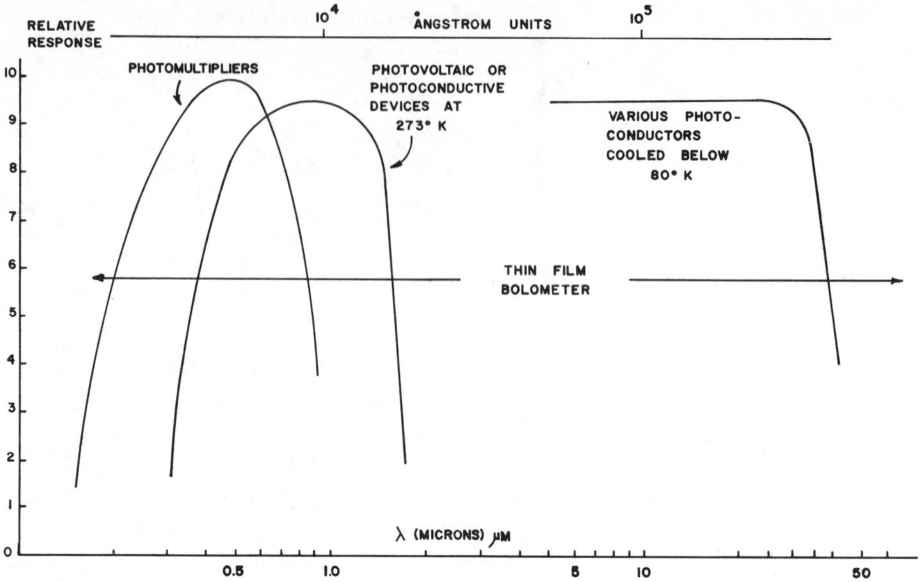

Figure 7-33. Response curves of various photodetectors.

Another point to remember is that various types of photodevices may have current, voltage, or resistance changes as their output. With op-amps, you can convert current or resistance signals to voltage signals. Then away you go!

Photoconductive Devices

This category covers devices that change their internal resistance when illuminated by a light source. Once again, there is a bewildering variety of devices available, but the *RCA Electro-Optics Handbook* is a useful guide to the products available from RCA.

In general, these devices are packaged in plastic, so their response is limited to the visible and near IR regions. They are low in cost, are available in large and small sizes, and have a fairly constant output over their range of operating wavelengths.

Their response is not as rapid as the photoelectric devices, and this may or may not be of importance to you. The main application of photoconductors is in the far IR, out at 3 or 4 or even 30 microns.

PHOTOTRANSISTORS

These devices look like a regular bipolar transistor, except that there isn't any base connection, but the base has a window over it. You hook it up just like a regular transistor, with reverse bias on the collector. When the base is illuminated, current flows from the collector to the emitter. These devices are usually low in cost but their gain and response speed are limited. Their best spectral region is about 0.8 micron (for silicon based units).

PHOTODIODES

These devices are essentially reverse-biased diodes with a transparent envelope. When the diode is illuminated with light of the proper wavelength (again about 0.8 micron), current carriers are produced and diffuse across the junction, thereby producing a current. Photodiodes generally have fast response, and they can be made in vast arrays in which each individual device is quite small. The small sizes involved can be appreciated when you realize that arrays of photodiodes have been used to generate television-type images with almost unnoticeable grain. This may be something to keep in mind.

You may hear the term *avalanche photodiode* now and then. These devices operate somewhat differently and have a higher output than regular photodiodes, but they can be evaluated in much the same way.

BOLOMETERS – THE ULTIMATE DETECTOR

All the gadgets discussed so far have their place, but *none* of them responds with uniform output from the far IR to the far UV, and the day may come when you will want a device to do just that. Such devices are called *bolometers,* and until recently they were slow in response, difficult to use, and expensive. Now they are only expensive. A bolometer such as the Hewlett-Packard gadget works with a blackened foil that has a string of thin-film thermocouples on the back. The incident light warms the foil, and the thermocouples develop a voltage accordingly (for details, contact your local Hewlett-Packard office and ask about the Model 8334A thermopile detector, or write to the Hewlett-Packard Corp., 1501 Page Mill Rd., Palo Alto, CA 94304). From 0.2 to 40 microns, it is precise to ±3% with a response time (90%) of about 0.7 second. If you are really going to do an absolute calibration on an optical detector, this is the only way to go.

LIGHT AMPLIFIERS AND STARLIGHT SCOPES

Here you are really getting up into high cost devices, and you should plan to investigate twice before you invest once. These gadgets were developed during the recent fiasco in Southeast Asia, and they may be the only good result of the whole business. They usually operate by turning the light into electrons using an array of devices, and multiplying the number of electrons. Other devices have optical emitters that are triggered by the incident light from the "target" and give off more light than they get. Some of these gadgets are coming up on the surplus market (check with Edmund Scientific Co., Barrington, NJ 08007), and may be useful for the observation of animals at night.

CHOOSING A PHOTOCONDUCTIVE DEVICE

Now we finally get back again to the nitty-gritty of the problem: what detector do we pick? The first consideration should be the frequency range of your light. The second should be the light intensity available. The third, of course, is how much money you have.

Having settled these three questions, you go to the RCA books referred to earlier (p. 339) and try to pick a device. This doesn't mean that you have to buy from RCA, but their books are a good guide. If you don't have catalogs from any other manufacturer, you can go to the library and look at a freebie magazine called *Optical Spectra* (The Optical Publishing Co., Inc., Berkshire Common, P.O. Box 1146, Pittsfield, MA 01201). Another good one is *Electro-Optical Systems Design* (222 West Adams St., Chicago, IL 60606). Write to a few of their advertisers and the first thing you know, you too can be a part of the Post Office deficit problem. Good luck.

OPTICAL-ELECTRONIC CIRCUITS: POSTCRIPT

If you have to design and build electro-optical circuits, we suggest you begin by looking at the article by M. G. Leonard, "Controlling Electrical Circuits With Light," *Machine Design,* July 23, 1970. Mr. Leonard discusses all sorts of neat circuits, from coupling optics to electronics to mechanical devices.

If you need optically isolated switches for high voltage or medical electronics, you might write to Motorola Semiconductor Products, Inc., P.O. Box 20912, Phoenix, AZ 85036, and ask for literature on their optical-isolation devices.

Applications of optical devices in op-amp and transistor circuits can be found in *Electronics* and *Popular Electronics.* M. H. Loughnane's article, "Light-Emitting Diode Pair Forms Null Indicator" (*Electronics,* August 2, 1971, p. 58) will be helpful. Reading *Popular Electronics* every month will give you a good collection of useful circuits. Some of the articles we suggest reading are: D. C. Conner's "Light-Operated Bistable Switch (March, 1961, p. 70); H. Olson's "Infrared Intrusion Alarm (March, 1972, p. 51); and N. P. Huffnagle's "H.E.L. Panel Driver" (May, 1971, p. 43). Optical electronic devices data can be obtained from National Semiconductor Ltd., 331 Cornelia St., Plattsburgh, NY 12901.

TEST YOURSELF PROBLEMS AND SUGGESTED EXPERIMENTS

Suggested Experiments

A. TRANSISTOR CURVE TRACER AND TESTER

Test the following transistors on both the Heathkit and Tektronix testers:

1. TI 7228 β Heath = ____________ β Tek = ____________
2. TI 7437 β Heath = ____________ β Tek = ____________
3. Motorola A20 β Heath = ____________ β Tek = ____________

B. SEVEN STEPS TO TRANSISTOR DESIGN

Design a common emitter using the following seven steps. Fill in the necessary information at each step. Use the MPS-A20 Transistor and data sheet in your design (*not* the transistor data from the book). Also, use the same conservative estimates as the book does (i.e., $R_b = 15\ R_c$, $V_e = 0.1\ V_{ce}$, etc.).

1. Draw a schematic of the common emitter amp.
   Step 1: ____________
   Step 2: given $I_c = 10$ mA
   Step 3: given $V_{ce} = 10$ volts
   Step 4: ____________
   Step 5: given $R_c = 1$ kΩ
   Step 6: ____________
   Step 7: ____________

2. Construct the circuit and check how closely the measured values matched the calculated values.

	Measured	*Calculated*
V_e	____________	____________
V_{ce}	____________	____________
V_{be}	____________	____________
V_b	____________	____________
$V_c/R_c = I_c =$	____________	____________

3. Connect the function generator to the amplifier input and observe the output.
   What is the gain?
   Why would you use transistors instead of op-amps?

C. TRANSISTOR CURVE TRACER AND TESTER

Define the following in words and by diagram:

β	____________	V_{ce}	____________
I_{ceo}	____________	V_e	____________
I_{cbo}	____________	V_c	____________
I_c	____________	P_d	____________
I_b	____________	h_{fe}	____________

Draw an I_c versus V_{ce} plot.

8 Conclusions

As one of my friends would say, "What's to conclude? By now you either hate electronics or love it forever." All we can say is we hope you have learned that electronics is fun and that it doesn't require a BS in electrical engineering to do useful things. We also hope that you will want to continue to learn about new devices and how to use them. It might not be long before you will be contributing to the literature yourself.

You might have wondered about how to get things for nothing from various corporations. The process first involves swallowing any false pride you might have and accepting the fact that *you want something for nothing.* The second thing to do is write a good letter to the company involved. If you don't know any individual name, write to the "Director of Public Relations."* With big outfits like G.E. or RCA, this might get you a form letter about their corporate gifts program. Smaller companies may respond to your first letter, but in any case it is better to write to *someone.* If you are at a university, find out if the company recruits at your school. If they do, write to the recruiter; he will send your letter to Public Relations for possible action. Don't be afraid to make a pest of yourself if the company is slow in responding. What can they do besides ignore you?

A couple of typical and *successful* letters are included here to give you an example of how to do it. From here on, you are on your own!

FACULTY VERSION

Dear Mr. Smith, Jones, Brown, or whoever:

We are running a course in electronics for graduate students in the biological, zoological, and medical sciences. Most of the course is built around operational amplifiers, and we are trying to give every student an opportunity to work with these new devices. We feel that the application of op-amps will expand *only* as fast as students who have worked with them go into the industry.

A major part of our present laboratory is built around your model *XYZ* op-amps and power supplies. The student response to this laboratory course has been excellent, and we have had to turn people away for lack of equipment.

In view of this problem, I am writing in the hope that you can help us obtain

*The "home office" addresses of all major American corporations are given in *Thomas Register* (available at your local or university library).

more model *XYZ* op-amps and power supplies. We have some very limited state funds that we would like to stretch by purchasing any obsolete, out-of-specification, or demonstrator op-amps that you might have available. Your cooperation in this matter would serve a pressing educational need and help introduce op-amps to a wider group of users.

Yours truly,

If this letter, which would bring tears to the eyes of a stone image, doesn't work, then you may have to break down and pay cash. A list of corporations and surplus sales outlets is given in the "Where to Buy It" section in the Appendix.

STUDENT VERSION

The attitude you must take is that of a *humble student* in awe of the Mighty, but Beneficent Corporation. Address your letter to the Head of the Professional Recruitment Department. If there is no answer, try again in a month or so. Sometimes it takes three letters to get a corporation to lay some freebie merchandise on you. A typical letter is given below (feel free to modify it if you wish).

Dear Sirs:

I am a student in the Department of at the University of I am working on a project to and, as part of this work, we are using your product. These devices are working quite well and we are very pleased with them, but we are hampered by a lack of this equipment. Many other students who would like to make use of your product are being turned away because of our limited supplies of this type of apparatus.

In view of this problem, ...

Yours truly,

HOW TO FIND OUT WHO SELLS WHAT AND TO WHOM

If you have come this far, you must be interested in how and where to buy things. The best general source is *Thomas' Register:*

Thomas Publishing Company
1 Penn Plaza
New York, NY 10001

Thomas' Register lists all American companies by name, products, and industry. All libraries have it, and if it weren't so expensive, we would have it in the lab.

For the electrical and electronics industry, *EEM (Electronic Engineers Master)* is very good, and best of all it is free. Write to *EEM* magazine and ask to be put on their free mailing list. They have great op-amp circuits that you can steal, and they publish *EEM.* They do try to limit the freebie list to those readers who buy or specify electronic devices. Even if they were to charge you $20 per year, its worth it.

EEM
Herman Publishing Service, Inc.
755 Boylston Street
Boston, MA 02116

Another good freebie magazine with lots of op-amp circuits is

EDN
Cahners Publishing Company
221 Columbus Avenue
Boston, MA 02116

This is the only way to keep up with new developments in this industry. Things change so fast in this field that one must run hard just to stay in place, and to gain you must run twice as fast!

Appendixes

Where to Buy It

MISCELLANEOUS ELECTRONIC AND LABORATORY SUPPLY HOUSES

(Catalogs usually available)

Allied Electronics
3160 Alfred Street
Santa Clara, CA 95050
(800) 538-8030

American Design Components
39 Lispenard Street
New York, NY 10013

Analog Devices, Inc.
Rte. 1, Industrial Park
P.O. Box 280
Norwood, MA 02062

Burr-Brown Research Co.
P.O. Box 11400
Tucson, AZ 85734

Continental Resources, Inc.
175 Middlesex Turnpike
Bedford, MA 01730

Edmund Scientific Co.
101 E. Gloucester Pike
Barrington, NJ 08007

EICO Electronic Instrument Co.
283-T Malta Street
Brooklyn, NY 11207

Electronic Associates, Inc.
185 Monmouth Parkway
West Long Branch, NJ 07764

Electronic Supermarket
P.O. Box 619
Lynnfield, MA 01940

Fairchild Camera & Instrument Corp.
464 Ellis Street
Mountain View, CA 94042

John Fluke Mfg. Co., Inc.
P.O. Box 43210-T
Mountlake Terrace, WA 98043

General Electric Corp.
Electronic Components Sales Dept.
Electronics Park
Syracuse, NY 13201

General Electric Corp.
Lamp Dept.
21888 Tungsten Road
Cleveland, OH 44117

Gould, Inc.
Instrument Systems Division
3631 Perkins Avenue
Cleveland, OH 44114

Heath Company
Benton Harbor, MI 49022

Herbach and Rademan, Inc.
401 E. Erie Avenue
Philadelphia, PA 19134

Hewlett-Packard Co.
1501 Page Mill Road
Palo Alto, CA 94304

Instrument Systems Division
IBM
1000 Westchester Avenue
White Plains, NY 10604

Kepco, Inc.
131-44 Sanford Ave.
Flushing, NY 11352

Lafayette Radio
111 Jericho Turnpike
Syosett, NY 11791

Lambda Electronics
Division of Veeco Instruments, Inc.
515 Broad Hollow Rd.
Melville, NY 11746

Leasametric
A Trans Union Co.
1164 Triton Drive
Foster City, CA 94404

John J. Meshna, Jr., Inc.
P.O. Box 62
19 Allerton Street
E. Lynn, MA 01904

Minco Products, Inc.
7302 Commerce Lane
Minneapolis, MN 55432

Motorola Semiconductor Products, Inc.
P.O. Box 20912
Phoenix, AZ 85036
Attn: Sales Dept.

Newark Electronics Corp.
500-T N. Pulaski Road
Chicago, IL 60624

Olson Electronics
260 S. Forge Street
Akron, OH 44308

Poly Paks
P.O. Box 942
S. Lynnfield, MA 01940

Power/Mate Corp.
514 S. River Street
Hackensack, NJ 07601

Princeton Applied Research
P.O. Box 2565
Princeton, NJ 08540

Radio Shack
(A Division of Tandy Corp.)
500 One Tandy Center
Fort Worth, TX 76102

RCA Commercial Electronics
Systems Division
Front & Cooper Streets
Camden, NY 08102

Sorensen Power Supplies
676 Island Pond Road
Manchester, NH 03103

Teledyne Philbrick
Allied Dr. at Rt. 128
Dedham, MA 02026

Westinghouse Electric Corp.
Semiconductor Division
Armbrust Rd.
Youngwood, PA 15697

MISCELLANEOUS SURPLUS MATERIALS*

ADC
39 Lispenard Street
New York, NY 10013

Airborne Electronics Co.
6813 1/2 Troost Avenue
North Hollywood, CA 91605

AST/Servo Systems
930 Broadway
Newark, NJ 07104

Barry Electronics
512 Broadway
New York, NY 10012

BEC
2709 N. Broad Street
Philadelphia, PA 19132

C & H Sales
2176 E. Colorado Boulevard
Pasadena, CA 91107

Kelvin Electronics Inc.
1900 New Highway
P.O. Box 8
Farmingdale, NY 11735

Leasametric
Division of Metric Resources Corp.
822 Airport Boulevard
Burlingame, CA 94010

Lee Lab Supply
13714 S. Normandie Avenue
Gardena, CA 90249

Palley Supply Co.
11630 Burke Street
Los Nietos, CA 90606

Phoenix Design
P.O. Box 357
Douglassville, PA 19518

Scientific Leasing Services, Inc.
3 Delaware Drive
Lake Success, NY 11040

The Surplus Center
P.O. Box 82209
Lincoln, NE 68501

Tucker Electronics Co.
P.O. Box 1050
Garland, TX 75040

*For miscellaneous catalogs and manuals see Selected Reading List.

Selected Reading List

BOOKS AND ARTICLES

Abrecht, R. L., et al. *BASIC 2nd Edition Self Teaching Guide.* New York: Wiley, 1978.

Ackermann, P. G. *Electronic Instrumentation in the Clinical Laboratory.* Boston: Little, Brown, 1972.

Bennett, W. R., Jr. *Scientific and Engineering Problem Solving with the Computer.* Englewood Cliffs, N.J.: Prentice-Hall, 1976.

Diefenderfer, A. J. *Basic Techniques in Electronic Instrumentation.* Philadelphia: Saunders, 1972.

Finkel, J. *Computer-Aided Experimentation.* New York: Wiley, 1975.

Geddes, L., et al. *Principles of Applied Biomedical Instrumentation.* New York: Wiley, 1968.

Geddes, L. *Electrodes and the Measurement of Bioelectric Events.* New York: Wiley, 1972.

Geffe, P. R. *Simplified Modern Filter Design.* New York: Ryder, 1963.

Graeme, J. G., Tobey, J. E., and Huelsman, L. P. (eds.). *Operational Amplifiers.* New York: McGraw-Hill, 1971.

Granville, W. A., et al. *Elements of the Differential and Integral Calculus.* New York: Ginn, 1934.

Hilburn, J. L. *Manual of Active Filter Design.* New York: McGraw-Hill, 1973.

Kleppner, D., and Ramsey, N. *Quick Calculus.* New York: Wiley, 1966.

MacKay, S. *Biomedical Telemetry.* New York: Wiley, 1968.

Meyers, G. H. *Engineering in the Heart and Blood Vessels.* New York: Wiley, 1969.

Morrison, R. *Grounding and Shielding Techniques in Instrumentation.* New York: Wiley, 1967.

Plonsey, R. *Bioelectric Phenomena.* New York: McGraw-Hill, 1969.

Sallen, R. P., and Key, E. L. A practical method of designing RC active filters. In *IRE Transactions, Circuit Theory,* Vol. CT-2, pp. 74–85, March 1955.

Smith, J. I. *Modern Operational Circuit Design.* New York: Wiley, 1971.

Smith, R. J. *Electronics: Circuit Devices and Systems.* New York: Wiley, 1973.

Strong, C. L. Little radio transmitters for short range telemetry. *Scientific American,* p. 128, March 1968.

Tektronix Corporation. *Biophysical Measurements.* P.O. Box 500, Beaverton, OR 97077: Tektronix Corporation, 1970.

Tektronix Corporation. *Transducer Measurements.* P.O. Box 500, Beaverton, OR 97077: Tektronix Corporation, 1970.

Wilcox, G. *Electronics for Engineering Technology.* Boston: Allyn and Bacon, 1969.

Zatzick, M. R. How to make every photon count. *Electro-Optical Systems Design,* p. 20, June 1972.

MAGAZINES AND JOURNALS

Applied Optics
American Institute of Physics
335 E. 45th Street
New York, NY 10017

Electrical Design News
Computer Center
P.O. Box 5563
Cenver, CO 80217

Electronics Magazine
McGraw-Hill, Inc.
1221 Avenue of the Americas
New York, NY 10020
(Note: This journal distributes *Circuit Designers' Casebook.*)

Electro-Optical Systems Design
222 W. Adams Street
Chicago, IL 60606

Optical Spectra
The Optical Publishing Co., Inc.
Berkshire Common, P.O. Box 1146
Pittsfield, MA 01201

Popular Electronics
Ziff-Davis Publishing Co.
1 Park Avenue
New York, NY 10016

MANUALS AND MISCELLANEOUS PUBLICATIONS

(If you check the catalogs obtained from our *Where to Buy It* section, you can often find a section of *Data Books* or *Electronic Books and Manuals* sold by that particular supplier. You may also obtain such publications from your local electronics store at nominal expense.)

American Radio Relay League
225 Main Street
Newington, CT 06111
(Publications on amateur radio)

Brush Instruments Co.
(A division of Gould Inc.)
Belmont & City Line Avenues
Bala-cynwyd, PA 19004
Signal Conditioning

Burr-Brown Research Co.
P.O. Box 11400
Tucson, AZ 85734
(Various publications)

Emergency Care Research Inst.
5200 Butler Pike
Plymouth Meeting, PA 19462
Health Devices

General Electric Corp.
Electronics Park
Syracuse, NY 13201
SCR Manual
Transistor Substitution Book
Design Manual on Varistors

General Electric Corp.
Lamp Dept.
21888 Tungsten Road
Cleveland, OH 44117
Manual on Solid State Lamps
Glow Lamp Manual

Howard W. Sams and Co.
4300 W. 62nd Street
Indianapolis, IN 46206
Basic Electricity/Electronics, Volumes 1 and 2

Motorola Semiconductor Products, Inc.
P.O. Box 20912
Phoenix, AZ 85036
(Publications on Semiconductors)

National Fire Protection Assn.
470 Atlantic Avenue
Boston, MA 02110
(Handbooks and regulations on industrial safety)

Radio Shack
(A Division of Tandy Corp.)
500 One Tandy Center
Forth Worth, TX 76102
#62-2016 *BASIC Computer Language*
Dictionary of Electronics

RCA Commercial Engineering
Front & Cooper Streets
Camden, NJ 08102
Electro-Optics Handbook
Linear Integrated Circuit Manual

Schaevitz Engineering Co.
P.O. Box 505
Camden, NJ 08101
Handbook of Measurement and Control

Tektronix Corp.
P.O. Box 500
Beaverton, OR 97077
(Miscellaneous manuals on measurement techniques; also books mentioned above)

Teledyne Philbrick
Allied Dr. at Rt. 128
Dedham, MA 02026
Applications Manual for Operational Amplifiers

Texas Instruments Co.
P.O. Box 5012
Dallas, TX 75222
Transistor Design Manual

Westinghouse Electric Corp.
Semiconductor Div.
Armbrust Road
Youngwood, PA 15697
SCR Designer Handbook

An Informal Glossary of Technical Terms

AC voltage or current A voltage or current signal that varies symmetrically about the zero voltage axis.

A_v The DC voltage gain of a transistor circuit. Note that the value of this parameter depends on the value of h_{fe} for the transistor in the circuit, but the two symbols usually have different numerical values.

AM radio Transmission of audio signals by a system that uses the audio signal to change the amplitude of a fixed-frequency carrier wave.

Ammeter A device or system for measuring the rate of flow of electricity in coulombs per second or amperes.

Amplifier A device whose output current or voltage is larger than its input current or voltage. In this book, the term is used only for linear amplifiers; in such amplifiers, the input-output ratio may be adjustable, but it is constant once set.

Analog In general, an analog system behaves in the same way as some other systems we are interested in. In a mechanical "real world" system, for example, we have parameters of force, mass, speed, displacement, and so forth. In the analog computer system, we have a voltage for every "real world" variable. By observing how the analogous voltages behave, we can see how the "real world" system will respond.

Antenna An array of wires, rods, or plates used to transmit or receive electromagnetic signals. Antennas are available from the very low frequencies of 3 to 5 Hz up to super high frequencies of 10,000 MHz.

Assembler A program that prepares a program in machine language from a program in symbolic language by substituting absolute operation codes for symbolic operation codes and absolute or relocatable addresses for symbolic addresses.

Astable circuit A circuit that oscillates between two states without a trigger signal. Used for generating square waves.

Attenuator A circuit used for reducing a given voltage to some lower value. (See Voltage divider.)

Audio frequency (AF) Any signal in the audio range, i.e., from 10 Hz to 25 kHz.

Autotransformer A special type of transformer having a variable output voltage. It is often called by the trade name "Variac." An autotransformer does *not* isolate its output from ground.

Average Given a series of values for a quantity, one adds them and divides the sum by the number of values taken.

BASIC A simplified computer language intended for use in engineering applications.

Battery A device for storing electrical energy and delivering it upon demand.

β ratio This symbol is used for the ratio of transistor collector-to-base current in the older literature. (See h_{fe}.)

Binary-coded decimal (bcd) A system of number representation in which each decimal digit of a number is expressed by binary numbers.

Bistable (flip-flop) circuit) A circuit that will remain in either of two output states. It switches from one state to the other upon receiving the proper input signal.

Bit Abbreviation for binary digit. A unit of information equal to one binary decision, or the designation of one of two possible and equally likely values or states (such as 1 or 0) of anything used to store or convey information. It may also mean "yes" or "no."

Bode plot A graph of the open loop gain of an op-amp vs frequency. The Bode plot is obtained with "small signal swing." (See Roll-off, Open loop, Slew rate, Large vs small signal operation.) Filter characteristics are often presented as Bode plots.

"Bootstrapping" This term is used for two very separate and distinct situations: (1) In analog computer applications, we assume that the highest derivative of a differential equation exists and use it as the input to an integrator. This is called "bootstrapping." (2) In some situations, two op-amps are connected in a circuit that has a high effective input impedance because any current taken from a signal source by the first op-amp is replaced by current from the second op-amp. This, too, is often called "bootstrapping" in the literature.

Boundary condition Condition defining how a system will behave at some point in its life (usually the beginning or the end). Normally, the boundary conditions set the maximum or minimum values of the system; in a sense, they set the "boundaries" that the system cannot exceed.

Bounding Limiting the maximum voltage to which the op-amp output may go. This is usually done to increase the speed of operation and prevent saturation in switching circuits.

Bridge rectifier Several diodes connected together to turn AC signals into pulsating DC.

Butterworth, Chebyshev, and Bessel filters These filters are analyzed in terms of specific mathematical functions that were first investigated by the scientists after whom the filters are named. A Bessel filter is one that can be described in terms of Bessel functions.

Byte (pronounced "bite") A computer group that is eight bits long.

Capacitive load A circuit in which the capacitive reactance exceeds the resistance. Long electrical cables may have significant capacitance.

Capacitor or condenser This consists of two pieces of metal foil separated by an insulator which are formed into a solid mass. It is used for storing electric charge. A capacitor (measured in farads) offers a high resistance to DC voltages and a comparatively low resistance to AC voltages. The quality of the insulator determines the internal leakage and the cost; as leakage goes down, cost goes up.

Carrier signal A radio frequency (RF) signal that is used for transmission of information by AM or FM techniques.

Charge An excess or lack of electrons. It is measured in coulombs.

Chopping Periodic interruption of the input signal to a device by electrical or mechanical techniques. This allows low frequency signals to pass through an AC amplifier that is free of drift.

Clipper circuit A circuit that either removes or amplifies all signals above a given level.

COBOL Acronym for *CO*mmon *B*usiness *O*riented *L*anguage. Used to express problems of data manipulation and processing in English narrative form.

Common emitter circuit A circuit in which the emitter of a transistor is part of both the input and the output loop.

Common mode impedance This is measured by connecting both op-amp inputs together and then applying a voltage with respect to ground. The common mode impedance is determined from the ratio of current to voltage in this condition.

Common mode input A circuit in which the signal is applied to both op-amp inputs (neither input is grounded). It is used primarily for noise rejection applications.

Common mode rejection (CMR) The ability of an op-amp to reject any signal presented to both inputs at the same time. This parameter is important for noise rejection. For best results, the noise signal to be rejected should be presented to the inputs via resistors of the same value. In this case, we can speak of optimum rejection because the noise signal is coming from a "balanced source."

Common mode rejection ratio (CMRR) The numerical measure of the ability of an op-amp to reject signals presented in the common mode. The larger the CMRR value, the better the op-amp.

Common mode voltage limit The maximum voltage that can be applied to the op-amp input mode when both inputs are connected together. In such a case, the voltage is applied to both inputs with respect to op-amp ground.

Comparison detector A circuit that measures the difference between two voltages or currents and switches its output at some pre-set difference value. See gate circuit.

Compensation Changing the characteristics of a device by connecting various resistors and capacitors between its terminals. With op-amps, discrete op-amps are usually internally compensated at the factory, whereas integrated circuit op-amps must generally be compensated externally by the user.

Computer reference voltage The maximum voltage that an analog computer can produce; usually 10 volts on modern computers and 100 volts on older models.

Conductivity A measure of the ability of a material to conduct electrical current. (See also Resistivity.)

Crowbar protection circuit A circuit that protects a delicate device against the consequences of an accidental short circuit by putting a fast acting switch across the power supply output terminals.

Current The electric fluid that moves through the circuit, measured in I (amps). (See Ohm's law.)

Current divider Two or more elements in a parallel across a voltage source. The current from the source is divided between the two elements.

Current regulator diode A special type of diode so designed that the forward bias current remains constant over a wide range of applied voltages.

Current-to-voltage circuit A circuit that has zero input impedance to an applied current and provides an output voltage proportionate to the input current.

Darlington circuit Any of a variety of circuits in which the output of one transistor is used as the input signal for the next.

DC voltage or current Any voltage or current signal that does not pass through the zero voltage point. Any signal, however variable, that *does not change sign* is considered DC.

Decade A change in the value of a parameter by a factor of 10 is one decade. For example, the changes from 10 to 100 or from 1000 to 10,000 are each one decade.

Decade box A metal box containing an assortment of resistors, connectors, and switches such that it can provide a wide range of resistance values for experimental purposes. There are also decade inductor and capacitor boxes.

Decibel A mathematical system for expressing large numbers, usually used for the ratio of two voltages. The gain of a circuit in terms of the input-output voltage ratio would be $\text{dB} = 20 \log_{10} (V_{out}/V_{in})$.

Derate To reduce the voltage, current, or power rating of a device to improve its reliability or to permit operation at high ambient temperatures.

DIAC A device that acts like a reverse-biased diode until the applied voltage reaches some pre-chosen value, at which point the DIAC resistance drops almost to zero.

Differential input resistance On an op-amp, this parameter is measured by applying a voltage difference between the two inputs and observing the current under open loop conditions. The differential input resistance is the ratio of voltage to current in this test.

Differential mode voltage limit The largest voltage that can be applied between the two inputs of an op-amp.

Digital computer An electronic calculator that operates with numbers expressed directly as digits, as opposed to the directly measurable quantities (voltages, resistance, etc.) in an analog computer. In other words, the digital computer counts (as does an adding machine); the analog computer measures a quantity (as does a voltmeter).

Diode A device that has more resistance to current flow in one direction than in the other. The polarity of easy current flow is called "forward bias," the other polarity is referred to as "reverse bias." In forward bias, a small voltage is required to induce current flow. This is called the "forward voltage drop."

Discrete circuit A device built of individual resistors, transistors, diodes, and so forth. The cost is generally high because of the hand assembly operations required, but post-construction adjustments can produce a very high quality device.

Discrete device A term used for a single electronic component – e.g., a transistor, resistor, or diode – as distinguished from an assembly of such devices in an integrated circuit.

Drift A slow change in the output signal from a device or system without any corresponding change in the input signal. It is usually due to temperature changes.

Dynode A special type of cathode used in electron multipliers; when a single electron hits the cathode, the cathode responds by emitting anywhere from 2 to 10 secondary electrons. This is why the system is called an electron multiplier.

Effective input impedance The actual impedance of an op-amp circuit as distinguished from the "rated" impedance of the op-amp.

"Eith" point or time A measure of the response of a system to a step change in the input signal. The eith time is the interval required for the output signal to reach 0.63 of its final value.

Electrical leakage Any flow of electricity through paths other than those designed into the device. Faulty insulation and broken wires, for example, can result in electrical leakage.

Electrocardiography (ECG) Measurement of the electrical potentials induced on the body by the action of the heart.

Electroencephalography (EEG) Measurement of the electrical potentials induced on the scalp by the activity of the brain.

Electromotive force (emf) In electrical engineering, it can be considered as another term for voltage.

Equivalent circuit A circuit having the same electrical characteristics as some other, more complicated circuit.

Extinguishing voltage The voltage at which the device turns off or current flow ceases.

Federal Communications Commission (FCC) The people in Washington that control who uses the radio frequency spectrum and what power level they can operate on.

Feedback If some or all of the output signal from a device is returned to the input, that is called "feedback." Feedback can be negative or positive; negative feedback tends to reduce the output signal, positive feedback tends to increase it.

Feedback factor (β) A term used to indicate the fraction of the output voltage that is fed back to the input.

Field effect transistor A three-terminal device used for detection and amplification of signals from high impedance sources or for controlling current flow. A small voltage signal to the gate induces a large current flow from the source to the drain.

Field effect transistor characteristics A graph of the drain current versus the drain-to-source voltage for various values of the gate voltage of a field effect transistor.

Filter A circuit or device that can attenuate or amplify either high, low, or intermediate frequencies depending upon the design involved.

Filter order A term used to describe the ability of a filter to separate signals differing only slightly in frequency. A second-order filter is better than a first-order filter, and so on.

Firing voltage The voltage at which the device turns on or begins to conduct current.

Floating circuit or device A circuit or device that is not connected or referenced to power ground.

FM radio A transmission system in which the amplitude of the carrier wave remains constant but the frequency is modulated by the audio signal.

FORTRAN FORmula TRANslation, a procedure-oriented computer language designed to be used with problems that are expressible in algebraic notation. There are several forms: FORTRAN II, FORTRAN IV, etc.

Forward voltage drop The forward bias voltage that must be applied to a diode to induce current flow. For silicon diodes it is 0.7 volt; for germanium diodes it is 0.2 volt.

Four-point probe A test system in which current is put into a system at two points and the resultant voltage drop is measured at two other points. It allows the experimenter to separate the high current power inputs from the low voltage signal outputs.

Frequency (*f*) The number of occurrences of a periodic phenomenon in a unit of time. Electrical frequency is usually expressed in hertz; while radio frequency is expressed in kilohertz or megahertz.

Full-wave control A system where control is effective during the entire AC cycle. (See Half-cycle control.)

Gain-bandwidth product An expression of the general rule that a high gain circuit cannot operate over a wide frequency, and vice versa. As the gain goes up, the bandwidth must go down to keep the gain bandwidth product constant.

Galvanic skin resistance (GSR) This term is usually used for measurements of the change in skin resistance that follows the application of a stimulus to the subject.

Gate circuit A circuit which changes its output at some pre-chosen voltage. See Comparison detector.

GIGO The term *GIGO* simply means that if the data you put into a computer is bad, the output that you get will be useless; hence the term, *Garbage In, Garbage Out.*

Glow lamp or neon lamp A glass envelope that is filled with neon or other inert gas. There are two internal wires which are connected to external leads by a glass seal. When the proper voltage is applied, the gas becomes an electrical conductor and the device emits light.

Grounded receptable A standard receptable having three openings: "hot," "neutral," and "ground." The ground connection provides an important safety feature and permits the use of three-wire cords.

Grounding and shielding A two-step process in which conducting metal is placed around wires, signal sources, and amplifiers. This conducting metal is referred to as the shield because it keeps stray signals out of the system. The connection of the shield to "ground" is called grounding.

h_{fe} A symbol for the current gain of a transistor. In essence, this is the ratio of collector current to base current.

Half-cycle control A system where control is exercised only during one half (the positive or the negative part) of the AC cycle. (See Full-wave control.)

Hardware 1. Mechanical, magnetic, electrical, or electronic devices; physical equipment (in contrast with software). 2. Particular circuits or functions built into a system.

Heterodyne Two signals, each of a different frequency, are mixed to produce signals having frequencies given by the sum and difference of the original signals. The lower or higher frequency signals are then separated by filters for amplification or recording.

High frequency noise, Johnson noise, or white noise These terms are used rather loosely for noise signals whose frequency is above a few thousand hertz. True white noise extends over the whole frequency spectrum. The noise that is observed at a particular instant is essentially random in both amplitude and frequency.

Hum rejection filter A filter that is designed specifically to reject only 60 to 120 Hz signals. It is generally used in audio amplifier output circuits.

Ideal voltage source An ideal voltage source has zero internal resistance and therefore does not change in value as current is drawn from the source.

Impedance When we have only resistors in a circuit, we can speak of the impedance as resistive. However, when there are inductive or capacitive elements whose effective resistance depends upon the *frequency* of the applied signal, we must use the term impedance. Impedance is taken to include all capacitive, inductive, and resistive effects and is denoted by Z.

Impedance (or RLC) bridge A device for measuring the combined resistance and reactance of a component part of a circuit.

Inductive load A circuit in which the inductive reactance exceeds the resistance. Relay coils and induction motors are typical inductive loads.

Inductor or coil An iron or plastic core would with many turns of wire. An inductor (measured in henrys) offers a low resistance to DC voltages and a high resistance to AC voltages. For this reason, it is sometimes called a "choke" because it "chokes" down the flow of AC.

Initial condition A term used in analog computer operation for the voltage applied to an integrator before the computation begins.

Input-Output A general term applied to the equipment used in communicating with a computer and the data involved in the computer.

Instrumentation amplifier A device which is somewhat similar to the isolation amplifier in that the common mode rejection is high, even in the presence of signals from unbalanced sources. Gains up to 5000 are available. The thermal drift is very small, and the unit is ideally suited to floating operation in instrument circuits.

Integrated circuit A device built upon a single chip of silicon or sapphire with all components deposited during manufacturing by chemical processes. The cost is low because of large volume production, but manufacturing variations may lower system capabilities.

Integration for noise removal Application of a series of integrators to a signal that is masked by truly random noise. Over a period of time, the random noise will integrate out to zero, leaving the original signal available to the experimenter.

Internal resistance The electrical resistance of a device, usually inaccessible to the user and not subject to adjustment.

Inverting amplifier An op-amp circuit that inverts the sign of any signal applied to the input. For example, a positive input signal will yield a negative output signal.

Isolation A term used for many different situations. In a transformer circuit, the output is isolated because there is no voltage between the transformer outputs and ground. For biomedical experiments we isolate the subject to prevent injury.

Isolation amplifier (IA) A system used to pass signals from an easily injured source, e.g., a human subject, to an amplifying and recording system. Isolation amplifiers decouple the source from all stray AC or DC voltages that might endanger the patient, and they have high common mode rejection ratios that allow them to accept signals from unbalanced sources. In general, their gain is low and their frequency response may be limited.

Joulean heat The rate at which electrical energy is turned into heat in some part of a circuit depends upon the product of the square of the current in the element and the resistance of the element as I^2R. If the element is an impedance rather than a simple resistor, it is more convenient to define the heat dissipation or the power loss as V^2/Z. Here V is the DC or rms voltage in the element and Z is the impedance of the element.

Key 1. A hand-operated switching device for switching one or more parts of a circuit. It ordinarily consists of concealed spring contacts and an exposed handle or push-button. 2. A projection which slides into a mating slot or groove so as to guide two parts being assembled and assure proper polarization.

Kirchoff's law The mathematical statement of the fact that the sum of all the voltage drops around a closed circuit must be zero. Another way of stating this law is that electricity is a quantity that is neither created nor destroyed – all the electricity that goes into a system must come out. It is important to note that when current flows through a resistor, the lost energy appears as heat, and what has been changed is the voltage, not the current. The current out of a resistor is the same as that coming in but the voltage drop across the resistor has been lost as heat. The expression I^2R or V^2/R is a measure of the rate at which this heat dissipation is occurring.

Large-signal vs small-signal operation This term was taken over from transistor and vacuum tube designers. If the maximum output value of a device is ±10 volts and we design a circuit whose output varies only ±1 or 2 volts, we can speak of "small signal swing." In this case, the important op-amp parameter is the frequency response, as given by the Bode plot. However, if we design a circuit that "swings" ±9 volts, we must then be concerned about the slew rate and settling time of the op-amp.

Lenz's law A mathematical statement of the fact that if a current is passed through an inductor, the applied voltage depends on the magnitude of the inductance multiplied by the rate of change of the inductor current.

Light-emitting diode (LED) A device with diode characteristics that emits light when conducting current in the forward bias condition.

Linearity A measure of the change in output for a given change in input signal. In a linear device, a 10% change in input changes the output by 10% at any DC signal level.

Line or load protection A general term for a variety of systems that are used to protect a power supply, or the load on the supply, from the effects of a change in the line or load conditions. Other types of protection may include short circuit protection for the load, constant voltage or current systems that hold the proper output even when the line or load varies.

Lissajous figures The patterns obtained on an oscilloscope screen when sine or cosine signals of various frequency are used to drive the horizontal and vertical inputs.

Load The element that receives the signal generated by the circuit or device in question.

Low-pass or high-pass filters Terms used to designate filters that attenuate signals above or below some particular frequency.

Mixer A device that accepts two input signals and produces an output that contains sums and differences or products of the two signals.

Modulation Change of one parameter by another. In AM radio, for example, the amplitude of the carrier signal is modulated (changed) by the audio signal.

Monostable multivibrator A circuit that has only one stable output state. Upon receiving the proper signal, it switches to its other output; then it switches back to its original state after some fixed length of time.

MOS-FET (metal-oxide semiconductor field effect transistor) A special type of field effect transistor used for control of very small currents at high impedance levels.

Multivibrator A circuit that goes from one output voltage to the other at a fixed frequency. It is often used as a square wave generator.

Negative resistance A voltage-current relationship in which the current increases as the applied voltage decreases. Since the voltage-current curve decreases with increased voltage, the slope is negative, hence the term "negative resistance."

Network An electrical circuit.

Neutral or current-carrying neutral The wire through which electricity flows back to the power plant. The neutral wire is not at "ground potential" inside a building and should never be referred to as "ground" neutral. A wire that carries current cannot be at ground potential.

Ninety percent time This is defined as the time required for the output to reach 90% of its final values. (See "Eith" time.) Some manufacturers refer to the 90% time as the response time.

Nodal point See Summing point.

Node current law All the electricity that flows into a junction must flow out of it.

Noise A generic term for any unwanted signal that appears at the output of a detector or a system. It may be subdivided into drift noise and short or high frequency noise. (See Noise rejection.)

Noise rejection Removal of unwanted signals. A variety of techniques may be used. (See Chopping, Phase-locked loop, Synchronous detection, Common mode rejection.)

Noninverting amplifier An op-amp circuit that produces a positive output signal when the input signal is positive, and a negative output signal when the input is negative.

Notch filter See Slot filter.

Octave A change in the value of a parameter by a factor of two is one octave. For example, the changes from 10 to 20 or from 2000 to 4000 are each one octave.

Offset current The difference between the two input bias currents of an op-amp.

Ohms law A mathematical relationship between the voltage, current, and resistance in an electrical circuit ($V = IR$).

Ohms-per-volt The system used for specifying the current required for a DC ammeter to deflect full scale. For example, a 20 kilohms-per-volt rating indicates that a current of 50 microamperes will drive the meter full scale.

Op-amp (operational amplifier) The ideal op-amp is a high gain, stable, linear, DC amplifier. The input impedance is taken to be infinite, and the output impedance is considered to be zero. These characteristics are usually used for circuit design. Under some conditions, the ideal op-amp characteristics do not fully describe the way the device behaves in a circuit. In that event, consideration is given to the "real" values of input impedance, gain, and so forth as they effect circuit operation.

Op-amp buffer The application of an op-amp in a circuit that isolates one part of a circuit from the other.

Op-amp ground If the op-amp is being supplied directly from a power source that is driven by the 115 volt mains (line) coming from the power company, the op-amp ground is power ground. If the op-amp is floating (i.e., if it is supplied by batteries or an isolated power source), op-amp ground is just the neutral point on the supply. For example, if the op-amp is running on a ±15 volt supply, ground is the reference point for the ±15 volts.

Open circuit voltage The voltage of a source when no current is being drawn from it.

Open Loop Operation of an op-amp without feedback from the output to the input.

Open loop gain The input-output voltage ratio of an op-amp circuit without feedback.

Optical-electronic device A device that controls a current or produces a signal when there is a change in its optical input or the illumination of its detector changes.

Oscillator A device or circuit that changes regularly and repeatedly from one output value to another. (See Monostable multivibrator.)

Oscillograph A device for recording electrical signals.

Oscilloscope A device for viewing electrical signals by allowing the signal to move a stream of electrons that impact on a phosphor screen.

Overshoot A condition in which the output of a circuit or system goes above its correct final value. It must then settle down by a series of decaying oscillations.

Partial concentration If we have x liters of fluid in which are dissolved y grams of one compound and z grams of another, the concentrations are y/x and z/x. The partial concentrations are

$$c_y = \frac{y}{y + z}$$

$$c_z = \frac{z}{y + z}$$

Passive vs. active filters A passive filter is made up of resistors, inductors, and capacitors. As such there is some loss of signal in the filter, and the output signal can never be as large as the input signal. An active filter may contain one or more op-amps or transistors, and the output signal at some given frequency may be larger than the input signal.

Phase-locked loop A special electronic system that activates the receiver only when a signal from the source is expected. In a sense the source and the receiver are locked in phase; one opens as the other provides a signal. Sometimes called synchronous detection.

Phase shift An unwanted or unexpected change in the input-output signal characteristics of a device. For example, if a sine wave is applied to the (–) input of an op-amp, the output will be an inverted sine wave but the node points (where the signal goes through the zero voltage line) will be unchanged. This is 180 degrees of phase shift, but it is usually considered to be "normal" or zero phase shift in op-amp circuits. As the input signal frequency increases, more phase shift will occur and the output nodal points will shift with respect to the input. This can induce oscillation or distortion and is referred to as "phase shift." In some circuits – e.g., the phase shift oscillator – the phase shift is induced deliberately.

Phototransistor A two-terminal device that has a window instead of a base contact. Illumination of the base window turns on the device.

Pickup noise Stray signals at 60 or 120 Hz, usually due to fluorescent lights or power machinery.

Pinch-off voltage A term used in design of field effect transistor circuits. It is the voltage at which the drain-to-source current is zero.

Potentiometer The term is used in two different ways. A measuring potentiometer is used for measuring a voltage without drawing a current from the circuit being measured. A variable resistance is often called a potentiometer or "pot."

Power ground Power ground is the earth itself, a water pipe, or anything connected directly to the ground cable provided by the power company.

Processor 1. In hardware, a data processor. 2. In software, a computer program that includes the compiling, assembling, translating, and related functions for a particular programming language, including logic, memory, arithmetic, and control.

Program 1. A sequence of instructions that tells a computer how to receive, store, process, and deliver information. 2. A plan for solving a problem, including instructions that cause the computer to perform the desired operations and such necessary information as data description and tables.

Pulsed oscillator A radio frequency (RF) oscillator that is turned on and off in a controlled cycle.

***Q* factor or quality** A measure of the ability of a filter to separate signals that differ only slightly in frequency.

Quantum efficiency A measure of the response of an optical-electronic device to light. In the case of a photoelectric tube, it is the number of electrons generated per incident photon.

Quiescent point (*Q* point) In linear transistor circuit design, the *Q* point is the DC condition of the circuit. It is that point in the characteristic diagram that the circuit remains at when there is no AC input signal.

Radio frequency (RF) Any signal with a frequency from 50 kHz to 2MHz.

Ramp generator A signal source having an output that increases in a regular way with time. A voltage source that changed a fixed rate of 1 volt per minute would produce a ramp-shaped signal on an oscilloscope screen.

Reactance For a pure capacitor or inductor the reactance is another term for impedance. If there is both resistance and capacitance (or inductance) in a circuit the reactance is the imaginary part of the total impedance of the circuit.

Real voltage source A voltage source with significant internal resistance such that the output voltage is a function of the current drawn from the source.

Real world The system or equation that we are trying to model on an analog computer. Real world variables are feet, miles, gallons, or whatever. Computer world variables are always volts.

Rectifier See Diode.

Reference voltage or current source A voltage or current generator that holds its output very close to a fixed value despite variations in ambient temperature or load.

Register 1. A short-term storage circuit the capacity of which is usually one computer word. Variations may include provisions for shifting, calculating, etc. such as Static Shift Register and Dynamic Shift Register. 2. The relative position of all or part of the conductive pattern with respect to a mechanical feature of the board or to another pattern of the obverse side of the printed circuit board (e.g., pattern-to-hole register or pattern-front-to-pattern-back register). 3. Also called registration. The accurate matching of two or more patterns such as the three images in color television. 4. A range of notes used for playing a particular piece or part of it (e.g., melody or harmony), particularly the range covered by a clavier or manual. 5. In an automatic switching telephone system, the part of the system that receives and stores the dialing pulses that control the additional operations necessary to establish a telephone connection.

Resistance A measure of the degree to which a circuit or element resists or impedes the flow of electricity. Measured in R (ohms). (See Ohm's law.)

Resistivity A measure of the ability of a material to resist the flow of electrical current. The resistivity of a material is independent of the shape of the material involved.

Resistor tolerance The possible variation of a resistor from the value marked upon it. On carbon resistors, no band means ±20%, a silver band ±10%, and a gold band ±5%.

Response time A measure of the ability of a device to change its output signal when the input signal level changes. Response time specifications must include the time for the device to achieve the correct output value after all transients have subsided.

Ripple A small AC voltage, usually 60 or 120 Hz, in the output of a power supply or amplifier.

Roll-off curve The decrease in op-amp open loop gain with frequency on a Bode plot. This decrease is usually expressed in decibels per octave or per decade. The term "roll-off" is often used in discussion of filter characteristics.

Root-mean-square (RMS) In DC circuits there is no question of what value to use in formulas. We use the value of the voltage or current at the instant involved. If the signal is varying rapidly, it is possible to time-average the signal over some convenient period and use the average value in formulas. With AC signals, however, the time average is zero and it is customary to use the rms average, which is defined as

$$V_{rms} = \left(\frac{1}{T} \int_0^T V^2 \, dt \right)^{1/2}$$

$$I_{rms} = \left(\frac{1}{T} \int_0^T I^2 \, dt \right)^{1/2}$$

An advantage of this definition is that formulas such as I^2R yield the same heating value for AC or DC current as long as the AC current is rms.

Saturation A condition in which an op-amp goes to its maximum plus or minus output value and holds in this condition despite changes in the input signal.

Scaled equation An equation that has been transformed from the real world to the analog computer world by means of scaling factors. It is sometimes called the "computer-ready equation."

Scaling factor A constant used to translate "real world" variables into computer world variables, i.e., voltages. The scaling factor has units of whatever the real world variable is, divided by volts.

Schmidt trigger circuit A circuit that accepts a variety of pulses (which may be poorly shaped) and delivers pulses of uniform size and duration at its output.

Settling time The period required for an amplifier or system to reach the correct output value after a step change in the input signal.

Signal averaging A term used for several different types of signal processing. In many cases the objective is reduction of noise. Circuits are given for simple averaging, root-mean-square averaging, and the like.

Silicon controlled rectifier (SCR) A special type of diode that can pass a substantial current in forward bias when it receives a signal at its third, or gate, terminal. It is used as a unidirectional controlled switch.

Silicon controlled switch (SCS) A three-terminal device used for switching in a manner similar to the SCR or TRIAC. Unlike SCRs and TRIACs, however, the SCS can be switched off *without* first reducing the applied voltage across the device to zero.

Single-ended, balanced, and driven sources Typical signal sources observed in experiments with live subjects.

Sixty-three percent time See "Eith" time.

Skin contact A device used to apply or receive electrical signals without breaking the skin of a living subject.

Slew rate The rate of change of the amplifier output signal when a step function change is applied to the input. A large slew rate is an advantage in circuits that must switch rapidly from one state to another. Op-amps with high slew rates, however, may have overshoot problems. (See Overshoot.)

Slope The change in the value of a parameter in a given period of time divided by that period of time. It is, for our purposes, identical with the first derivative of a function with respect to time.

Slot filter or notch filter A slot filter stops all frequency signals except those in a small range called its "pass band." A notch filter passes all frequencies except those in a narrow range called its "stop band."

Software Programs, routines, codes, and other written information for use with digital computers, as distinguished from the equipment itself, which is referred to as "hardware."

Spectral response A measure of the change in device output with the wavelength of the incident light.

Square wave test Application of a square wave signal to a detection and recording system to check its response to AC and DC signals.

Stability A measure of the tendency of an op-amp to go into oscillation in a circuit where oscillation is not desired. Usually this is due to phase shift or poor compensation.

Standard cell A special type of battery used as a voltage standard, which is often called a "Weston cell."

Standard receptable A receptacle that provides only two openings: "hot" and "neutral." This type of installation can be very dangerous if any electrical leakage exists.

Step function A change in voltage or current level in a very short interval of time. This is sometimes called an "infinite step" because the ratio of the voltage change to the time required for the change to occur is very large (e.g., volts per microsecond).

Summing point Any point where we apply the current conservation law (the sum of all currents in and out of a summing point must be zero).

Synchronous detection A noise rejection technique that involves disconnecting the output of a device at all times except those times at which a signal might be expected.

Tachometer A device or system used for measurement of rotational speed (revolutions per minute).

Thermistor A device whose resistance varies with temperature in a fixed and reproducible way.

Thermocouple A device used for measurement of temperature. Two dissimilar metals are brought into contact, and a voltage occurs at the contact that is a function of the metals involved and the temperature of the contact.

Time average The value of a parameter is integrated over some period of time and the integral is divided by the total time of integration.

Time scaling Adjusting the analog computer so that it runs either faster or more slowly than the real world system that it represents.

Transfer function See Scaling factor.

Transformer A device used to change AC voltages from one value (rms) to another (rms).

Transient A sudden change in voltage or current in a circuit, usually without previous warning. Often caused by line or load variations.

Transistor (bipolar) A three-terminal device used as a controllable valve for current. A small current into the base produces a large current flow between the collector and the emitter.

Transistor characteristics (bipolar) A graph of the collector current vs the collector-to-emitter voltage for various values of the base current of a bipolar transistor.

TRIAC A device that can pass current in either direction upon receiving a signal at its gate terminal. It is used as a bidirectional controlled switch. (*Note:* SCR and TRIAC devices cannot be turned off unless the applied voltage [not the gate voltage] is reduced to zero; see the Silicon controlled switch.)

Tuned filter A slot filter designed to reject all signals except those at the frequency generated by a chopper circuit.

Tungsten lamp (incandescent lamp) A device that emits light when an internal filament is heated by an electrical current.

Two- and three-wire cords A two-wire cord provides one lead for the electricity to flow to the device and another lead for the current to flow back to the receptacle. A three-wire cord provides a third ground wire to prevent injury to the operator in case the device has electrical leakage.

Unbalanced source A signal source with two outputs, both of which are isolated from ground, but having different output impedances.

Variable resistor (pot or potentiometer) A resistor whose value can be changed by turning a knob or moving a contact.

Varistor A trade name for a series of devices that are normally high in resistance until the applied voltage rises above some rated level, at which point they become conductors. They are used for elimination of voltage transients.

Voltage The voltage in an electrical circuit may be thought of as the electrical pressure that pushes a fluid (electricity) through a resistance (the wires) measured in V (volts).

Voltage or current regulator A device that holds the output voltage or current from a source to some pre-set value.

Voltage divider An electrical circuit, usually using resistors, that is used to split a fixed voltage source into two (or more) sources whose sum must equal the magnitude of the original voltage.

Voltage-time or controlled-curve generator A term describing a wide variety of devices used to generate specific voltage-versus-time signals. These devices are often driven by a linear ramp voltage signal generated by an op-amp.

Voltmeter A device or system for measuring electrical voltage.

Volt-ohmmeter (VOM) A low-cost battery-driven meter and associated control system that is used for the measurement of resistance, voltage, and current at various levels.

VTVM, FET-VOM These terms are used as abbreviations for *V*acuum *T*ube *V*oltmeter and *F*ield *E*ffect *T*ransistorized *V*olt-*O*hm *M*eter. They are measurement systems with very high input impedance. They are used for measurements where a simple VOM might draw too much current from the circuit that is being tested.

Watt The unit used for measurement of electrical power. The flow of power in an electrical circuit is the instantaneous product of the voltage and the current: volts $\times$ amps = watts. The rate at which electricity is turned into heat in a resistor is given by $P = I^2R$ = watts. (See Joulean heat.)

Wheatstone bridge A system used for accurate measurement of resistance. If it is capable of measuring inductance and capacitance as well, it is called an "*RLC* bridge."

WWV The U.S. government station run by the Bureau of Standards out of Boulder, Colorado. They broadcast a variety of standard frequencies and time signals for world use.

X-Y recorder A recorder for which the independent variable is a parameter other than time.

Zener diode A special type of diode having an almost constant voltage drop under reverse bias conditions. It is used as a fixed voltage source in constant voltage or current supplies.

Zero-crossing detector A circuit that switches its output when the input signal or its first derivative passes through zero.

Index

Abbreviations, used in electrical engineering, 2-3
AC (alternating current), 16-17
  power receptacles for, 5-8
  root-mean-square (rms) voltage (V_{rms}), 16-17
  skin resistance measurement with, 230
  voltage-versus-time plot, 16
AC meter, high impedance, 112
Accounting, computer, 277
Addition circuit, 132, 134-136
  gain law for, 134
Agricultural applications of computers, 278
Airline check-in procedures, computer, 263-264
Ammeters, 37-38. *See also* Galvanometers
  cathode ray oscilloscopes as, 48, 53-54
  clip-on type, 37-38
  ideal, 37
  measurement error caused by, 37
  multi-range, 38-39
  op-amp feedback circuit and, 111-112
  resistance of, 37, 44
  shunts and, 38-39
Amplifiers
  for audio system, 119-121
  with bass or treble control, 107
  for cathode ray oscilloscope, 49, 50
  common emitter-common collector. *See* Darlington amplifier
  Darlington amplifier, 117-121
    as power supply current booster, 181
    as power supply regulator, 185
  inverting, 73-75, 78, 290-291, 296
  light, 343
  linear, 73, 114
  noninverting, 75-77, 83
  open loop, 73
  operational. *See* Op-amps
  roll-off in, 74
  thermal drift in, 182
  transistor circuits for, 118-119, 312-313
Analog computation, 133-141. *See also* Analog computers
  addition with, 134-136
  differentiation with, 136, 137-140
  integration with, 136-137
  multiplication, division, and square root with, 140-141
  subtraction with, 136
Analog computers, 249-282. *See also* Analog computation
  accounting applications of, 277
  agricultural applications of, 278
  airline check-in procedures with, 263-264
  animal movement monitoring with, 272-273
  applications of, 251-271
  assembly and shipping systems with, 255-256
  auto crash simulation with, 259, 266
  cement plant problems and, 270-271
  conversational therapy and, 266
  counting and classifying operations with, 255, 266
  courtroom applications of, 280-281
  data reduction with, 253
  definitions in, 250-251
  design engineering problems in, 259-261
  drilling process control with, 270
  drug information in, 277-278
  educational trend analysis with, 252-253
  electrical power problem detection with, 267-268
  emergency medical network with, 276-277
  graphics with, 258-263
  hospital instrument and apparatus inventory control and exchange with, 276
  human engineering in product design with, 261-262
  identification and verification with, 251-252
  industrial safety and, 268-269
  instrument connections with, 253-254
  laboratory instrumentation and, 271-273
  languages for, 250, 262-263
  library applications of, 254, 275, 281
  as lie detector, 279-280
  machine control and minitoring with, 262, 269
  medical applications of, 273-278
  memory applications of, 252-254

Analog computers – *Continued*
mileage recorder and shock load sensor system with, 270
multi-phasic health testing centers (MHTC) with, 273–274
municipal services and, 258
op-amps and, 133
patient evaluation with, 274–275
people matching with, 279
price coding with, 257–258
professional sports and, 278–279
security applications of, 268, 277
software vs. hardware distinctions in, 256
solid models of two-dimensional drawings with, 258–259
store checkout systems with, 256–258
three-dimensional stress analysis with, 259
three-dimensional views represented by, 258–259
traffic control with, 267
troubleshooting and repair applications of, 254–255
verbal communication and response systems with, 264–265
visual recognition applications with, 266–267
voice commands for, 264
wire routing with, 263
Anemometer resistance thermometer, 107–110
Animal monitoring by computer, 272–273
Animation, computer, 259
Antennas, 208
Assembling systems, computer, 255–256
Assembly, computer, 250
Audio system amplifiers, 119–121
Auto crash simulation, computer, 259, 266
Autotransformers, 30–31
Avalanche photodiode, 343
Averaging. *See* Signal averaging

Band-pass filters, 23–24, 146–148, 157
Bode plot for, 147
Butterworth vs. Chebyshev types, 152
op-amp active types, 150, 151
quality (Q) factor for, 147–148
BASIC (computer language), 250
Batteries
circuit symbol for, 4
as ideal and real power sources, 15
internal resistance of, 15–16
mercury, as reference voltage, 124
as op-amp power supply, 185, 186, 244
testing of, 15
zinc-carbon dry-cell, as reference voltage, 124–127
Battery charger, 33–34
Beat, 56
Bessel filters, 151–153
Bias current, 284
Bibliographies, computer-generated, 254
Binary coded decimal (bcd) computer outputs, 253–254
Biomedical applications, of computer, 229–247
AC or DC current usage in, 230–231
in ECG and EEG work, 233–236
electrical safety and patient resistance in, 236–238
four-point probes in, 232–233
grounded three-wire cords and receptacles in, 241–242
in intensive care (IC) facilities, 240–241
isolation amplifiers in, 230, 239–245
maximum current used in, 229–230
signal sources and shielding in, 236–238
skin contact problems in, 245–247
skin resistance usage in, 229–232
Bits, computer, 251
Bode plots, 22, 74, 294
Bolometers, 343
Bootstrapping, 295–298
"Bounding the circuit" technique, 129
Bridge, impedance, 156, 246
Bus-bar ground, 174
Buffer, 84
Butterworth filters, 148, 151–153
Bytes, computer, 251

Calculators, hand, 251
Cameras, for recorders, 64
Candela, 334
Capacitors
circuit symbol for, 4
to eliminate transients, 331
feedback, 106–107
flashlamps with, 336, 337
phase shift in, 28–29
reactance of, 17–18
Carbon resistors, 1
Carrier frequency (RF), 209
Cathode ray oscillograph, 64
Cathode ray oscilloscope, 48–60
applications of, 48, 56–60
frequency measurements, 56–57
phase measurements, 57–59
beat notes in, 56
cathode ray tube for, 49
controls for, 49–53
horizontal deflection, 51
intensity and focus, 52
sweep or trigger voltage, 51–52
vertical deflection, 49–51
display of waveforms in, 52–53
grounding of, 53, 240–241
Lissajous figures and, 56–57
operation of, 53–56
cathode ray beam controls, 53

multiple trace, 55
positioning, 54
sampling, 55
storage, 55
time base, 54–55
vertical amplifier, 53–54
X-Y input, 55
operator safety in use of, 240–241
as recorder, 64
with retrace blanking, 53
saw-tooth wave generator for, 162
testing frequency-dependent phase shift with, 235–236
Cathode ray tube, 49
Cautery machines, 246
Cement plant problems, and computers, 270–271
Chamois skin, 232
Change-of-slope detector, 139, 140
Chebyshev filters, 148, 151–153
Check systems, computer-controlled, 256–258
Chemical burns of skin, 246
Choppers. *See also* Chopping
DC noise and drift removal with, 199
field effect transistor type, 202
op-amp stabilization with, 195–197, 284
phase-locked system in, 198
in synchronous detection system, 199, 200, 201
synchronous rectification and amplification with, 195, 196–197
Chopping, 192. *See also* Choppers
for detector drift reduction, 197–198
of high frequency signals, 202
of low frequency signals, 196
op-amp stabilization with, 194–197
in synchronous detection, 198–202
Circuit symbols, 4
Classification, computer, 255
Clipper circuits, 131
Coaxial connector, 53
COBOL (computer language), 250
Coefficient of resistivity, 10
Coil. *See* Inductors
Common emitter, 115
Common mode impedance, 84, 285
Common mode rejection (CMR), 84, 85–86, 90, 288
Communication systems, and computers, 264–266
Complex numbers, 24
Computers. *See* Analog computers
Condensor discharge lamps, 336
Condensors. *See* Capacitors
Conductivity (σ), 9, 10
skin, 236
Constant current sources, 48, 172, 183–184
battery-resistor type, 107
op-amp type, 82, 107–110, 183–184
Constant voltage sources, 48, 172, 182–183
op-amp type, 82, 182–183
Constantan, 88
Control circuits for inductive loads, 325–328
Copper-constantan thermocouple pair, 88–89
Counting operations, computer, 255
Counting techniques for single photons, 202, 203
Court applications of computers, 280–281
Crossover distortion, 117
Crowbar protection circuits, 175–176
Current booster circuits, 113, 114–117
Current conservation law, 78
Current-crossing detector, 132. *See also* Gate circuits
Current density (J), 9
Current divider equation, 14
Current gain. *See* Gain
Current-limiting modules, 230
Current output, 172
Current rating, for diodes, 32
Current reference sources, 127
Current regulator diode, 127, 128
as current source, 127
V-I curve for, 127
zener diode voltage source, stabilization of, 127, 128
Current-to-voltage circuit, 129–130

Darlington amplifier, 117–121
as power supply current booster, 181
as power supply regulator, 185
Data reduction by computer, 253, 272
DC (direct current)
peak voltage (V_{peak}), 16, 17
power sources, ideal and real, 15–16
pulsating, 32
skin resistance measurement and, 230, 236, 245–246
transformers for, 30–31
voltage-versus-time plot for, 16
DC pulse circuit, 162
Decibel (dB), 22–23, 74
Derating factor, 188–190
Derivative. *See* Differentiation
Design engineering, 261, 262
Design point, 318, 319
Detector drift reduction, 197–198
Detectivity (D*), 340
DIAC, 177–178
power supply protection circuit with, 178
V-I curve for, 177
Difference circuit, 87, 88, 90
Differential circuit, 87, 88, 90, 203–204
Differential mode impedance, 85
Differential signals, 207–208
Differentiation, 137

Differentiation circuits, 137–138, 285
Digital computers, 250
Diodes, 31–34
  bias on, 32
  circuit symbol for, 4
  ideal vs. real, 31–32
  instrument or switching types, 132
  light-emitting (LED)
    with isolation op-amp, 239–240
    as light source in photometry, 336, 337, 338
  magnetic tape recorder and, 219
  in mixer circuit, 212, 213, 217
  nonlinearity of, 164
  peak inverse voltage (PIV) of, 32
  as rectifiers, 32, 33
  as transistor model, 301, 302
  V-I curve for, 32, 301, 302
  zener. *See* Zener diodes
Discrete amplifiers, 289, 290
Discriminator, 203
Division circuit, 140–141
Drift reduction. *See also* Noise, reduction
  by chopper stabilization of op-amp, 194–197
  for detectors, 197–198
  for photodetectors, 339
  for synchronous detection, 198–202
  two op-amp circuit for, 298–299
Drift temperature in op-amps, 290
Drilling operations, computer, 270
Drug reaction monitoring, computer, 277–278
Dry-reed relay, 163

Earth ground, 46
Educational trends, and computers, 252–253
Eith time, 92
Electrical engineering abbreviations, 2–3
Electrical noise. *See* Noise
Electrical safety. *See* Grounding; Patient safety
Electrocardiography (ECG)
  computer evaluation of, 274–275
  frequency-dependent phase shift and, 235–236
  op-amp circuits in, 233–236, 246
  testing of circuits in, 235
  typical signal in, 234
Electroencephalography (EEG)
  frequency-dependent phase shift in, 235–236
  op-amp circuits in, 233–236
  testing of circuits in, 235
Electrometers, 35. *See also* Voltmeters
Electromotive force (emf), 15
Engineering problems, and computers, 259–261
Envelope, 195–196

Experiments, 66–68, 97–103, 223–228, 344–345

Farad (F), 4, 18
Fatigue, photodetector, 339–340
Federal Communications Commission (FCC), 213, 214
Feedback
  in op-amp circuits, 73, 80–81, 94, 106–107, 110–111
  stabilization of op-amps and, 158, 159–160
Feedback factor (β), 92, 94
Field effect diodes, 230
Field effect transistors (FET), 295, 315–323. *See also* Field effect transistor (FET) circuits
  in amplifiers, 322
  bias line for, 318, 322
  vs. bipolar transistors, 315–316, 317
  characteristic curve of, 316
  circuit design for, 317–322
  circuit symbols for, 316, 317
  common mode impedance of, 285
  design point (quiescent point) of, 318, 319
  and device variations, 318, 319, 322
  drain terminal of, 316
  forward transfer admittance of, 320
  gate leakage current of, 320
  gate terminal of, 316, 317
  gate-to-source breakdown voltage of, 319–320
  junction type
    N-channel, 316, 317
    P-channel, 316, 317
  metal-oxide semiconductor type (MOS-FET), 316–317
  in op-amps, 290
  parameters for, 319–320
  and pinch-off, 317–318
  source terminal of, 315
  temperature variation and, 318, 319, 322
  voltmeters with, 37
Field effect transistor (FET) circuits. *See also* Field effect transistors (FET)
  biasing of, 317, 318, 322–323
  chopper, 202
  design of, 317–322
Filters, 17–27, 146–157
  amplitude, 146
  band attenuation, 23, 146
  band-pass, 23–24, 146–148, 157
    Bode plot for, 147
    Butterworth vs. Chebyshev types, 152
    op-amp active types, 150, 151
    quality (Q) factor for, 147–148
  basic design of, 17–24
  Bessel type, 151–153

Bode plots for, 22, 147, 148, 150
Butterworth type, 148, 151–153
Chebyshev type, 148, 151–153
complementary pairs in, 153
for frequency rejection, 155–157
high-pass, 18, 21, 25, 146, 153
active vs. passive types, 150
design of, 21
op-amp active types, 150, 151
high-quality, 153, 154
inductor type, 21–22
low-pass, 18, 25, 146, 148, 154
design of, 18–20
op-amp active type, 149–150, 154
RC passive type, 148, 149
mathematical analysis of circuits of, 24–27
notch, 153, 154
op-amp active type, 149–154
order of, 22–23, 153
quality (*Q*) factor for, 27, 147–148
RC (resistor-and-capacitor) type, 18–21, 148–149, 151, 152
resonance and, 26–27
ringing in, 152
rise time of, 152
roll-off of, 22, 148, 153
slot, 23, 146, 147, 148, 150–151
testing of, 21
transient response and, 153
twin-tee, 155–157
Fire departments, 258
Flashlamps, 336–337
Flip-flop circuit, 167, 168
Fluorescent lamps, 336, 337
Flow chart, computer, 251
FORTRAN (computer language), 250
Four-point probes, 232–233, 247
Frequency, standard and radio station WWV, 218
Frequency analyzer, heterodyne type, 217–218
Frequency doubler, 140
Frequency meter, 48
Function generators. *See* Ramp generators
Fuses, 34

Gain, 73
closed loop, 74
current, of transistors, 115
op-amp
ideal, 78, 79, 84
real, 285
open loop, 73
unity, 74–75
Gain-bandwidth product, 75
Gain law. *See* Gain
Galvanometers. *See also* Ammeters
multi-element oscillographs and, 63
potentiometers and, 41
Gas chromatographic data, 253
Gate circuits, 131–133
addition circuits and, 132
diode feedback network for, 132, 133
negative type, 132–133
zener diodes in, 133
Generators. *See* Ramp generators; Saw-tooth wave generators; Signal generators; Square wave generators
Globar lamps, 335–336
Glow lamps, 328–331
oscillator, 329–330
V-I characteristics of, 329
voltage regulator, 330–331
Graphics, computer, 258–263
animation with, 259
computer language in, 262–263
general engineering problems with, 259–261
human engineering with, 261–262
machine control in, 262
solid models of two-dimensional drawings in, 258–259
stress analysis with, 259
three-dimensional views with, 258–259
Ground, 5–8. *See also* Grounding
circuit symbol for, 4
ground neutral, 5, 6–7
safety, 5–8
three-phase power and, 6
three-wire cord and, 5, 6–8, 241–242
two-wire cord and, 5, 6
Grounded device, 207
Grounding. *See also* Ground
balanced sources and, 237
biomedical measurements and, 236, 237–238, 239, 240–241, 244
floating sources vs., 47, 237–238
ground loops and, 238
human body potentials and, 207
isolation amplifiers and, 241
op-amps and, 121, 122
oscilloscopes and, 241
of power supplies, 48, 174, 175
of recorders, 206
shielding and, 204–205, 236–238
of signal sources, 46–47, 236–238
single-ended signal source and, 237–238

h_{fe} (transistor current gain), 302–304, 306
minimum/maximum values of, 309
temperature dependence of, 306–307, 309
Hardware, computer, 256
Hartley oscillator circuit, 216
Heating, joulean, 10, 242
Henry (H), 18
Heterodyning, 217–218
High-pass filters, 18, 21, 25, 146, 153

High-pass filters – *Continued*
active vs. passive types, 150
design of, 21
op-amp active types, 150, 151
Hospitals. *See* Medical applications
House wiring, 6–8
Human engineering, 261–262
Hybrid amplifiers, 289–290
Hysteresis loop, 169

Identification, computer, 251–252
Impedance
common mode, 84–85
matching of source and detector in, 77, 83
measurement with VOM or VTVM, 76–77
in measuring voltage, 76–77, 81
op-amp
ideal, 78, 80–81, 83
real, 84, 283–284, 285–287
in transferring maximum power to load, 81
Impedance bridge, 156, 246
Inductance-capacitance (*LC*) circuit, 211–212. *See also* Resonant circuits
Inductors
circuit symbol for, 4
in filters, 21
phase shift in, 28–29
reactance of, 18
transformers and, 29–31
Industrial safety devices, 268–269
Infinite step voltage, 93
Infrared light, 273, 334
Initial condition (IC) circuits, 163
Input/output devices, computer, 251
Instrumentation
amplifier, 90–91
computer, 271–273
Instrument diodes, 132
Instrument Society of America, 271
Integration, 136–137
Integration circuits, 137, 187
for saw-tooth generators, 162
for signal averaging, 143–144
Intensive care (IC) facilities, 240–241
Intermediate frequency (IF), 218
Internal compensation of op-amp, 188, 189, 289, 294
Inventory control, computer, 276
Isolation op-amps (IA), 230, 239–245
application of electrical stimulus and, 242–243
fetal heartbeat detection with, 243–244
with optical system, 239–240
patient safety and, 238, 239–245
hospital intensive care (IC) facilities, 240–241
pacemaker output measurement, 241, 242
shielding of, 240
with transformer coupling, 240

Isolation techniques, 238
using conventional op-amps, 244, 245
using isolation op-amps, 239–244

Joulean heating, 10, 242
Joule's law, 10

Kirchoff's laws, 11–15, 303, 304, 321
loop voltage law, 12–13
node current law, 13–15

Languages, computer, 250, 262–263
Latch-up, 216–217
Leakage current, 119, 306
in field effect transistors, 320
Lenz's Law, 29
Library applications of computers, 254, 275, 281
Lie detectors, computer, 279–280
Light amplifiers, 343
Light-emitting diodes (LED)
with isolation op-amp, 239–240
as light source in photometry, 336, 337, 338
in optical detectors, 337
Light sources
in computer monitoring, 273
for optical detectors, 333, 335–337
choice of, 337
condensor discharge lamp, 337
electroluminescence, 336
flashlamps, 336–337
gas discharge or spectral sources, 336
light-emitting diode (LED), 239–240, 336, 337, 338
pulsed sources, 336–337
quartz iodine, 335, 337
thermal sources, 335–336
Linear resistance, and potentiometer, 42
Linearity
and diodes, 164
and op-amps, 114, 135
resistance, 10
and transistors, 114
Lissajous figures, 56–57, 235
Low-pass filters, 18, 25, 146, 148, 154
design of, 18–20
op-amp active type, 149–150, 154
RC passive type, 148, 149
Lumen, 334

Machine control and monitoring, computer, 262, 269
Mail sorting, computer, 256
Mass spectrometric data, 253
Mean-square function generators, 141, 142
Medical applications, of computer, 273–278. *See also* Biomedical applications, of computer

diagnosis and treatment applications in, 275
drug reactions and monitoring in, 277–278
emergency medical network in, 276–277
history-taking in, 273–274
hospital instrumentation and apparatus
inventory control and exchange in, 276
hospital security and accounting in, 277
in multi-phasic health testing centers (MHTC), 273
patient evaluation in, 274–275
Memory applications of computers, 252–254
data reduction with, 253
educational trends and, 252–253
instrument connections with, 253–254
tracing applications in, 254
Mercury batteries, 124
Mercury cells, 59, 182
Metals, resistance of, 10–11
Meters, power supply, 172, 173, 273–278
Mhos, 319
Microfarads, 18
Microprocessors, 250, 271
Millihenrys, 18
Minicomputers, 250, 251
Mixer, 209, 210
diode circuit for, 212, 213, 217
Modulation, 209, 210
side bands and, 212–213
signals, 209
for tape recorder data collections, 219
Monolithic amplifiers, 289, 290
MOS-FET transistors, 316–317
Multi-phasic health testing centers (MHTC), 273
Multiplication circuits, 140–141
Multiplier, and voltmeters, 39–40
Multivibrator circuits, 166–169
bistable, 167, 168–169
flip-flop type, 167, 168
free-running or astable, 166, 167
monostable, 167, 168
as square wave generator, 167
Municipal services, 258
MYCIN (computer program), 275

National Technical Information Service (NTIS), 254
Neon bulbs
in glow lamps, 328–329
as pilot lights, 34
as pulse generators, 214, 215
Ninety-percent time, 92
Nodal point, 78
Noise
in biomedical measurements, 236–237, 242, 243–244
current in op-amps and, 288–289
electrical, defined, 18
in optical detectors, 339–340
pickup, 205, 208, 237
reduction, 192, 194–202, 203, 205
common mode rejection of, 87, 90
isolation op-amps and, 243–244
signal averaging and, 199–200
single-ended signals and, 207–208
signal sources and, 237
voltage in op-amps and, 288–289
white, 200, 288
Non-ohmic contact, 246
Notch filters, 153, 154
NPN transistors, 301
circuit symbol for, 301, 303
common emitter transistor circuit and, 302–304
complementary pair with, 118
in Darlington amplifiers, 117–118
determination of type in, 314
PNP type vs., 115–116
power supply regulation and, 181
push-pull circuit and, 117
two-diode model of, 302

Offset voltage
of ideal op-amp, 78
of real op-amp, 283–284
Ohmic contact, 246
Ohm's law, 8–9, 19
derivation of, 9–10
One-half power points, 147
Op-amps (operational amplifiers), 69–228. *See also* Op-amp circuits
amplifier roll-off and, 74
analog computation with, 133–141
basic types of, 69–72, 289–290
bias current of, 284
compensation for, 292–293
error caused by, 291
Bode plot for, 74, 294
bootstrapping in, 295–298
can, type of, 69
chopper-stabilized, 194, 196, 197–198, 290
closed loop frequency response of, 293–294
closed loop impedance of, 286
common mode impedance of, 84–90, 285
common mode rejection (CMR) of, 85–86, 90, 288
common mode voltage limit in, 288
communications systems and, 208–221
as constant current generators, 82
as constant voltage generators, 82
current delivered by, 81–82
differential input to, 85, 90
differential mode impedance of, 84–90, 285
differential offset current in, 284

Op-amps – *Continued*
discrete, type of, 289, 290
drift in, 290
chopping in, 197–198
reduction of, 194–197, 298–299
dual inline, type of, 69, 70, 299
dual output, type of, 207
feedback and, 79, 94
error calculations in, 290–295
resistor for, 73, 106
feedback factor (β) in, 94
field effect transistors in, 285, 290, 295, 323
finite gain error of, 291–292
flat pack, type of, 69
frequency compensation for, 294
gain of, 73, 75, 285
gain-bandwidth product for, 75
grounding of, 121, 122
heterodyning for amplification with, 217–218
how to buy, 95–96
hybrid type, 289, 290
ideal, 78–84, 283
feedback and, 79, 94
gain of, 78, 79, 83
impedance of, 78, 80, 83
offset voltage of, 78
phase shift of, 78
impedance of
input, 76, 77, 80–81, 83, 85, 90–91, 283–284, 295–298
output, 77, 78, 81, 83, 288
inputs in, 71–72
instrumental amplifier in, 90–91
integrated circuit (IC) type, 187–192, 193, 194
compensation of, 188–190
equivalent circuit for, 188, 189
manufacturer's specifications for, 190–192, 193, 194
maximum ratings for, 188
internally compensated, 289, 294
internally shielded, 238
isolation. *See* Isolation op-amps
latch-up of, 216–217
levels of sophistication in, 105–106
limitations as current or voltage sources with, 110–111
linearity of, 114, 135
low drift unit and, 298–299
magnetic tape recording and, 218–221
models
of bias current, 284
of common and differential mode impedance, 286
of noise voltage and current, 289
of output impedance, 285
module, type of, 70, 71
monolithic type, 289, 290
noise current in, 288–289
noise voltage in, 288–289
notation systems for, 72
offset current of, 284
offset voltage of, 283–284
compensation for, 292–293
error caused by, 291, 292
open-loop gain frequency response and, 287, 289, 293-294
oscillation in, 76, 289, 293–294
output impedance of, 285
packages in, 70, 71
parameters of, 283–289
phase shift in, 78, 93–94, 289
power supplies and, 184–186
connections for, 71
improvement of, 172
radio receivers and, 211–212
rated vs. effective output impedance of, 285–287
recorders and, 206
response time of, 92
eith time, 92
ninety-percent time, 92
roll-off of, 74, 289
saturation of, 158
settling time of, 93
shielding of, 203–207, 238
slew rate of, 93, 287
stability criterion for, 293–294
step function and, 93
step voltage and, 93
temperature coefficients for, 290
terminal notation for, 70–71, 72
transistors vs., 315
unity gain in, 74–75
white noise in, 288
Op-amp circuits, 72–77
for addition, 132, 134–136, 220
band pass filter and, 146–148
Bessel filter and, 152, 153
biomedical applications of, 229–247
bootstrapping techniques in, 295–298
bounding the source technique in, 129
Butterworth filter and, 152, 153
change-of-slope detector and, 139, 140
Chebyshev filter and, 152
chopper-stabilized, 196, 197–198, 284
clipper, 131
closed loop gain in, 74
constant current supply and, 82, 107–110, 183–184
constant voltage supply and, 182–183
converting single-ended to differential signals with, 207–208
crossover distortion in, 117

current booster in, 113, 114–117
current regulator diode in, 127
current-to-voltage, 129–130
Darlington amplifiers and, 117–121
difference, 87, 90
differential, 87, 89, 90
division and, 140–141
ECG and EEG studies and, 233–236
feedback in, 73, 80–81, 110–111, 158
capacitor for, 106–107
resistor for, 73, 106–107
field effect transistor chopper system and, 202
as filters, 146–157
band-pass, 146–148
Bessel, 152, 153
Butterworth, 152, 153
Chebyshev, 152
high-pass, 150
high quality, 153, 154
low-pass, 149–150
quality factor for, 147–148
RC filters, 148–149
slot, 150–151
twin-tee, 155–157
gate circuits and, 131–133
high-impedance AC motor and, 112
high-pass filters and, 150, 151
high quality, 153, 154
human body potential measurement with, 207
inductance-capacitance, 211
initial condition or zero-reset, 163
integrated circuit op-amp with compensation and, 187
for integration, 136–140
inverting amplifier and, 73–75, 78–80, 123, 129, 295–298
isolation system using conventional op-amps, 121–122, 244, 245
low drift, using two op-amps, 298–299
low-pass filter, 149–150, 154
mean-square functional generator and, 141, 142
modulator, 220–221
for multiplication, 140–141
multivibrator, 166–169
bistable, 168–169
free-running, 167
monostable, 168
neon-bulb probe generator and, 214, 215
noninverting amplifier and, 75–77, 82–83
open loop amplifier and, 73
oscillation in, 76
oscillator, 158–161
phase-shift, 158–160
for telemetry, 215–216
Wien bridge, 159, 160–161
as peak detector, 139–140
power supply hook-up and, 174, 175
for power supply protection
DIAC crowbar type, 177–178
SCR crowbar type, 175–177
for power supply regulation, 178–182, 184–186
programmed power supply and, 187
pulsed oscillator circuit, 215–216, 217
pulse-rate-to-voltage converter, 234
push-pull, 116
ramp generators and, 165–166
reference voltage sources in, 122–127
regulated voltage source in, 124–125
relay control, 113–114
resistance isolation, 121–122
resonant circuits and, 211–212
reverse biased, 116
Schmidt trigger circuit and, 169–171
shielding of, 205–207
signal averaging and, 141–146
of noisy signal, 141–144, 199–200
for rms average of pure AC signal, 144–146
signal conditioning and, 84
skin resistance measurement with, 230–232
slot filter and, 148, 150–151
for square root computation, 141, 142
special function generators and, 161–166
ramp, 163–166
saw-tooth wave, 162–163
square wave, 161–162
stabilization of, and feedback, 159–161
for subtraction, 136–140
synchronous detection with, 195, 196–197
synchronous rectification and amplification with, 199–202
tape recorder data collection and, 218–221
thermal runaway in, 117
thermocouples in, 88, 89
transistor oscillator circuit, 215, 216
twin-tee filters and, 155–157
unity gain and, 84
voltage booster, 113–114
voltage comparison for, 128–129
with Western cell as reference voltage, 123–124
zener diode and, 124–125, 127, 133
zero-reset, 163
Optical detectors, 333–344
cost of, 340
detectivity or D* of, 340
light sources for, 333, 335–337
choice of, 337
condensor discharge lamp, 337
electroluminescence, 336
flashlamps, 336–337

Optical detectors, light sources for – *Continued*
fluorescent lamp, 336, 337
gas discharge or spectral sources, 336
light-emitting diode (LED), 239–240, 336, 337, 338
pulsed sources, 336–337
quartz iodine, 335, 337
thermal sources, 335–336
linearity of, 338
noise and, 272, 339–340
dark current, 339
drift, 339
fatigue, 339–340
noise-equivalent power (NEP), 339
shot, 339
photoconductive devices in, 342–344
bolometers, 343
choice of, 343–344
light amplifiers, 343
photodiodes, 343
phototransistors, 342
starlight scopes, 343
photoelectric devices in, 340–342
photometry of, 334–335
photomultiplier, 341
phototube, 341
quantum efficiency of, 338
response time of, 338–339, 342
spectral response of, 337–338
transmission characteristics of various materials in, 333–334
window materials for, 341
Order of filters, 22–23, 153
Oscillation in op-amp circuits, 76
Oscillators, 158–161
glow-lamp type, 329–331
Hartley, for telemetry, 215, 216
op-amp, for field effect transistor chopper circuit, 202
phase-shift, 158–160
relaxation type, 330
unijunction type, 316, 317
Wien bridge, 159, 160–161
Oscillographs
cathode-ray type, 64
multi-element recording type, 63

Pacemakers, 241, 242
Paper, recorder, 61–62
Patient safety, 236–247
battery-powered op-amps and, 244
conductivity of human tissue and, 236
grounding of equipment and, 236, 237–238, 241–242
in hospitals, 240–241
isolation op-amps and, 239–245
levels of hazard protection in, 239
measurement of skin resistance and, 229–230
shielding of equipment and, 236–237
skin-electrode contact problems and, 245–247
Peak detector circuits, 139–140
Peak inverse voltage (PIV), 32
Peak voltage (V_{peak}), 16, 17
Pens, recorder, 61, 63
Phase difference meter, 48
Phase-locked system, 198
Phase shift, 27–29
in capacitors, inductors, and resistors, 28–29
frequency-dependent, 235–236
in op-amps
ideal, 78
real, 93–94, 289, 293–294
and oscillators, 158–160
Photocells, 200–202, 273
Photoconductive devices, 342–343
bolometers, 343
choice of, 343–344
light amplifiers, 343
photodiodes, 343
phototransistors, 342
starlight scopes, 343
Photodectors. *See* Optical detectors
Photodiodes, 340, 343
Photoelectric devices, 340–342
Photographic recorders, 62–63, 64
Photometry, 334–335
photometric (PM) units for, 334
radiometric (RM) units for, 334
spectral units for, 334, 337–338
Photomultipliers, 341
Photomultiplier (PM) tubes, 202, 203
Photons, counting of, 202, 203
Phototransistors, 240, 342
Phototubes, 341
Pilot lights, neon-bulb type, 34
Pinch-off regions, 317
Pinch-off voltage, 318
Platinum, resistance of, 11
Plato, 77
PNP transistors
circuit symbols for, 301, 303
complementary pair with, 118
in Darlington amplifiers, 117–118
determination of type in, 314
power dissipation limit of, 309
power supply regulation with, 179, 181
push-pull circuit and, 116
Polaroid Land camera, 64
Police departments, 258
Potentiometers, 41–43
calibrated type, 42–43
linear resistance wire for, 42

measuring current with, 43
measuring resistance with, 43
self-balancing recorder, 43, 59–60, 64
slide-wire type of, 43
standardization of, 42–43
thermocouples and, 41
as voltmeters, 41
Power (*P*), defined, 10
Power dissipation by resistance, 9, 10
Power receptacles, 5–8
Power supplies, 15–16, 47–48, 171–187. *See also* Batteries; Constant current sources; Transformers
buying vs. building, 173–174
computer detection of problems in, 267–268
constant current type, 48, 172, 183–184
constant voltage type, 48, 172, 182–183
DC from AC, 30, 32
dual output type, 49
evaluation of, 172–173
grounding of, 47
hooking up, 174, 175
line-load changes, response to, 172, 173, 175, 178
meters with, 172, 173
for op-amps, 184–186
battery-resistor type, 185, 186
ground connections and, 174, 175
op-amp regulated, 184–185
transformer-rectifier type, 184
zener diode regulated, 185–186
outputs, types of, 172, 173
programed, 187
protection circuits for, 174–178
cascaded SCR, 176–177
DIAC type, 177–178
SCR crowbar, 175–176
ripple in, 172
short-circuit protection for, 172, 173, 174–178
signal generators vs., 47
three-phase, 6, 7
transformer-diode type for DC, 32
transformers in, 30–31
two- and three-wire cords in, 47–48, 241–242
voltage regulation circuits for
Darlington amplifier, 181, 185
op-amp type, 179, 184–185
transistor type, 179–180
zener-diode type, 185–186
Price codes, computer, 257–258
Problems, 65–66, 96–97, 221–223
Processor, computer, 250
Product engineering, 261, 262
Program, computer, 251
Programed power supplies, 187
Programmable signals, 161, 164, 166
Pulse generators, 46, 162
neon-bulb type, 214, 215
Pulse-width modulator, 220–221
Push-pull circuit, 116

Quality (*Q*) factor of filters, 27, 147–148
Quarter-square multiplier, 140
Quartz iodine lamps, 335, 337
Quick Cast, 279
Quiescent point (design point), 318, 319

Radio
AM broadcast band for, 209
antennas, length of, 208–209
carrier frequency (RF) for, 209
heterodyning, 217–218
latch-up in, 216–217
mixer device in, 209, 210, 212–213
modulated carrier, 209, 210
modulation in, 209, 210, 212–213
neon-bulb pulse RF circuit in, 214, 215
oscillators for transmission in, 215, 216
reception in, 210–213
resonant circuit for, 211–212
side-band frequencies in, 212–213
transmission in, 208–209
WWV, 218
Ramp generators, 46, 138, 161, 163–166. *See also* Signal generators; Staircase voltage generators
output of, 165–166
programmable signals and, 164
voltage-time generator circuit and, 165–166
RC filters, 18–21, 148–149, 151, 152
Reactance, 17
of capacitors, 18
of inductors, 18
Recorders, 60–64
cameras for, 64
cathode ray oscillograph, 64
chart types in, 60
circular, 60
strip chart, 60
X-Y type, 60, 62
Z-fold, 60, 62
direct writing, 61–62
chopper bar and typewriter, 62
heated stylus on temperature-sensitive paper, 61–62
ink pen in, 61
stylus on carbon-coated paper, 62
stylus on electrosensitive paper, 62
EEG or ECG work with, 235
grounding of, 206
multi-element oscillograph, 63
photographic, 62–63
self-balancing, 43, 44, 59–60

Recorders – *Continued*
tape for data collection, 218–221
X-Y type, 63–64
Rectifiers. *See* Diodes; Semiconductor-controlled rectifiers (SCR)
Reed switches, 325
Reference voltage sources, 122–127
mercury battery, 124
Weston cell in, 123–124
zinc-carbon dry-cell battery with zener diode in, 124–125
Register, computer, 250
Regulator circuits, 124–125
Relaxation oscillators, 330
Relay control circuits, 113–114
Repair operations, computer, 254–255
Reset circuits, 163
Resistance, 9–10
of different metals, 10, 11
reactance and, 18
temperature variations in, 10–11
Resistance isolation circuit, 121–122
Resistance thermometers, 11
Resistivity, 9–10
coefficient of, 10
Resistors, 1-4
carbon, 1
circuit symbol for, 4
color code for, 1, 2, 3
feedback, in op-amp circuits, 73, 80–81, 94, 106–107, 110–111
parallel, and Kirchoff's laws, 13–15
phase shift in, 28–29
pilot lights and, 34
reactance of, 18
series, and Kirchoff's laws, 12–13
variable, 73
wire-wound, 1
Resonant circuits
diodes in, 212
inductance-capacitance (*LC*) type, 211–212
with neon-bulb pulse generator, 214, 215
op-amp amplifier for, 212
Response time, in op-amps, 92
Ribbons, recorder, 61, 62
Ripple, 172
RLC bridge, 156, 246
Roll-off
of filters, 22–23, 148, 153
of op-amps, 74, 289, 293–295
Root-mean-square voltage (V_{rms}), 16–17

Safety devices, 268–269
Sampling oscilloscopes, 56
Saturation, 144, 158
Saw-tooth wave generators, 161, 162–163
dry-reed relay in, 163
initial condition or zero-reset circuit for, 163
Scales, and computers, 271–272
Schmidt trigger circuit, 167–171
input-output curve for, 169,170
for pulse-rate-to-voltage converter, 234
as square wave generator, 169
SCR. *See* Semiconductor-controlled rectifiers (SCR)
SCS (Silicon-controlled switch), 325
Security, computer, 268, 277
Semiconductor-controlled rectifiers (SCR), 175, 323–328
circuit design for, 324–325
circuit symbol for, 324
conduction angle of, 325, 326
Darlington amplifiers and, 118, 119
latching current of, 326
in power supply protection (crowbar) circuits, 175–176
time of turn-on of, 325, 326
V-I curve for, 176, 324
Semiconductor-controlled rectifier (SCR) circuits
applications of, 325–328
cascaded, 176–177
for control
of electric motor speed, 325
full-wave type, 326, 327
half-wave type, 327
waveform of, 326
crowbar protection circuit with, 175–176
design of, 324–325
V-I curve for, 175, 177
Semiconductor materials
germanium, 306
N-type, in transistors, 301
P-type, in transistors, 301
silicon, 306
Servo systems, 43
Settling time of op-amps, 93
Shielded wires, circuit symbol for, 4
Shielding, 203–207
for op-amps, 204–207, 236–238, 240
patient safety and, 236–238
of signal sources, 236–238
Shipping systems, computer, 255–256
Shot noise, 339
Shunts, and ammeters, 38–39
Side bands, 212–213
Signal averaging, 141–146, 199–200
rms average, 142
short-term, 142
true average, 143
Signal averaging circuits
for noisy signals, 142–143, 199–200
for rms average of pure AC signals, 144–146
Signal conditioning, 84
Signal generators, 46–47. *See also* Signal sources

audio frequency type, 46
floating system, 47
grounding of, 46–47
output impedance of, 46
output waveforms of, 46
Signal sources, 236–238
balanced, 236
circuit symbol for, 4
floating, 237, 238
shielding, 236–238
single-ended, 237, 238, 244
types of, 237
Silicon-controlled switch (SCS), 325
Sine wave generators. *See also* Oscillators
phase-shift oscillator, 158–160
Wien bridge oscillators, 160–161
Single-ended device, 207
Single-ended signals, converting to differential, 207–208
Skin conductivity, 236
Skin-electrode contact problems, 245–247
cautery work and, 246
conductive grease and, 246
four-point probe and, 247
low voltage DC and, 245–246
non-ohmic contact and, 240
skin-contact testing in, 246–247
skin-drilling technique in, 246
Skin resistance
measurement of, 229–232
AC vs. DC in, 230–231
chamois skin experiment in, 232
current-limiting modules for, 230
field effect diodes for, 230
four-point probes for, 233
isolation amplifiers for, 230
op-amp circuits for, 230–232
surface resistance in, 229
Slew rate of op-amps, 93, 287–288
Slot filters, 23, 146, 147, 148, 150–151
Society for Computer Medicine, 273
Software, computer, 256
Spacecraft engineering, 259–261
Spectrometric Oil Analysis Program (SOAP), 269
Speed control of electric motors, 325–328
Sports, and computer applications, 278–279
Square root circuit, 141, 142
Square wave generators, 161–162. *See also* Multivibrator circuits
for amplifier testing, 161–162
multivibrators as, 167
outputs of, 162
Schmidt trigger as, 169
Square wave test, 161, 235
Staircase voltage generators, 138, 242
Standard cells for potentiometers, 42
Starlight scopes, 343
Step function, 93
Step voltage, 93
Stress analysis, computer, 259
Strip charts, 60
Stylus, recorder, 61–62
Subtraction circuit, 136
Summing point, 78
Switches
circuit symbol for, 4
reed, 325
SCS, 325
Switching diodes, 132
Synchronous detection, 195, 196–197, 198–202

Tape recorders for data collection, 218–221
Telemetry, 208, 213–217. *See also* Radio
heterodyning in, 217–218
neon-bulb pulse generator for, 214, 215
op-amp oscillator circuit for, 216
transistor oscillator circuit for, 215, 216
Television scanners, and computers, 268
Temperature
field effect transistors (FET) variations with, 318, 319, 322
op-amps and, 290
resistance and, 10–11
zener diode circuit sensitivity to, 126
Therapeutic applications of computers, 266
Thermal drift in DC amplifiers, 182
Thermal runaway in transistors, 117
Thermistors, in isolation systems, 244, 245
Thermocouples
op-amp circuits and, 88–89
potentiometers and, 41
Thermometer, resistance, 11
Three-wire cords and ground, 5, 6–8, 241–242
Traffic control, by computer, 267
Transfer function, 200
Transformers, 29–31
autotransformer (Variac) and, 30–31
center-tapped, 30, 31
circuit symbol for, 4, 30
DC power supply and, 30–31, 184
inductors and, 29–31
isolation amplifiers and, 240
primary and secondary coils in, 30
Transient elimination, 331–332
capacitor for, 331
varistor for, 331–332
Transient response of filters, 153
Transient test circuit, 331
Transistors, bipolar, 301–315. *See also* Transistor circuits
AC voltage gain of, 313
base-emitter voltage
of germanium type, 306
of silicon type, 306
base terminal of, 115, 301, 302, 314

Transistors, bipolar – *Continued*
biasing for stabilization in, 307–309
characteristic curves for, 304–307
collector terminal of, 301, 302, 314
collector-to-base bias in, 301
common emitter configuration in, 115
current gain of, 115, 301, 302, 313
temperature dependence in, 306–307, 309
DC load line of, 305
DC voltage gain of, 305–306, 313
diode model of, 302
emitter-to-base bias in, 301
emitter terminal of, 115, 301, 302, 314
field effect transistors (FET) vs., 315–316
germanium type, 306
h_{fe} ratio (current gain) for, 302–304, 306
minimum/maximum values of, 309
h_{ie} ratio (AC emitter-to-base impedance) in, 313
ideal characteristics of, 304
junction of, 301
leakage current of, 306
linearity of, 114
NPN type, 301
circuit symbol for, 301, 303
common emitter transistor circuit and, 302–304
complementary pair with, 118
in Darlington amplifiers, 117–118
determination of type in, 314
PNP type vs., 115–116
power supply regulation and, 181
push-pull circuit and, 117
two-diode model of, 302
N-type semiconductor in, 301
op-amp current booster circuit and, 114–117
op-amps vs., 315
packages in, 314
PNP type
circuit symbols for, 301, 303
complementary pair with, 118
in Darlington amplifiers, 117–118
determination of type in, 314
power dissipation limit of, 309
power supply regulation with, 179, 181
push-pull circuit and, 116
power dissipation limit of, 309
power supply regulation with, 179–180
P-type semiconductor and, 301
reverse biasing of, 116, 301
silicon type, 306
terminals of, 301
Transistor circuits. *See also* Transistors, bipolar
amplifier, 312–313
biasing for stabilization of, 307–309
common base, 314
common collector, 314
common emitter, 302–304, 314
crossover distortion in, 117
Darlington amplifier and, 117–121
emitter follower, 314
leakage current in, 119
oscillator (Hartley) for telemetry and, 215, 216
push-pull type, 116
Seven Steps to Design of, 309–312
thermal runaway in, 117
TRIACs, 323–328
applications of, 325–328
circuit design and, 325
conduction angle of, 325, 326
Darlington amplifiers and, 118, 119
latching current of, 326
time of turn-on of, 325, 326
TRIAC circuits
applications of, 325–328
for control
of electric motor speed, 325, 328
waveform of, 327
design of, 325
Tungsten lamps, 333–334
Twin-tee filters, 155–157
Two-wire cords and ground, 5, 6
Typewriter, and recorders, 62

Unijunction, 316, 317
Unity gain, 74–75
Unity gain amplifier, 84
Unity gain circuits, 84
Used equipment, 95–96

Vacuum tube voltmeters (VTVM), 35, 37. *See also* Voltmeters; Volt-ohmmeters (VOM)
matching with source output in, 77
in measuring high impedance source, 76–77
Variac, 30–31
Varistors, 331–332
Verbal communication, and computers, 264–265
Verification, computer, 251–252
Visual applications of computers, 266–267
Voice recognition systems, computer, 264, 279, 280
Voltage
alternating current (V_{rms}), 16–17
direct current (V_{peak}), 16–17
Voltage booster circuits, 113–114
Voltage comparison circuits, 128–129
Voltage-crossing detector, 132–133
Voltage divider circuit, 8–9
Voltage follower circuit, 84
Voltage gain. *See* Gain
Voltage gradient (E), 9
Voltage, offset. *See* Offset voltage

Voltage range, 172
Voltage regulators
glow-lamp type, 330–331
zener-diode type, 124–125
Voltage (V_{rms}), root-mean-square (rms), 16–17
Voltage-time generator. *See* Ramp generators
Voltmeters, 34–37, 145. *See also* Vacuum tube voltmeters (VTVM); Volt-ohmmeters (VOM)
cathode ray oscilloscope as, 48
electrostatic, 35
field effect transistor type, 37
high impedance AC, 112
ideal, 37
measurement error caused by, 36
multipliers and, 39–40
multi-range, 39–40
op-amp feedback circuit and, 111–112
potentiometers as, 41
resistance of, 35–36
Volt-ohmmeters (VOM), 1, 4–5. *See also* Vacuum tube voltmeters (VTVM); Voltmeters
matching with source output, 77
measuring high impedance source and, 76–77
reading, 5
true rms reading in, 145

Wave numbers, 334
Weighing, computer, 271–272
Weston cells, 42, 123–124
Wheatstone bridge, 44–46, 246
White noise, 200, 288
Wien bridge oscillators, 159, 160–161
Wire routing systems, 263
WWV, 218

X-Y charts, 60, 62
X-Y recorders, 63–64

Zener diodes, 124–127
constant voltage supply and, 182
gate circuits and, 133
op-amp latch-up prevention and, 216–217
potentiometric recorder and, 59
power supply regulation and, 185–186
reference voltage source regulation and, 59, 124–125
skin resistance measurement and, 230
temperature sensitivity and, 126
temperature stabilization of, 126, 127, 128
V-I curve for, 124
Zero-crossing detectors, 132–133. *See also* Gate circuits
Zero-reset circuits, 163
Z-fold paper, 60, 62